Applied Mathematics
for the Petroleum and Other Industries

Fourth Edition

by

Will L. McNair, P. E.

Published by
THE UNIVERSITY OF TEXAS AT AUSTIN
PETROLEUM EXTENSION SERVICE
Continuing & Extended Education
Austin, Texas

2004

Library of Congress Cataloging-in-Publication Data

Applied mathematics for the petroleum and other industries.— 4th ed. / Will L. McNair.
   p. cm.
 Includes index.
 ISBN 0-88698-211-1 (alk. paper)
 1. Mathematics. I. McNair, Will L.
 QA39.3.A56 2004
 510—dc22

                                                                    2004004386

CONTENTS

FIGURES

TABLES

FOREWORD

Employees, students and others who are involved in the petroleum and other industries are discovering that operation of their facilities includes equipment, products, and systems requiring knowledge of mathematics. From the simplest valve to the automated operation of sophisticated production systems, mathematics is an integral part of understanding how they function. To paraphrase Lord Kelvin from an earlier era, "full understanding of anything requires a number…"

The original intent of this work was to provide a basic reference for field employees in the oil and gas industry who had little or no mathematical background. It has now grown to the point that personnel in many industries require knowledge of math as a means of understanding system operation as well as designing, maintaining, and troubleshooting equipment.

This new edition is designed to provide both a text and reference source in one book that covers basic mathematics as well as advanced technology. Handheld calculators are available with basic mathematic functions as well as specialized functions (such as exponents, logic functions, trigonometric functions, etc.) and use of these is suggested as the book is studied.

When used as a refresher course in elementary mathematics, it is assumed that the user will refer to those chapters where review is needed. Practice problems and self-tests are provided to assist the learners in testing their skills in the appropriate areas.

When used as a technical reference, the book gives basic guidelines and examples in a number of mechanical, hydraulic, and electrical areas that are commonly found in the field.

PETEX encourages those studying this book to communicate to us any suggestions or improvements you may find for consideration to be included in future revisions.

Will L. McNair, P. E.
Petroleum Extension Service

ACKNOWLEDGMENTS

This work is the result of many people who have contributed to *Applied Mathematics* in previous editions. Since the early 1930s, the American Petroleum Institute (API) sponsored the first oil field mathematics books and sponsored subsequent publications by PETEX by providing resources and contributors.

Among these earlier contributors are:

Dois D. Dallas, Tanana Valley Community College

Bruce Whalen, PETEX Consultant

Tommy Bicknell, Sedco, Inc.

R. G. Carson, Exxon Co.

Charlie Smith, Conoco, Inc.

Howard Swaim, Conoco, Inc.

Lejeune Wilson, Sedco, Inc.

Contributors to this current work include TRAINCO staff member Jeanette Maddox and PETEX staff members Ron Baker, Kathryn Roberts, Doris Dickey, and Deborah Caples.

1
The Number System

OBJECTIVES

Upon completion of chapter one, the student, without using a calculator, will be able to—

1. Convert numbers that are written out into digits in their proper places.
2. Convert numbers from one unit into another given unit.
3. Add and subtract whole numbers.
4. Multiply and divide whole numbers.
5. Determine whether a word problem requires addition, subtraction, multiplication, or division.
6. Find the least common denominator of a group of fractions.
7. Reduce a common fraction to its lowest terms.
8. Add and subtract common fractions.
9. Multiply and divide common fractions.
10. Add and subtract mixed numbers.
11. Multiply and divide mixed numbers.
12. Add and subtract decimal fractions.
13. Multiply and divide decimal fractions.
14. Convert common fractions into decimal fractions and decimal fractions into common fractions.
15. Find the square root of a number.
16. Calculate the quantity of a number squared or raised to another power.

INTRODUCTION

Imagine that you are a lone hunter for your tribe on the plains of Africa, thousands of years ago. You have spotted a herd of gazelles that will, if you can successfully bring a few of them down, provide food for your family and your tribe for several days. Since you cannot effectively hunt the gazelles alone, you need to find your fellow hunters and get their help. You realize that your friends will want to know how many gazelles are in the herd, so you begin counting them. You count to ten using ten fingers; then, you start over and count to ten again. Finally, you count three more on your fingers. You can now tell your fellow hunters that the herd is made up of two tens and three gazelles, which is large enough to bring them running to help.

This very short story illustrates that early humans could easily count to ten using ten fingers. Then, they could start over and count to ten again. And, when they counted to ten, ten times, they had counted to one hundred. Besides remembering the count, they could also scratch straight lines into the dirt or onto the wall of a shelter to record a count. But, as time passed and humans became more sophisticated, they found that they had to keep track of large numbers. Indeed, these numbers were so large that using fingers and scratching straight lines was not adequate. So, they came up with special symbols, or figures, to represent numbers. A group of people who lived in the Middle East—the Arabs—cleverly invented symbols to represent the numbers from one to nine and for zero. Having a figure for zero was a great step forward, as you will see in a moment. Because the Arabs were the first to create symbols for numbers, everyone began calling the symbols *Arabic numerals*.

Later, the Romans also developed a system of symbols, which, logically enough, were called Roman numerals. Today, we mostly use Arabic numerals; however, we sometimes use Roman numerals for such items as chapter headings in books, hours on clocks, and copyright dates for movies.

The Arabic figures are 0, 1, 2, 3, 4, 5, 6, 7, 8, and 9. Letters of the alphabet stand for Roman numerals. For example, the Roman numeral I represents the Arabic numeral 1. Similarly, the Roman V is equivalent to the Arabic 5, X is 10, L is 50, C is 100, D is 500, and M is 1,000. Table 1.1 lists several Arabic numerals and their Roman equivalents.

Notice that the Roman numeral system sometimes places a symbol of lower value before the symbol for the next higher value. This placement indicates that you should subtract the lower value from the higher value. For example, the Roman numeral for 4 is IV, instead of IIII. The I before the V means to subtract 1 from 5. Similarly, the Roman numeral for 9 is IX and the Roman numeral for 49 is XLIX. For the equivalent of 9, the I before the X means subtract 1 from ten. For 49, the X before the L means to subtract 10 from 50, which is L, and the I means subtract 1 from 10, which is 9. Therefore, Arabic 49 is Roman XLIX. Notice, too, that the Roman system does not have a symbol for zero.

Suppose a movie was copyrighted in 1969. The Roman numeral equivalent is MCMLXIX. M equals 1,000; C (100) less M (1,000) is 900; LX is 60; and IX is 9. On the other hand, the Roman numeral equivalent for a movie copyrighted in 2000 is simply MM.

Because the Arabic system contains the figures 0 through 9, it certainly saves time in writing—that is, Arabic 3 is easier to write than Roman III. But this advantage is only a minor part of the system's usefulness. The truly revolutionary aspect of the Arabic system is that the placement of each symbol determines its value. That is, Arabic numerals are written one after the other on a single line and their place in that line indicates the number's value.

Take, for example, the number 692. Because we read from left to right, six starts the number, nine follows it, and two follows the nine. The number's position, or place, in the line determines the value of that number. In this case, the position of the six represents 600, or six hundreds; the nine represents 90 or ten nines; and the two represents two ones. Consequently, we read it as six hundred ninety-two—that is, the number contains six hundreds, nine tens, and two ones.

Another feature of Arabic numbers is that we can put different numbers in columns and manipulate them in many ways. For example, as you will soon learn, several numbers can be added together, subtracted from each other, multiplied by each other, and divided by each other.

Table 1.2 shows four Arabic numerals arranged in columns. The numbers are 5, 50, 500, and 5,000. Reading from left to right and starting in the table's first row, the table displays these numbers as 0005, 0050, 0500, and 5,000 to

TABLE 1.1
Arabic Numerals and Roman Equivalents

Arabic	Roman
1	I
2	II
3	III
4	IV
5	V
9	IX
10	X
19	XIX
20	XX
40	XL
44	XLIV
45	XLV
49	XLIX
50	L
90	XC
100	C
500	D
700	DCC
900	CM
999	CMIX
1,000	M
1,500	MD

TABLE 1.2
Number Value and Placement

Thousands	Hundreds	Tens	Ones	Number
0	0	0	5	five, or 5 ones
0	0	5	0	fifty or 5 tens
0	5	0	0	five hundred or 5 hundreds
5	0	0	0	five thousand or 5 thousands

show their value. You can always place a zero in front of a number and not change the value of the number itself. For instance, 0005 is the same as 5. However, a zero after a number is a different story. Zeros after a number give the number a different value. So, in this case, 0005 is only five; however, 5,000 is five thousand—quite a difference!

Table 1.2 shows that the place of the number in each column represents the number's value. Thus, the table has a ones column, a tens column, a hundreds column, and a thousands column. (Often, the ones column is called the units column.) Number values increase to the left, so if the table was expanded beyond the thousands column, the next column to the left of the thousands column would be the ten thousands column, the next would be the hundred thousands column, the next the millions column, and so on.

Using Arabic numerals means that writing a single-digit number—five in the table—puts it in the ones, or units, column. Thus, the numbers 1 through 9 all fall in the ones column. Similarly, two-digit numbers, 10 through 99—50 in the table—fall in the tens column. Three-digit numbers, 100 through 999—500 in the table—fall in the hundreds column. Four-digit numbers, 1,000 through 9,999—5,000 in the table—are in the thousands column.

As mentioned before, although table 1.2 goes no higher than the thousands column, larger numbers can be shown by adding columns to the left of the thousands column—for example, five-digit numbers fall in the ten thousands column, six-digit numbers in the hundred thousands column, seven-digit numbers in the millions column, and so on. Thus, the five-digit number 50,000 reads fifty thousand and has 5 ten thousands. Similarly, the six-digit number 500,000 reads five hundred thousand and has 5 hundred thousands; and the seven-digit number 5,000,000 reads five million and has 5 millions.

With this ingenious number system, we can write any whole number by combining the ten symbols of 0, 1, 2, 3, 4, 5, 6, 7, 8, and 9. Incidentally, a whole number is a number (also called an integer) equal to or greater than zero. Put another way, a whole number is one of the numbers from 0 to 9 or any number made up of the numbers from 0 to 9.

The Arabic system is such a convenient way to express numbers that virtually everyone uses it. And, because Arabic numbers are based on ten numbers including 0, it is called a decimal system from the Latin word for ten, *decem*. So, we may refer to this base-10 system not only as Arabic, but also as the *decimal system*. The decimal system is simply a number system based on units of ten.

WHOLE NUMBERS

Table 1.3 shows large and small Arabic figures. The Arabic number system is divided from right to left into units (or ones), thousands, millions, billions, and trillions. Each of the larger divisions is further divided from right to left into ones, tens, and hundreds (1, 10, 100). The divisions indicate position value and make it easier to read large numbers.

TABLE 1.3
Decimal System of Numbers

Number Written in Arabic Symbols	Trillions 100	10	1	Billions 100	10	1	Millions 100	10	1	Thousands 100	10	1	Unit/One 100	10	1	Number Spoken or Spelled Out
5															5	five
42														4	2	forty-two
340													3	4	0	three hundred forty
6,088												6	0	8	8	six thousand eighty-eight
44,307											4	4	3	0	7	forty-four thousand, three hundred seven
529,600										5	2	9	6	0	0	five hundred twenty-nine thousand, six hundred
1,000,000									1	0	0	0	0	0	0	one million
25,202,019								2	5	2	0	2	0	1	9	twenty-five million, two hundred two thousand, nineteen
555,555,555							5	5	5	5	5	5	5	5	5	five hundred fifty-five million, five hundred fifty-five thousand, five hundred fifty five
2,250,000,000						2	2	5	0	0	0	0	0	0	0	two billion, two hundred fifty million
30,000,000,000					3	0	0	0	0	0	0	0	0	0	0	thirty billion
300,000,000,000				3	0	0	0	0	0	0	0	0	0	0	0	three hundred billion
1,064,038,000,000			1	0	6	4	0	3	8	0	0	0	0	0	0	one trillion, sixty-four billion, thirty-eight million
10,454,300,000,000		1	0	4	5	4	3	0	0	0	0	0	0	0	0	ten trillion, four hundred fifty-four billion, three hundred million
254,100,000,000,000	2	5	4	1	0	0	0	0	0	0	0	0	0	0	0	two hundred fifty-four trillion, one hundred billion

Also, note that 1 ten is the same as 10 ones, and 1 hundred is the same as 10 tens. Continuing this process, 1 thousand is the same as 10 hundreds and 1 million is the same as 1,000 thousands.

Normally, we use commas to indicate the divisions between numbers in the thousands, ten thousands, hundred thousands, and so on. For example, fourteen thousand, three hundred fifty is written as 14,350. However, two thousand, four hundred twenty-three may be written either with a comma—2,423—or without a comma—2423. It is mainly a matter of choice. For the most part, this manual uses a comma except in the case of denoting a year, such as 1982, for example. The comma indicates the division between the thousands and the hundreds.

Also, the Système International d'Unités (the International System of Units, or SI for short, which is a measurement system based on the metric system) separates values having more than four digits on either side of the decimal into groups of three with a small space between the numbers. So, for example, where people in the U.S. write 14,568.3394, SI convention calls for it to be written as 14 568.339 4. SI does not use commas to separate numbers into groups of three; instead, it uses spaces.

Going back to table 1.3, note that it shows how whole numbers are read as well as how they are written. For example, the number 555,555,555 in the table reads "five hundred fifty-five million, five hundred fifty-five thousand, five hundred fifty-five." Note that each 5 in this large number is ten times as great as the 5 to its immediate right. For example, reading from right to left in the number 555,555,555, the second 5 is ten times as great as the first 5, the third 5 is ten times as great as the second 5, and so on. Hence, moving any number one place to the left is the same as multiplying it by 10.

Addition

Addition puts together, or unites, two or more numbers or groups of objects of the same kind. The number obtained by adding is the *sum*. The plus symbol (+) indicates that the numbers or objects are to be added. Another way to say it is the plus sign (+) is the operational sign of addition.

Keep in mind that only like quantities can be added. To state it as the old cliché does, "you cannot add apples and oranges." Likewise, you cannot add 5 dollars and 3 bolts to get 8 dollars or 8 bolts. Nor can you add 6 pounds of grease and 90 gallons of oil to get 96 pounds or 96 gallons of lubricant. The unit of measurement must be the same.

The first step in addition is to arrange the numbers correctly in vertical columns. For example, suppose that we wish to add two numbers, each of which has several digits. For the first number, place the units (or ones) digit in the right-hand, or units, column. Then, put the tens digit of the number in the next column to the left of the units column, which is the tens column. Next, put the hundreds digit of the number in the next column to the left of the tens column, which is the hundreds column, and so forth. Now, repeat this action for the second number, and start by placing the ones, or units, digit for the second number directly under the units digit of the first number. Regardless of the sizes and lengths of the numbers, the rightmost digit of each number (the units digit) should be in a vertical column, one number directly under the other. Put another way, the units digits of each number are flush right.

Example Problem: arrange 31,542 and 14 properly for addition and add them.

Solution:

```
  31,542
+     14
  31,556.
```

Start by writing 31,542. Then, place the 4 in 14 directly under the 2 in 31,542—that is, make the two units columns flush right. Next, put the 1 in 14 directly under the 4 in 31,542. Now, draw a line under the second row of numbers, and write the plus sign (+). Add the numbers by starting with the right-hand, or units, column. In this case, 2 and 4 are 6. Then, move to the column (the tens column) immediately to the left of the units column and add them. So, 4 and 1 are 5. Finally, write the rest of the numbers (3, 1, and 5) under the line. Simply writing these numbers is proper because zeros are implied and not shown to the left of 14. (Refer to table 1.2.) So, 0 + 3 = 3, 0 + 1 = 1, and 0 + 5 = 5. The final answer is 31,556.

In the preceding problem, the sums can be expressed by one digit—that is, 2 + 4 = 6 and 4 + 1 = 5. So, they are simply written under the columns. No other steps are needed.

However, when adding several columns and a sum contains more than one digit, such as 6 + 7 = 13, write the right-hand digit of the sum under the column and add the remaining digit or digits to the top of the next column to the left. This operation is called carrying.

Example Problem: Add the following numbers: 8,145; 234; and 56.

Solution:

```
   1 1
  8,145
    234
+    56
  8,435.
```

First, write down the numbers so that they are in columns that line up on the right—make them flush right as the example shows. Then, add the right-hand column. So, 5 and 4 are 9, and 9 and 6 are 15. Thus, the three numbers add up to 15. Write, or put down, 5 and carry the 1 to add to the next column. Add the numbers in the next column, which is 13, because 4, 3, and 5 are 12 and 1 more (which is the carried number) is 13; so, put down 3 and carry 1. Add the third column, which is 4. Finally, add the last column and put down 8. The sum is 8,435.

You can check addition by either adding the column numbers in reverse sequence or by adding column totals separately. To check by adding the column numbers in reverse, the arrangement becomes—

```
    1 1
    56
   234
+ 8,145
  8,435.
```

This method merely changes the addition sequence. If a difference occurs in the total, add both sequences again and locate the error.

The second checking method involves moving the total of each column one place to the left of the preceding total and then adding the totals. This method is often used when adding long columns of figures.

Example Problem: Check the addition of the previous problem by adding the column totals.

Solution:

```
   8,145
     234
 +   56
 ─────────
      15
      12
       3
 +8
 ─────────
   8,435.
```

In this case, the units column totals 15, the tens column totals 12, the hundreds column totals 3, and the thousands column totals 8. When added, the answer is 8,435, which agrees with the original answer.

Subtraction

Subtraction takes an amount away from another amount, or takes one number away from another number, to arrive at a third number. The number that is subtracted is the *subtrahend*. The number from which it is taken is the *minuend*, and the number that represents the results, or answer, is the *remainder*, or *difference*. The minus symbol (−) is the operational sign for subtraction. As with addition, only like quantities can be subtracted. Also, numbers should be arranged uniformly in vertical columns with the units digit of each number in the right-hand column, the tens digits in the next column, and so on.

Example Problem: Subtract 83 from 597.

Solution:

```
    597   (minuend)
 −   83   (subtrahend)
 ─────────
    514   (difference, or remainder).
```

Beginning with the right-hand column, subtract 3 from 7, which is 4. (Another way to state it is to decide what number must be added to the 3 in the subtrahend to result in 7 in the minuend. The number is 4.) In either case, write 4 under the line and continue with each column until the remainder is complete. So, next, take 8 from 9 and write down 1 under the line. Finally, simply write 5 under the line because 0 from 5 is 5. (As previously stated, the 0 is implied—that is, we normally do not write a 0 before another number; we know that the 0 is there, but we do not write it.)

Example Problem: 273 minus 186, or 273 − 186 = ?

Solution:

$$\begin{array}{r} {\scriptstyle 1\ 16} \\ 2\!\!\!/7\!\!\!/^13 \\ -\ 186 \\ \hline 87. \end{array}$$

The first step is to subtract 6 from 3; however, 6 is bigger than 3. So, an operation called borrowing must be done. In this case, borrow a 10 from the tens place, which reduces the 7 tens to only 6 tens. Now bring the borrowed 10 over to the ones place. When you bring the 10 to the ones place, you add it to what is already there, making the 3 a 13. Now subtract the 6 from 13 to get 7. Again, to subtract the 8 from the remaining 6, you have to enlarge the 6 by borrowing a 100 from the hundreds place and adding it to the 6 to make it 160, which leaves 1 instead of 2 in the hundreds column. Now, 8 from 16 leaves 8, and 1 from 1 leaves 0, resulting in the answer of 87.

It is helpful when subtracting to keep track of borrowing by running slashes through changed numbers and writing the new number above the changed number as shown.

You can check subtraction by adding the subtrahend and the difference:

$$\begin{array}{r} {\scriptstyle 1\ 1} \\ 186 \\ +\ 87 \\ \hline 273. \end{array}$$

In this case, the subtrahend of 186 and the difference of 87 add up to the minuend of 273, which means that the subtraction was done correctly.

Multiplication

Multiplication is adding one number to itself a specified number of times. For example, when we multiply 5 times 6, we actually add 5 to itself 6 times. Put another way, 5 times 6 is the same as 5 plus 5 plus 5 plus 5 plus 5 plus 5. (The solution to both operations is 30.) The number that is multiplied, or increased in multiples, is the *multiplicand*. The *multiplier* is the number of times the multiplicand is added to itself. Thus, 451 times 25 means that 451 (the multiplicand) is added to itself 25 (the multiplier) times. The result obtained by multiplying one number by another is the *product*.

The sign of multiplication is × or • and is called *times*. For example, 3 × 4 or 3 • 4 is read as "three times four." The symbol • is often used when unknown quantities are being multiplied and an *x* represents an unknown quantity. In such cases, the times sign × is similar to the unknown quantity *x*. So, to avoid confusion, • is used instead of ×, as in 3 • *x*, rather than 3 × *x*. (This fact is particularly true when one is writing the operation by hand rather than typing or entering it into a computer's word processing program.)

Multiplication shortens the process of addition. For instance, 3 + 3 + 3 + 3 is the same as 3 × 4, and both equal 12. A multiplication table (table 1.4) aids in the shortening process. For example, to find the answer of 3 × 4 in the table, locate the multiplicand, 3, in the horizontal row across the top of the table and the multiplier, 4, in the left-hand vertical column. Follow these two numbers horizontally and vertically until they meet as 12, the product. To successfully multiply without referring to the table or using a calculator, you have to memorize the multiplication table.

TABLE 1.4
Multiplication Table

1	2	3	4	5	6	7	8	9	10	11	12
2	4	6	8	10	12	14	16	18	20	22	24
3	6	9	12	15	18	21	24	27	30	33	36
4	8	12	16	20	24	28	32	36	40	44	48
5	10	15	20	25	30	35	40	45	50	55	60
6	12	18	24	30	36	42	48	54	60	66	72
7	14	21	28	35	42	49	56	63	70	77	84
8	16	24	32	40	48	56	64	72	80	88	96
9	18	27	36	45	54	63	72	81	90	99	108
10	20	30	40	50	60	70	80	90	100	110	120
11	22	33	44	55	66	77	88	99	110	121	132
12	24	36	48	60	72	84	96	108	120	132	144

Keep in mind that the multiplier is always an abstract number while the multiplicand may be either abstract or concrete. An abstract number does not refer to an actual thing or object. Instead it denotes a quality or an idea. A concrete number, on the other hand, does refer to an actual thing or object. For example, an hour is abstract whereas a joint of pipe is concrete.

The multiplicand and the multiplier can be switched—that is, the multiplicand can become the multiplier and the multiplier can become the multiplicand without affecting the results. For example, $3 \times 4 = 12$ is the same as $4 \times 3 = 12$. In the first case, 3 is the multiplicand and 4 is the multiplier; in the second case, 4 is the multiplicand and 3 is the multiplier. In both cases, 12 is the correct product.

However, keep in mind that the product must be the same quantitative unit as the multiplicand. Thus, you can multiply 8 horses by 4 horses to get a correct answer of 32 horses. But you cannot correctly multiply 8 horses by 4 goats because both are concrete. That is, 12 horse-goats cannot exist, because no such animal exists. In like manner, it is impossible to multiply dollars by persons. However, it is possible to multiply the dollar rate of pay per person (which is concrete) by the number of persons working (which is abstract) and get as the product total dollars paid (concrete). Thus, 46×3 persons = $138.

Example Problem: Multiply 232 by 3, or $232 \times 3 = ?$

Solution:

    232 (multiplicand)
  ×   3 (multiplier)
    ---
    696 (product).

Start with the right-hand number 2 in the multiplicand, and multiply it by 3, which is the multiplier, to get the product 6. Place the 6 under the line and even with (flush with) the 3. Next, multiply 3 in the multiplicand by the 3 in the multiplier. Place the answer, 9, under the line. Finally, multiply 2 by 3 and write down the answer, 6, under the line. The resulting product is 696.

As the multiplier becomes larger, multiplication gets more involved and placement becomes very important. As you will see shortly, multiplying large numbers requires determining partial products, which are numbers that result during each step of multiplication. And, when the multiplier contains a zero, proper placement of the partial products is vital. It is also important to remember that any number multiplied by 0 is 0, and any number multiplied by 1 is the same as the number. Thus, 234,892,784 × 0 = 0 and 234,892,784 × 1 = 234,892,784.

Example Problem: Multiply 25 × 16.

Solution:

```
      3
     25  (multiplicand)
  ×  16  (multiplier)
    150  (partial product)
  +  25  (partial product)
    400  (product).
```

First, multiply 6 × 5, which gives the two-digit number of 30. Enter the 0 below the line and carry the 3 to the top of the next column. Then multiply 6 × 2, which gives 12; then, add the carried 3 to the 12, which results in 15. Enter 15 to the left of the zero. Thus, the first partial product is 150. Next, multiply 25 by the multiplier of 1 to get 25. Enter 25 under the partial product of 150 *one place to the left*, so that the fives line up one under the other. Finally, add the two partial products to obtain the product 400.

Example Problem: Your rig has 162 joints of 30-foot drill pipe stored on the rack. When all of the joints are made up (joined), what is the total length of the string?

Solution: To find the number of feet in the string of joined drill pipe, multiply 162 times 30.

```
    162  joints  (multiplicand)
  ×  30  feet    (multiplier)
  4,860  feet    (product).
```

Because the last figure in 30 is 0, for convenience, make 30 the multiplier and 162 the multiplicand, although, as mentioned earlier, either number could be the multiplier or the multiplicand. Because of the 0, it is more convenient to make 30 the multiplier. Here is why: when writing down 30, put the 0 one place to the right of the right-hand digit of the multiplicand. Placing the 0 in the multiplier to the right of the right-hand digit of the multiplicand allows you to simply bring the 0 down and multiply 162 by 3 for a result of 4,860 feet.

Since 0 times any number is 0, it is not necessary to multiply out and put a row of zeros in the answer. The important thing is to bring the 0 down as a placeholder. A placeholder simply indicates the proper position of a number.

When the multiplier contains more than one zero, apply the same principle. In this case, move the multiplier to the right two places instead of one for each 0.

Example Problem: Multiply 1,234 by 200.

Solution:

$$
\begin{array}{r}
1,234 \\
\times \quad\;\; 200 \\
\hline
246,800.
\end{array}
$$

Again, simply make 200 the multiplier, bring down the two zeros as place holders, and multiply 1,234 by 2 to obtain the result of 246,800.

Also, when one or more zeros occur between two other numbers, apply the same principle.

Example Problem: An oil company conducted a 105-hour school for 345 of its employees. How many person-hours of training did the company provide in this school?

Solution: Each person received 105 hours of training; thus, 345 × 105 gives the number of person-hours.

$$
\begin{array}{r}
345 \\
\times \;\; 105 \\
\hline
1,725 \\
3,450 \quad\; \\
\hline
36,225 \;\; \text{person-hours of training.}
\end{array}
$$

In this case, the zero occurs between 1 and 5. So, to obtain the second partial product, simply put the 0 under the 0 in the multiplier (which is also under the 2 in the first partial product). Then, multiply 345 by 1 and place 345 to the left of the 0. Finally, add the partial products to obtain 36,225 hours of training.

You can check multiplication by switching the multiplier and multiplicand and multiplying again. You should obtain the same answer. You can also check multiplication by dividing the product by the multiplier. If you multiplied correctly, the answer is equal to the multiplicand. To check the previous answer, for instance, 36,225 (the product) divided by 105 (the multiplier) equals 345 (the multiplicand). (See the following section on division.)

Division

Division is the reverse of multiplication. It determines how many times one number is contained in another number. For example, to divide 8 by 4 means to find how many times 4 is contained in 8; or, stated another way, what number is multiplied by 4 to result in 8? The answer to this problem is, of course, 2.

One symbol used to indicate division is ÷ and is read "divided by." When the division sign (÷) appears between two numbers, it indicates that the first number is to be divided by the second. For example, 16 ÷ 4 reads "sixteen divided by four" and means what number equals 16 if multiplied by 4?

The number to be divided is the *dividend*, and the number that does the dividing is the *divisor*. The result obtained by division is the *quotient*.

Two methods of division are short division and long division. Short division is used when the divisor contains only one digit.

Example Problem: Divide 345 by 3, or what number multiplied by 3 equals 345?

Solution: Write down the numbers in the following manner and use the short division method.

$$
\begin{array}{r}
115 \text{ (quotient)} \\
\text{(divisor) } 3\,\overline{)\,345} \text{ (dividend).}
\end{array}
$$

Divide the first, or left-most, figure, which is 3, in the dividend by 3, obtaining 1 (3 is contained in 3 one time). Write this 1 above the line over the 3. Then, divide the next number 4 by 3. Note that 3 is contained in 4 one time, but 1 is left over. So place 1 in the quotient above the 4 and mentally place the 1 remaining in front of the 5, making it 15. Then divide 15 by 3, obtaining 5 as the final figure in the quotient.

Since division is the reverse of multiplication, check it by multiplying the answer or quotient by the divisor. The product should be the same as the dividend. So, in the example, 115 (quotient) × 3 (divisor) = 345 (dividend).

Example Problem: 552 ÷ 23 = ?

Solution: Using long division, write down the numbers in the following manner.

$$
\begin{array}{r}
24 \\
23\,\overline{)\,552} \\
46 \\
\hline
92 \\
92 \\
\hline
0.
\end{array}
$$

Since 23 is not contained in 5, determine how many times it is contained in 55. Note that it is contained two times (2 × 23 = 46) but not three times (3 × 23 = 69). So, write 2 above the second 5. Then multiply this 2 times 23 to get 46. Write 46 under the 55 and subtract, getting 9. Bring down the remaining digit 2 from the dividend and place it beside the 9, making 92. Now divide 92 by 23. By trial it is found that 23 is contained exactly four times in 92, so 4 is placed to the right in the quotient. Multiply 4 times 23 to get 92; subtract 92 from 92 to get 0, or a remainder of nothing. The answer shows that the process is complete and that 552 divided by 23 is equal to 24, or that 23 is contained in 552 exactly 24 times.

Many problems involving division of whole numbers do not work out evenly. Instead, they have a number left over, which is a remainder. A remainder occurs when the divisor is not contained in the dividend an exact number of times. For instance, 8 is not contained in 45 an exact number of times:

$$45 \div 8 = 5 \text{ with a remainder of 5.}$$

In long division, the remainder is written to the right of the answer preceded by an *r*.

Example Problem: 4,868 ÷ 30 = ?

Solution:

```
          162 r 8
   30 ) 4,868
       −30
       ‾‾‾‾
        186
       −180
       ‾‾‾‾
         68
        −60
        ‾‾‾‾
          8.
```

Practice Problems

1. Convert the following numbers into the units given.

   a. 5 millions is the same as _____ thousands.

   b. 11 thousands is the same as _____ hundreds.

   c. 3 tens is the same as _____ ones.

   d. 1 hundred is the same as _____ tens.

2. Write the Arabic figures for the numbers written out.

   a. A barrel of oil is equal to *forty-two* gallons.

      _____

   b. The companies drilled *three hundred sixty-one million* feet of hole.

      _____

   c. Proved crude oil reserves total *twenty-seven billion, fifty-one million, two hundred eighty-nine thousand* barrels.

      _____

   d. Natural gas production in the United States was *twenty trillion, four hundred billion, five hundred two million* cubic feet.

      _____

3. Write out the numbers for the Arabic figures in the blanks.

   a. The average depth of wells drilled in the field is *4,600* feet.

      _____

   b. The number of wells drilled is *78,000*.

      _____

   c. Crude oil production averaged *8,600,000* barrels per day.

      _____

   d. Fossil fuel imports totaled *$80,740,000,000*.

      _____

4. Work the following problems.

    a. $352 + 569 + 32 =$ _____

    b. $2{,}832 \div 12 =$ _____

    c. $1{,}950 \times 657 =$ _____

    d. $7{,}408{,}732 - 923{,}351 =$ _____

5. Work the following word problems, using addition, subtraction, multiplication, or division as needed.

    a. A company has 36 welders working at the same rate per month. The total monthly payroll for these employees is \$126,720. How much does each welder make per month?

        _____

        How much per year? _____

    b. A truck averages 55 miles per hour. How many hours of driving time will be needed to make a 935-mile trip?

        _____

    c. A drilling crew has 35 joints of 30-foot drill pipe attached to a 40-foot kelly. What is the total length of this string of pipe, including the kelly?

        _____

    d. A well is 6,500 feet deep. Casing lines 563 yards of the well. How many more feet of casing are needed to line the well all the way to the bottom? (Remember: 3 feet = 1 yard)

        _____

COMMON FRACTIONS

A fraction is a number that is not a whole number. For example, ½ is a vulgar, or common, fraction and 0.5 is a decimal fraction. Although common fractions can be cumbersome and time-consuming to use in computations, you often encounter them because of the industrial practices followed by the United States. U.S. industries chiefly use fractions to give the size of manufactured products and the machine tools used in manufacturing. Also, stocks and bonds are quoted in the security markets in dollars and fractions of a dollar.

In the U.S., dimensions of bolts, nuts, tubular goods, machine parts, tanks, and similar equipment require the use of common fractions. Some specific uses of fractions follow.

1. Manufacturers make pipe that has diameters sized to the nearest half, fourth, and eighth of an inch—for example, nine and five-eighths-inch (9⅝-inch) casing and five and one-half-inch (5½-inch) drill pipe.

2. Manufacturers make flow beans (chokes) sized in thirty-seconds and occasionally sixty-fourths of an inch—for example, a four and nine-thirty-seconds-inch (4⁹⁄₃₂-inch) flow bean. (Oil producers install flow beans into the pipe out of which a well's production flows. The bean, or choke, restricts the

flow and allows the producer to control the volume and pressure of flow from the well.)

3. Wire-rope companies size rope in halves, fourths, eighths, and sixteenths of an inch—for example, one and three-eighths-inch (1⅜-inch) drilling line.

4. Technicians measure the depth of oil in a tank in halves and fourths of an inch—for example, a tank depth of twenty-nine feet, four and one-quarter inches (29 feet, 4¼ inches.)

5. Mechanics usually measure shaft diameters, vessel openings and wall thicknesses, cylinder bores, and the length of compressor and engine strokes in halves, fourths, or eighths of an inch—for example, an engine cylinder bore of three and one-half inches (3½ inches).

6. Royalty—that portion of the crude production due the landowner—is normally one-eighth (⅛) of the production but may be any fraction agreed upon.

As just mentioned, companies must often measure a quantity of oil that is inside a metal tank. One method of measurement calls for a technician (a gauger) to measure (gauge) the distance from the bottom of the tank to the top of the oil. The gauger measures this height in feet, inches, and fractions of an inch, and then converts the height to barrels and decimal fractions of a barrel (or to metric tons and decimal fractions of a metric ton for international commerce). For example, suppose a gauger measured the height of the oil in one of the tanks on a lease and determined that it was 25 feet, 6¼ inches. By looking at an appropriate table for this tank, the gauger can determine that this height is the equivalent of 35,459.09 barrels of oil.

Machinists make many tools and devices that require measurements to the nearest sixty-fourths of an inch. For more precise work, they may use a special tool called a micrometer, which is a device that measures small diameters, thicknesses, or distances with a high degree of accuracy. Indeed, most micrometers are calibrated in thousandths of an inch, which are decimal fractions. For example, a device may have a diameter of 1.016 inches, which reads one-point-zero-one-six inches, or one and sixteen thousandths of an inch.

On the other hand, field personnel often measure pipe, tools, and equipment in fractions of an inch. For example, a nut on a bolt may be ¾ inch, which reads three-quarters of an inch. Or, the inside diameter of a joint of drill pipe may be 4½ inches (read four-and-one-half inches).

To add, subtract, multiply, and divide common fractions, you need to understand what a fraction is and how to deal with it. Fractions describe parts of whole things or a whole thing and a part of the same whole thing—for example, ½ of a pie, ⅞ of an inch, or 5½ inches.

In a common fraction, the number on top of the slash is the numerator and the number below is the denominator. So, in the fraction ¾, 3 is the *numerator*, and 4 is the *denominator*. The denominator indicates how many parts something whole is divided into, or how many parts it takes to make a whole. The numerator indicates how many of the total number of parts are being considered. For example, 1 inch is 1/12 of a foot, or 1 of the 12 inches in a foot; 30 minutes is ½ of an hour because 30 minutes is one of the two parts of an hour.

The fraction 5/5 means that something is divided into five parts (a pie, for example) and that all five parts put together make one whole. Likewise, two halves equal one whole (2/2 = 1), four quarters (or fourths) equal one whole (4/4 = 1), and so forth. Any fraction that has the same number in its numerator and denominator is the same as the whole number 1.

When working with fractions, a key principle is that the numerator and denominator can be multiplied or divided by the same number without changing the value of the fraction. For example, ½ is the same as ²⁄₄, ⁴⁄₈, ⁸⁄₁₆, and so forth. In this case, the numerator 1 and the denominator 2 were both multiplied by 2 to obtain ²⁄₄, by 4 to obtain ⁴⁄₈, and by 8 to obtain ⁸⁄₁₆. In all cases, the fraction's value is still ½. Similarly, in the fraction ⁸⁄₁₆, the numerator of 8 and the denominator of 16 can be divided by 8 to obtain ½. Put another way, ½ = ⁸⁄₁₆.

Addition

Before fractions can be added, they must have a *common denominator*—that is, all fractions to be added must have the same denominator. To add two or more fractions with the same denominator, simply add the numerators and place the sum over the common denominator. The denominator does not change. For example:

$$\frac{1}{8} + \frac{5}{8} = \frac{6}{8}.$$

When fractions have different denominators, before you can add them, you must first find the *lowest common denominator* (LCD)—that is, the smallest number that all of the denominators will divide into evenly. Sometimes, the largest denominator of the fractions will also be the LCD. For example, to add ¼ and ³⁄₁₆, 16 is used as the LCD. To express ¼ in sixteenths, multiply both the numerator 1 and the denominator 4 by a number that makes the new denominator 16. Dividing 4 into 16 gives the number 4, so 4 is multiplied times 1 and 4 to give 4 and 16, respectively, making the fraction ⁴⁄₁₆. Hence,

$$\frac{1}{4} + \frac{3}{16} = \frac{4}{16} + \frac{3}{16} = \frac{7}{16}.$$

Remember: multiplying or dividing the numerator and denominator by the same number does not change the value of the fraction.

Example Problem: Add ¾, ⅞, and ¹³⁄₁₆.

Solution: Set up the problem and find the LCD.

$$\frac{3}{4} + \frac{7}{8} + \frac{13}{16} = ?$$

The LCD is 16, because 4 and 8 both divide evenly into 16. Converting ¾ into sixteenths,

$$\frac{3}{4} = \frac{3 \times (16 \div 4)}{4 \times (16 \div 4)} = \frac{3 \times 4}{4 \times 4} = \frac{12}{16}.$$

Similarly,

$$\frac{7}{8} = \frac{7 \times (16 \div 2)}{8 \times (16 \div 2)} = \frac{7 \times 2}{8 \times 2} = \frac{14}{16}.$$

Now the problem can be set up with like denominators and can be solved by adding the numerators as

$$\frac{12}{16} + \frac{14}{16} + \frac{13}{16} = \frac{39}{16}.$$

Since 39 is larger than 16, the final answer is more than 1 ($^{16}/_{16}$ = 1). To reduce this fraction to a whole number and a fraction, divide 39 by 16:

$$
\begin{array}{r}
2\,r\,7 \\
16 \overline{)\,39} \\
32 \\
\hline
7.
\end{array}
$$

The remainder 7 becomes the numerator, and 16 remains the denominator, making the fraction $^{7}/_{16}$. Thus, the answer is $2^{7}/_{16}$.

To find the lowest common denominator when the denominators do not divide evenly into each other, the first step is to multiply the denominators by each other. This step determines the LCD. Then, divide the denominators in each fraction into the LCD and multiply the result by the numerator. Finally, add the numerators and place the result over the LCD.

Example Problem: Add $^{4}/_{5}$ and $^{2}/_{7}$.

Solution: To find the lowest common denominator, multiply the numbers in the denominators by each other. In this example, $5 \times 7 = 35$. Then, for the fraction $^{4}/_{5}$, divide 35 by the denominator 5, the result of which is 7. Now, multiply 7 by 4 to obtain the fraction $^{28}/_{35}$. (Put another way, $^{28}/_{35}$ = $^{4}/_{5}$.) Similarly, in the fraction $^{2}/_{7}$, divide 35 by 7, which is 5. Multiply 5 by 2 to obtain the fraction $^{10}/_{35}$. To write it—

$$
\frac{4}{5} = \frac{28}{35} + \frac{2}{7} = \frac{10}{35}.
$$

Now, add the numerators 28 and 10 to obtain:

$$
\frac{28}{35} + \frac{10}{35} = \frac{38}{35}.
$$

Since 38 is larger than 35, divide 38 by 35 to obtain the answer $1^{3}/_{35}$.

Subtraction

Subtracting common fractions is based on the same principles as adding common fractions. The first step in subtraction is to find a common denominator. Then one numerator is subtracted from the other, and the difference is written over the common denominator.

Example Problem: Subtract $^{5}/_{8}$ from $^{3}/_{4}$.

Solution: Using 8 as the LCD, convert $^{3}/_{4}$ to eighths.

$$
\frac{3}{4} = \frac{3 \times (8 \div 4)}{4 \times (8 \div 4)} = \frac{3 \times 2}{4 \times 2} = \frac{6}{8}.
$$

Then subtract numerators,

$$
\frac{6}{8} - \frac{5}{8} = \frac{1}{8}.
$$

Example Problem: Find the difference between $^{15}/_{16}$ of an inch and $^{19}/_{64}$ of an inch.

Solution: Find the least common denominator.

$$\frac{15}{16} = \frac{15 \times (64 \div 16)}{16 \times (64 \div 16)} = \frac{15 \times 4}{16 \times 4} = \frac{60}{64}.$$

Then subtract numerators,

$$\frac{60}{64} - \frac{19}{64} = \frac{41}{64}.$$

Sometimes, the resulting number can be reduced to a smaller fraction. Reducing is done only if dividing both the numerator and denominator by the same number results in even numbers. For example, $^{60}\!/_{64}$ can be reduced to $^{15}\!/_{16}$ by dividing by 4. However, $^{41}\!/_{64}$ cannot be reduced, because no number besides 1 divides evenly into 41. The operation of reducing to the lowest terms is especially necessary in the multiplication of fractions.

Multiplication

To multiply one fraction by another, multiply the numerators by each other to get a new numerator, and multiply the denominators by each other to get a new denominator. Remember to reduce the new fraction, which is the product, to its lowest terms. That is, divide the fraction by numbers contained in both the numerator and denominator until these terms are as small as it is possible to reduce them. In the fraction $^{15}\!/_{25}$, for example, dividing both terms by 5 results in $^{3}\!/_{5}$. Since no number other than 1 divides evenly into 3 and 5, the fraction $^{3}\!/_{5}$ is the lowest term.

Example Problem: Multiply ¾ by $^{5}\!/_{7}$.

Solution:

$$\frac{3}{4} \times \frac{5}{7} = \frac{3 \times 5}{4 \times 7} = \frac{15}{28}.$$

In this example, $^{15}\!/_{28}$ cannot be reduced to any lower terms because no number other than 1 divides evenly into 15 and 28.

Example Problem: $^{3}\!/_{16} \times \frac{4}{5} \times \frac{1}{2} = ?$

Solution:

$$\frac{3 \times 4 \times 1}{16 \times 5 \times 2} = \frac{12}{160} = \frac{3}{40}.$$

In this example, $^{12}\!/_{160}$ is reduced to its lowest terms by dividing it by 4.

Cancellation is a method of shortening the multiplication of fractions. It entails dividing any numerator and any denominator by the same number, then multiplying the new, smaller numbers.

Example Problem: $^{5}\!/_{12} \times \frac{3}{15} \times \frac{3}{8} = ?$

Solution: Here, cancellation saves time because the same number can divide the numbers in the numerators and denominators.

$$\frac{\cancel{5}^{1}}{\cancel{12}_{4}} \times \frac{\cancel{3}^{1}}{\cancel{15}_{\cancel{3}_{1}}} \times \frac{\cancel{3}^{1}}{8} = \frac{1}{4} \times \frac{1}{1} \times \frac{1}{8} = \frac{1}{32}.$$

First, divide the numerator of 5 in the first fraction and the denominator of 15 in the second fraction by 5, which leaves 1 and 3, respectively. Then divide the new numerator of 3 in the second fraction and the denominator of 12 in the first fraction by 3, which leaves 1 and 4. The numerator of 3 in the third fraction and the new denominator of 3 in the second fraction cancel each other, leaving 1 and 1. Now multiply $1 \times 1 \times 1$ to get a numerator of 1 and $4 \times 1 \times 8$ to get a denominator of 32.

Division

Dividing 12 by 3 is the same as multiplying 12 by ⅓ because

$$\frac{12}{1} \times \frac{1}{3} = \frac{12}{3} = 4.$$

Put another way,

$$\frac{12}{1} \div \frac{3}{1} = \frac{12}{1} \times \frac{1}{3} = \frac{12}{3} = 4.$$

The example calculation shows that when dividing fractions, exchange the numerator with the denominator of the divisor and multiply it by the first fraction. In other words, the fraction doing the dividing is inverted and used as a multiplier.

Example Problem: Divide ³⁄₁₆ by ⁹⁄₆₄.

Solution:

$$\frac{3}{16} \div \frac{9}{64} = \frac{\overset{1}{\cancel{3}}}{\underset{1}{16}} \times \frac{\overset{4}{\cancel{64}}}{\underset{3}{\cancel{9}}} = \frac{4}{3} = 1\tfrac{1}{3}.$$

Practice Problems

1. Convert the following fractions to the denominators given.

    a. ³⁄₃₂ to 64ths _____

    b. ¾ to 16ths _____

    c. ⁵⁶⁄₆₄ to 16ths _____

    d. ¹⁴⁄₁₆ to 8ths _____

2. Reduce the following fraction groups to the lowest common denominators.

    a. ½, ⅛ , ¾ _____

    b. ⅛, ¹⁄₁₆, ⁷⁄₃₂ _____

    c. ⅞, ¹⁹⁄₆₄, ¹¹⁄₃₂, ¹⁷⁄₆₄, ¹⁄₁₆, ⁶²⁄₆₄ _____

    d. ³¹⁄₃₂, ¹⁵⁄₁₆, ³⁄₆₄ _____

3. Reduce the following fractions to their lowest terms.

    a. $^{32}/_{64}$ _____

    b. $^{56}/_{64}$ _____

    c. $^{4}/_{32}$ _____

    d. $^{10}/_{12}$ _____

4. Add or subtract the following fractions.

    a. ½ + ¾ + ¼ = _____

    b. $^{59}/_{64}$ + $^{3}/_{32}$ + $^{9}/_{16}$ + $^{7}/_{64}$ + ⅛ + $^{7}/_{16}$ + $^{59}/_{64}$ = _____

    c. $^{5}/_{16}$ − $^{3}/_{32}$ = _____

    d. $^{26}/_{32}$ − $^{9}/_{64}$ = _____

5. Multiply or divide the following fractions.

    a. ⅝ × 9 = _____

    b. $^{7}/_{16}$ × $^{5}/_{10}$ = _____

    c. $^{11}/_{12}$ ÷ ⅚ = _____

    d. (⅚ ÷ ¼) × ⅔ × ⅛ = _____

6. A well produces 640 barrels of oil and salt water per day. If three-eighths of the production is oil, and the lease owner receives one-eighth royalty on this production, how many barrels of oil will he or she receive a daily royalty on?

_____

7. Jones has a three-eighths interest in a lease, Smith owns three-sixteenths, and White owns seven-sixteenths. If they sell the lease for $75,000, how much will each receive?

Jones: _____

Smith: _____

White: _____

8. What is the difference in the diameter of two wire ropes that are ⅞ inches and $^{9}/_{16}$ inches in diameter?

    $^{5}/_{16}$"

_____

9. A storage tank is 7 feet, 6 inches high overall. The top plates are $^{3}/_{16}$ of an inch thick, and the bottom plates are ¼-inch thick. What is the inside height of the tank?

    7' 5 $^{9}/_{16}$"

_____

10. A machinist places a worn pump plunger that is $^{15}/_{16}$ of an inch in diameter in a lathe, and takes three cuts from it. On the first cut, $^{3}/_{16}$ of an inch is removed. On the second cut, $^{3}/_{64}$ of an inch is taken off, and on the third cut, ⅜ of an inch. What is the diameter of the plunger after these cuts?

    $^{21}/_{64}$"

_____

MIXED NUMBERS

A mixed number is a number composed of a whole number and a fraction. For example, $18\frac{3}{4}$, $272\frac{1}{2}$, and $5\frac{3}{4}$ are mixed numbers. A mixed number is read with the word *and* between the whole number and the fraction. For example, $18\frac{3}{4}$ is read as "eighteen and three-fourths."

Addition

To add mixed numbers, first change the fractions so that they have a common denominator. Next, add the whole numbers; finally, add the fractions.

Example Problem: Find the sum of $9\frac{1}{4}$ and $12\frac{1}{8}$.

Solution:

$$\begin{array}{rcr} 9\frac{1}{4} & = & 9\frac{2}{8} \\ +\ 12\frac{1}{8} & = & 12\frac{1}{8} \\ \hline & & 21\frac{3}{8}. \end{array}$$

Should the fractions add up to more than 1, the whole number is carried over to the ones column and added in as a whole number.

Example Problem: Add $13\frac{3}{8}$, $17\frac{1}{8}$, $8\frac{1}{2}$, $340\frac{1}{4}$.

Solution: First, set up the fractions with a common denominator and add the fractions. Then convert the sum of the fractions into its lowest terms.

$$\begin{array}{lcl} \frac{3}{8} & = & \frac{3}{8} \\ \frac{1}{8} & = & \frac{1}{8} \\ \frac{1}{2} & = & \frac{4}{8} \\ \frac{1}{4} & = & \frac{2}{8} \\ \hline & & \frac{10}{8} = 1\frac{2}{8} = 1\frac{1}{4}. \end{array}$$

Then add the whole numbers, including the $1\frac{1}{4}$ in the total.

$$\begin{array}{r} 1\frac{1}{4} \\ 13 \\ 17 \\ 8 \\ +\ 340 \\ \hline 379\frac{1}{4}. \end{array}$$

Subtraction

In subtracting one mixed number from another, consider the fractions and the whole numbers together. Consider the two together because the fraction in the subtrahend may be larger than the fraction of the minuend. In such a case, it is necessary to convert one unit in the whole number of the minuend into a fraction.

Example Problem: Subtract $3\frac{5}{8}$ from $12\frac{3}{4}$.

Solution: Change $\frac{5}{8}$ and $\frac{3}{4}$ so that both fractions have the same denominator; then subtract fractions and whole numbers.

$$\begin{array}{rcr} 12\frac{3}{4} & = & 12\frac{6}{8} \\ -\ 3\frac{5}{8} & = & -\ 3\frac{5}{8} \\ \hline & & 9\frac{1}{8}. \end{array}$$

Example Problem: Subtract 7⁵⁄₁₆ from 18⅛.

Solution: Find the LCD of the fractions.

⅛ = ²⁄₁₆

⁵⁄₁₆ = ⁵⁄₁₆.

Because ⁵⁄₁₆ cannot be taken from ²⁄₁₆, it is necessary to convert one unit of the whole number in the minuend into a fraction and add it to ²⁄₁₆. In this case, convert 1 from the 18 to ¹⁶⁄₁₆, which makes the number 17¹⁸⁄₁₆ (17¹⁸⁄₁₆ is the same as 18²⁄₁₆). Then

18²⁄₁₆ = 17¹⁸⁄₁₆ (minuend)

− 7⁵⁄₁₆ =   7⁵⁄₁₆ (subtrahend)

           10¹³⁄₁₆ (difference).

Multiplication

Often, it is necessary to multiply mixed numbers by whole numbers, fractions, or other mixed numbers. To multiply a mixed number, it has to be changed to an *improper fraction*. An improper fraction is a fraction in which the numerator is larger than the denominator.

To change a mixed number to an improper fraction, multiply the whole number by the denominator of the fraction and add the product to its numerator. Then, write this sum over the denominator to form the improper fraction. For example, to change the mixed number 3⅘ to an improper fraction, multiply the whole number 3 times the denominator 5 to get 15. Then, add the product 15 to the numerator 4, changing the numerator to 19. Thus, 3⅘ changes to the improper fraction ¹⁹⁄₅ and can be multiplied as a common fraction. To change an improper fraction back to a whole number or a mixed number, divide the numerator by the denominator. For example, to change ¹⁵⁄₃ to a whole number, divide 15 by 3 to get 5. To change ¹⁹⁄₅ back to a mixed number:

$$\begin{array}{r} 3\ r\,4 \\ 5\,\overline{)\,19} \end{array} = 3⅘.$$

Remember that whole numbers have a denominator of 1 when changed to a fraction. For example, the whole number 14 is the fraction ¹⁴⁄₁.

Example Problem: Multiply 18¼ times 6⅛.

Solution:

18¼ × 6⅛ = ?

18¼ = ⁷³⁄₄, and 6⅛ = ⁴⁹⁄₈.

Then, $\dfrac{73}{4} \times \dfrac{49}{8} = \dfrac{3{,}577}{32} = 111²⁵⁄₃₂.$

Example Problem: 18 × 6⅓ = ?

Solution: 6⅓ = $\dfrac{19}{3}$

$\dfrac{6}{1} \times \dfrac{19}{1} = 114$

or

$18 \times 6 = 108.$

$$\frac{18}{1} \times \frac{1}{3} = \frac{18}{3} = 6.$$

$108 + 6 = 114.$

Division

To divide mixed numbers, change the mixed numbers to improper fractions and divide them the same as with common fractions.

Example Problem: Divide 203⁷⁄₁₆ by 24⅛.

Solution:

203⁷⁄₁₆ = ³,²⁵⁵⁄₁₆

24⅛ = ¹⁹³⁄₈

$$\frac{3,255}{16} \div \frac{198}{3} = \frac{3,255}{16} \times \frac{8}{193} = \frac{26,040}{3,088} \doteq 8^{167}/_{386}.$$

Practice Problems

1. An employee's time on a job included 8¼, 5½, 9¾, 7¼, 6¾, and 2¼ hours. What was the total time spent?

    40 hr
   _____

2. Find the dimension of *A* in the sketch below. (Note: inches are abbreviated with the symbol ".")

   17 ¹⁵⁄₁₆"
   _____

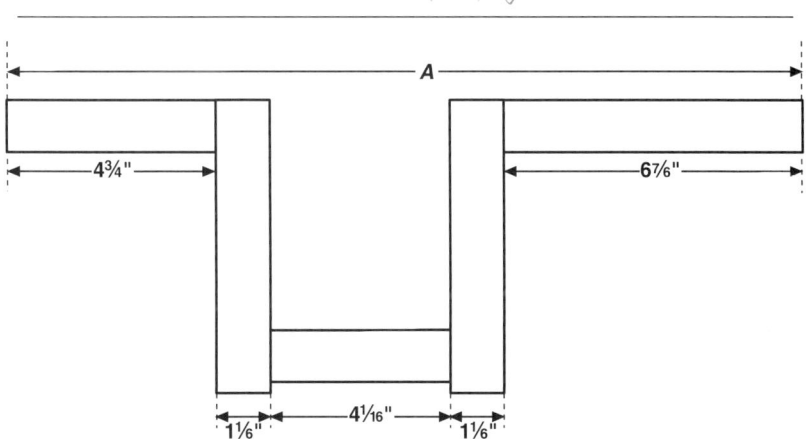

3. A worker cut off the following lengths of ¾-inch sucker rod for pins: 5³⁄₁₆, 4³⁄₃₂, 9⅛, and 2½ inches. What is the total length of sucker rod?

    20 ²⁹⁄₃₂"
   _____

4. A plant operates on a 40-hour week. The plant supervisor's weekly time sheets for June showed that he had worked 43½ hours, 42¼ hours, 45¾ hours, and 40¾ hours. How much overtime did the supervisor work during the month?

_____ 12¼ hrs _____

5. If the nominal inside diameter of a pipe is ⅝ of an inch and the outside diameter is ¹⁵⁄₁₆ of an inch, how thick is the pipe's wall? (Sketching a cross section will help you visualize this problem.)

_____ 5/16" _____

6. If an acetylene cutting machine is cutting at the rate of 1⅜ inches every 5 seconds, how far will it cut in 2¾ minutes?

_____ 45 ⅜" _____

7. On a production test, a well produced 103½ barrels of oil in 3 hours, 15 minutes. At this rate, what would the well produce in 24 hours?

_____ 764 ⁴⁄₁₃ bbl _____

8. A pipefitter is instructed to make a mark every 6⅞ inches, measuring from the end of a 6-foot pipe header. How many marks will he or she make on the header?

_____ 10 _____

9. A certain number of holes are spaced equally in a straight line. If the distance between the centers of the two end holes is 66¼ inches and if the spacing from center to center of two adjacent holes is 3⁵⁄₁₆ inches, how many holes are to be punched? (Note: be sure to count the beginning hole.)

_____ 21 _____

10. If 3 pints of chemical compound are required to treat 50 barrels of cut oil, how many barrels can be treated with 8¼ gallons of chemical? (Note: 8 pints = 1 gallon.)

_____ 1,100 bbl _____

DECIMAL FRACTIONS

Nowadays, we frequently encounter decimal numbers and fractions because calculators use decimal fractions. Once, operators measured casing, tubing, and drill pipe (tubulars) only in feet, inches, and fractions of inches. (In some cases, they still do because of established convention.) However, most operators today measure tubulars with measuring tapes divided into feet, tenths of a foot, and hundredths of a foot. So, when tallying (measuring the length of) a joint of casing, a joint could likely be recorded as being 39 feet, 1 and ²⁄₁₀ inches. But instead of using the fraction ²⁄₁₀, the person running the tally writes down 39 feet, 1.2 inches. In other words, ²⁄₁₀ is the same as 0.2.

What is more, rig personnel usually weigh drilling fluid with a mud balance that is calibrated in pounds per gallon and tenths of a pound per gallon. So, when derrickmen weigh mud, they do not write fractions; instead, they use decimal fractions. For example, if a mud weighs 9½ pounds per gallon (ppg), the derrickman records it as 9.5 ppg. The reason for using tenths and hundredths is that decimal fractions in any computation are easier to use and more precise than common fractions.

Converting common fractions to decimal fractions and vice versa is not difficult. A decimal fraction is a fraction whose denominator is 10 or some power of 10, such as 100 or 1,000. It may be written just as a common fraction, with the numerator over the denominator, as 6/10 or 93/100 (read as six tenths or ninety-three hundredths), or it may be written as 0.6 or 0.93. (When writing decimal fractions, convention calls for a zero to be placed in front of the decimal point. However, when reading them, normally, the zero is omitted.) To arrive at decimal fractions, omit the denominator and indicate the denominator by the number of places the numerator occupies to the right of the *decimal point*, which is a period (.) placed to the left of the first digit in the decimal fraction. In the case of a mixed number, the decimal point is placed between the whole number and the decimal fraction.

In reading decimal fractions, one figure to the right of the decimal point indicates the number of tenths, two figures to the right of the decimal point indicate hundredths, three figures indicate thousandths, four figures indicate ten thousandths, and so on. A mixed number may be read in two ways: with the word *and* between the whole number and fraction or the word *point* where the decimal appears. Thus, 6.25 is read "six and twenty-five hundredths" or as "six point two five."

To indicate decimal fractions of the proper denomination, the correct number of decimal places to the right of the decimal point is necessary. The same number of digits must be to the right of the decimal point as the number of zeros in the denominator if the decimal fraction is written as a common fraction. For example, to write 9/100 as a decimal, two digits must be placed to the right of the decimal point. In this example, the numerator has only one digit (9), so, place a zero between the decimal point and the number, which makes the decimal fraction 0.09.

Example Problem: Write the following decimal fractions in words: 0.25, 0.375, and 18.64.

Solution:

    0.25 = twenty-five hundredths
    0.375 = three hundred seventy-five thousandths
    18.64 = eighteen and sixty-four hundredths.

Because the U.S., Canada, and many other countries base their monetary systems on the decimal system, problems involving dollars and cents, pounds and pence, and so on are common decimal fraction problems. Instead of reading the fractional part of a dollar as so many hundredths, it is called cents. For example, $8.75 is read as "eight dollars and seventy-five cents," not as "eight and seventy-five hundredths dollars." Similarly, £3.50 is read as "three pounds and 50 pence," not as "three and fifty hundredths pounds."

Addition

To add or subtract decimal fractions or mixed decimals, arrange them in vertical columns and be certain to align the decimal points. Then, they can be added as whole numbers.

Example Problem: Add 875.3 and 6.05.

Solution: Write the numbers so that the decimal points fall in a vertical line.

$$
\begin{array}{r}
875.3 \\
+\ 6.05 \\
\hline
881.35.
\end{array}
$$

Note that the decimal point in the sum is directly below the decimal points in the two figures being added.

Example Problem: Add $35.42, $18.78, $19.25, and 35 cents.

Solution:

$$
\begin{array}{r}
\$35.42 \\
18.78 \\
19.25 \\
+\quad .35 \\
\hline
\$73.80.
\end{array}
$$

Subtraction

Subtracting decimal fractions is handled in the same manner as addition. Align the decimal points, and subtract the numbers as whole numbers. Sometimes, it is necessary to add zeros to the right of a fraction in the minuend to facilitate subtraction. Placing zeros to the right of a decimal fraction does not affect its value, just as placing zeros to the left of a whole number in no way affects the value of the whole number. For example, 0.27 and 0.270 have the same value, but 0.27 is its lowest term.

Example Problem: Subtract 125.873 from 340.27.

Solution:

$$
\begin{array}{r}
340.270 \\
-\ 125.873 \\
\hline
214.397.
\end{array}
$$

Multiplication

Multiplying decimal fractions, or mixed decimals, is the same as multiplying whole numbers. When the product has been obtained, locate the decimal point as many places from the right as there are decimal places in both the multiplicand and the multiplier. When multiplying decimal fractions, it is not necessary to align the decimal points.

Example Problem: Multiply 18.6 by 5.27.

Solution:

$$
\begin{array}{r}
18.6 \\
\times\ 5.27 \\
\hline
1,302 \\
372 \\
930 \\
\hline
98.022.
\end{array}
$$

Remember: because one decimal place is in the multiplicand and two are in the multiplier, the product contains three decimal places. In this case, the raw answer is 98022. But, a number with one decimal place is multiplied by a number with two decimal places. So, the answer has three decimal places and is 98.022. To determine where to place the decimal point, simply count from right to left the appropriate number of places.

Example Problem: Multiply 73.45 by 17.003.

Solution:

$$
\begin{array}{r}
73.45 \\
\times \quad 17.003 \\
\hline
22{,}035 \\
5{,}141{,}500 \\
7{,}345 \\
\hline
1{,}248.87035\,.
\end{array}
$$

In this problem, one number has two decimal places and the other three; so, the answer contains five decimal places. To place the decimal point, start at the answer's far right and count five numbers to the left.

Division

Decimals are divided the same as whole numbers but with one additional step, which involves the location of the decimal point in the quotient. Just as in multiplication, the decimal point must be accurately located for the answer to be correct. To locate the decimal in the quotient, subtract the number of decimal places in the divisor from those in the dividend. Also, relocate the decimal points before dividing.

Example Problem: Divide 0.875 by 0.5.

Solution:

$$
0.5\,\overline{)\,0.875} \;=\; 5\,\overline{)\,8.75}\,. \qquad \begin{array}{c}1.75\end{array}
$$

Move each decimal point one place to the right and eliminate the zeros, making the divisor the whole number 5 and the dividend 8.75. Place the decimal point in the quotient directly above the decimal point in the dividend before starting to divide. This action correctly positions the decimal point in the answer.

Example Problem: Divide 93.6404 by 3.71.

Solution:

$$
3.71\,\overline{)\,93.6404}\,.
$$

First, move the decimal point in the divisor, making it the whole number 371. Then move the decimal point in the dividend two places to the right, making it read 9,364.04. Place the decimal point in the quotient directly above the dividend and divide. The problem now reads

```
              25.24
     371 ) 9,364.04
             742
           1,944
           1,855
              890
              742
            1,484
            1,484.
```

When division results in remainders, using decimals makes them easy to handle. By simply adding zeros to the right of the decimal point in the dividend, division can be continued until no remainder is left, or it is so small that it becomes insignificant.

Example Problem: Divide 45 by 8.

Solution:

```
            5.625
     8 ) 45.000
          40
           50
           48
            20
            16
            40
            40.
```

In this case, carrying the division to three places makes the answer "come out even"—that is, this example's answer has no remainder after dividing to three decimal places.

If the division continues with a remainder after several decimal places have been recorded in the quotient, it may be desirable to round off the number. Rounding off an answer means to approximate the answer after a certain number of figures are known, rather than to complete the division until no remainder is left.

Example Problem: Find the quotient of 278 ÷ 34 to the nearest hundredth.

Solution:

```
             8.17
     34 ) 278.000
           272
            60
            34
           260
           238
            22.
```

Note that after finding two decimal places in the quotient, a remainder of 22 occurs. To determine the answer to the nearest hundredth, the division may be carried out to get another figure in the quotient. In this case, bring down the third zero, divide 34 into 220 to get 6, which is added to the quotient to make it 8.176. Since 8.176 is closer to 8.18 than it is to 8.17, the number is rounded off to 8.18.

A second method of determining the rounded off answer involves a short-cut. If the remainder 22 is more than half of the divisor 34, the quotient is rounded up to 8.18. If the remainder is less than half of the divisor, the answer stays 8.17.

Converting Fractions

For convenience in taking measurements, it may be necessary to change decimal fractions to common fractions, which can be read on a steel scale or a ruler. Likewise, the data needed to work problems might be given partly in common fractions and partly in decimal fractions, which makes it necessary to change the common fractions to decimal fractions.

Changing common fractions to decimals is finding their decimal equivalents. Table 1.5 gives decimal equivalents to common fractions frequently used in industry, including those used in the machine and general repair shops. Many machinists and mechanics memorize the equivalents they use regularly.

To change a decimal fraction to a common fraction, give the decimal its numerical denominator and then reduce to lowest terms.

Example Problem: Change 3.625 to a common fraction.

Solution:

$3.625 = 3\frac{625}{1,000} = 3\frac{5}{8}$.

Because decimal fractions are based on 10, 100, 1,000, etc., and because scales are graduated in ¼, ⅛, 1/16, etc., a decimal fraction often cannot be changed to a common fraction on the scale. Usually, however, the decimal can be changed to a common fraction close to one on the scale. In changing a decimal fraction to an approximate common fraction, multiply the decimal by the denominator of the fraction to which it is to be converted. The product is the numerator of the new fraction.

Example Problem: Convert 0.94 to the nearest sixteenth.

Solution:

$0.94 \times 16 = 15.04$.

Discard the .04 and write 15 as the numerator of the new fraction, making it 15/16. Thus, $0.94 \cong 15/16$. (The operational sign for approximately equals is $\cong$.) To confirm the approximation, divide 15 by 16. The result is 0.9375, which is pretty close.

As you can see, changing a common fraction to a decimal fraction is simple. Merely set down the numerator and divide it by the denominator.

Example Problem: Express 9/16 as a decimal fraction.

Solution:

```
         .5625
  16 ) 9.0000
       80
      ───────
       100
        96
      ───────
        40
        32
      ───────
        80.
```

TABLE 1.5
Decimal Equivalents

Fraction	Decimal Equivalent	Fraction	Decimal Equivalent
1/64	0.015625	33/64	0.515625
1/32	0.03125	17/32	0.53125
3/64	0.046875	35/64	0.546875
1/16	0.0625	9/16	0.5625
5/64	0.078125	37/64	0.578125
3/32	0.09375	19/32	0.59375
7/64	0.109375	39/64	0.609375
1/8	**0.125**	5/8	**0.625**
9/64	0.140625	41/64	0.640625
5/32	0.15625	21/32	0.65625
11/64	0.171875	43/64	0.671875
3/16	0.1875	11/16	0.6875
13/64	0.203125	45/64	0.703125
7/32	0.21875	23/32	0.71875
15/64	0.234375	47/64	0.734375
1/4	**0.25**	3/4	**0.75**
17/64	0.265625	49/64	0.765625
9/32	0.28125	25/32	0.78125
19/64	0.296875	51/64	0.796875
5/16	0.3125	13/16	0.8125
21/64	0.328125	53/64	0.828125
11/32	0.34375	27/32	0.84375
23/64	0.359375	55/64	0.859375
3/8	**0.375**	7/8	**0.875**
25/64	0.390625	57/64	0.890625
13/32	0.40625	29/32	0.90625
27/64	0.421875	59/64	0.921875
7/16	0.4375	15/16	0.9375
29/64	0.453125	61/64	0.953125
15/32	0.46875	31/32	0.96875
31/64	0.484375	63/64	0.984375
1/2	**0.5**	1	**1**

Practice Problems

1. Express the following numbers in words:

   a. 0.003 _____three thousandths_____

   b. 0.625 _____Six hundred and twenty five thousandths_____

   c. 120.04 _____One hundred twenty and four hundredths_____

   d. 8.3745 _____

2. Express the following fractions as decimal fractions and as common fractions:

   a. one hundred fifteen thousandths _____0.115   115/100_____

   b. seventy-six and seventy-six hundredths _____76.76   76 76/100 or 76 19/25_____

   c. five thousand and six tenths _____5000.6 , 5000 6/10 , 5000 3/5_____

   d. three thousand, one hundred twenty-five and eight tenths

   3125.8 , 3125 8/10 , 3125 4/5 _____

3. George received $316.89 on payday. He plans to pay the following bills out of his check: telephone, $24.50; charge card, $35.00; gas, $41.23. How much will he have left?

   _____$ 216.16_____

4. A contractor receives $1,543.75 for a job. His expenses on the job are: labor, $375.35; transportation, $53.65; materials, $544.60. How much is his profit?

   _____$570.15_____

5. Susan bought $10 worth of gasoline at $1.12 per gallon. How many gallons did she get?

   _____8.928   gal_____

6. Electric motors are going to be used to pump eighteen shallow wells. A test shows that 2.36 horsepower is required for each well. How much power will be required for pumping all of the wells at once?

   _____42.48 hp_____

7. One gallon of 45°API gravity crude oil weighs 6.675 pounds. What does 8,000 gallons of this oil weigh?

   _____53,400 lb_____

8. A group of fifteen employees decide to donate an equal amount each for the purchase of a microwave oven for their lunchroom. The price of the oven is $329.00 plus a tax of $16.45. How much will each person pay?

   _____$ 23.03_____

9. The depth of a groove in a wire-rope sheave should be 1.75 times the nominal diameter of the wire rope. How deep should the groove be for a ⅞-inch line?

   _____1.53 in_____

10. The distance across the flats (*W* in the following drawing) of a machine bolt is always equal to one and one half times the diameter of the bolt plus ⅛ inch. Using decimals, find the dimension of *W* in the drawing.

1.25 inch
_____

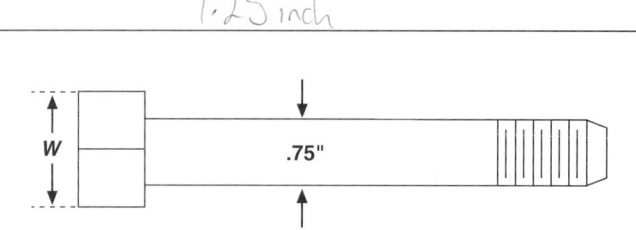

ROOTS AND POWERS

The *root*, or base, of a number or quantity is one of the equal factors which, when multiplied together, produce a given number or quantity. For example, the root of 4 is 2, because 2 × 2 = 4. However, a root number can be multiplied by itself any number of times. The number of times the root number is multiplied by itself is called the *power* and is indicated by an *exponent*. An exponent is the small number written to the right and a little above the root number or quantity. For example, the number 8^3 is read as "eight to the third power" or "eight cubed." The number 8 is the base, or root, number and 3 is the exponent, or power. Raising a number to a power means that the number is multiplied by itself the number of times indicated by the exponent. So, the value of 8^3 is determined by multiplying 8 by itself 3 times. Thus, 8 × 8 × 8 = 512. Another way of stating it is "eight cubed (or eight to the third power) is 512."

Example Problem: What is the value of 2^6 (two to the sixth power or 2 multiplied by itself six times)?

Solution: 2 × 2 × 2 × 2 × 2 × 2 = 64.

Squares and Square Roots

The product of a number multiplied by itself is the *square* of that number. Thus, the square of 10 (written as 10^2, and read as "ten squared") is 10 × 10, or 100. Conversely, the *square root* of a number is the number that when multiplied by itself gives the stated number. For example, the square root of 4 is 2 because 2 × 2 = 4. And, 2^2 is 4. In the same way, the square root of 100 is 10 because 10 × 10 = 100.

The common way to indicate that the square root of a quantity is to be found is to precede the quantity with a radical sign, $\sqrt{}$, and extend a line above all the numbers for which the square root is to be found. For example, the square root of 3 times 12 is written:

$$\sqrt{3 \times 12}.$$

The solution to the problem is:

$$\sqrt{3 \times 12} = \sqrt{36} = 6.$$

Note that the square root of 36 is 6 because 6 × 6 = 36.

Another way to indicate square root is to enclose the quantity in parentheses and use the exponent ½. Thus, $\sqrt{100}$ and $(100)^{1/2}$ both mean "the square root of 100."

Many math problems require extracting square roots of numbers. Examples include finding the radius of a circle when the area is known and finding the length of one side of a right triangle when the other sides are known. (The chapter on geometry covers these problems.)

Example Problem: Find the square root of 1,806.25.

Solution:

$$
\begin{array}{r}
4\ \ 2.\ 5 \\
\sqrt{18\ \ 06.25} \\
16 \\
\hline
80\,)\ \overline{2\ 06} \\
2\ \ \ 1\ 64 \\
\hline
82 \\
840\,)\ \overline{42\ 25} \\
5\ \ \ 42\ 25 \\
\hline
845.
\end{array}
$$

- First, divide the number into groups of two figures each, counting in each direction from the decimal point. In this case, 18 and 06 form a group of two, as does the 25 after the decimal point. The root has as many digits as the groups of numbers in the original number. In this case, the root has three numbers: two whole numbers and one decimal.

- Second, find the largest number which, when squared, is less than or equal to the first left-hand group. In this case, the number is 4 because 4^2 is 16.

- Third, record this number (4) above the first group. It is the first number in the root.

- Fourth, square the number 4 in the root and subtract the product 16 from 18, just as in long division.

- Fifth, bring down the remainder 2 and the next group (06) for a new dividend.

- Sixth, multiply the number 4 in the root by 20 (a trial number) and use the product 80 as a trial divisor.

- Seventh, find the number of times the trial divisor 80 is contained in the dividend 206, and place this number (2) as the second number in the root.

- Eighth, multiply the divisor thus obtained (82) by the second digit in the root (2) and subtract the product 164 from the dividend 206.

- Ninth, bring down the next group of figures (25) beside the remainder of 42 for a new dividend 4,225.

- Tenth, multiply the two figures in the root (42) by 20 and use the product 840 as a trial divisor.

- Eleventh, add the number of times the trial divisor of 840 is contained in the dividend 4,225 to the trial divisor, giving a divisor of 845.

- Twelfth, record the quotient 5 as the third digit in the root and multiply by the divisor 845, and obtain the product of 4,225. Subtraction leaves no remainder, so the square root of 1,806.25 is 42.5.

To check the results, simply multiply 42.5 times 42.5.

$$
\begin{array}{r}
42.5 \\
\times\ 42.5 \\
\hline
2{,}125 \\
850 \\
1{,}700 \\
\hline
1{,}806.25. \\
\end{array}
$$

You can see that working out the square root of a number on paper is not very easy. Using a calculator that can solve square roots greatly simplifies the operation. On such a calculator, you simply enter the number you wish the square root of, press the square root key, and the calculator gives it to you. However, to adhere to one of the objectives in this chapter—solving problems without a calculator—the preceding example is given.

Other Powers

A number or symbol may be raised to any power. For example, 5^4 (read as "five to the fourth power") means

$$5 \times 5 \times 5 \times 5, \text{ or } 625.$$

The fourth root of 625 is 5. The square root of 625 is 25 because 25^2, or 25×25, is 625. Letters or other symbols are often used to represent the root quantity, especially in algebraic equations such as

$$a^2 + b^2 = c^2.$$

Example Problem: $23^3 + 5^5 = ?$

Solution: Solve each expression, then add together.

$$
\begin{array}{rrr}
23 & 5 & 12{,}167 \\
\times\ 23 & \times\ 5 & +\ 3{,}125 \\
\hline
69 & 25 & 15{,}292 \\
46 & \times\ 5 & \\
\hline
529 & 125 & \\
\times\ 23 & \times\ 5 & \\
\hline
1{,}587 & 625 & \\
1{,}058 & \times\ 5 & \\
\hline
12{,}167 & 1{,}325 & \text{The answer is } 15{,}292.\\
\end{array}
$$

A simpler way to multiply like base numbers with exponents is to add the exponents of the numbers and solve the final term. For example:

$$10^5 \times 10^2 = 10^7 = 10{,}000{,}000.$$

Notice that 10^7 is 10 followed by six zeros. Similarly, 10^3 is 10 followed by two zeros, or 1,000 ($10 \times 10 \times 10 = 1{,}000$). Being aware of this relationship helps when calculating large numbers.

To divide like base numbers with exponents, subtract the exponents. For example:

$$85^4 \div 85^3 = 85^1 = 85.$$

Logarithms

Mathematics also involves the use of special exponents and exponential functions called logarithms, antilogarithms, and natural logarithms. Details about such exponents are beyond the scope of this text. However, in general, a *logarithm* (commonly called *log*) of a number is the exponent of the base 10 that makes the 10 equal to that number. For example, 100 equals 10^2, making 2 the logarithm of 100 (written as log 100 = 2). Not as simple is the log of 532.

$$\log 532 = 2.7259$$

since

$$532 = 10^{2.7259}.$$

Tables are available to find the log of such numbers. Also, some calculators offer the log function—you simply enter the number you wish the log of, press the log key, and the calculator reveals the number.

Reversing the log operation gives the *antilogarithm*. For example, the antilogarithm of 2.7259 is 532. Another kind of logarithm is the *natural logarithm*. The natural logarithm is so-called because it comes about naturally in a mathematical process. The base of the system of natural logarithms is designated by *e*, and its numerical value is approximately 2.71828.

Although logarithms, antilogarithms, and natural logarithms are advanced concepts, they are vitally important in many applications of higher mathematics. For one thing, logarithms make calculations faster. For another, logarithms are needed when solving problems with numbers raised to unknown powers.

Practice Problems

1. Find the square root of 784.

2. What is 14^3?

3. A square plot of ground measures 6 miles on each side. If the area of this plot is equal to one of its sides squared, how many square miles are contained in the plot of ground?

4. Extract the square root of 10.24.

5. $62^2 + 10^4 - \sqrt{25} = ?$

SELF-TEST

1. The Number System

Multiply each question or problem answered correctly by five to arrive at your percentage of competency.

1. In 1961, the estimated world population was two billion, two hundred fifty million. In 1982, it was four billion, four hundred ninety-two million. How much did the population grow in this 21-year period?

   2,242,000,000

2. If an operator owned a 3/8 interest in a lease and later sold 1/8 interest, what is his remaining interest?

   1/4

3. Three metal sheets are 3/16, 3/8, and 9/32 inches in thickness. What is their total thickness?

   27/32

4. Ignoring the collar diameter, how much clearance exists between a string of 7-inch casing with an inside diameter of 6¼ inches and a string of seamless tubing with an outside diameter of 2⅛ inches?

   4 1/8

5. Find the outside diameter of the pipe shown in the sketch below. (Note: the symbol ' is sometimes used to indicate feet and the symbol " is sometimes used to indicate inches.)

   27/32

6. A lease produced 7,952 barrels of oil in a 30-day period. The oil sold for $20 a barrel. The royalty owner received 1/8 and the operator 7/8. How much money did the royalty owner receive for the period? ________

   $19,880

   The operator? $139,160

7. A simplex, double-acting pump pumps 7/8 of a gallon per stroke. At 35 strokes per minute, how many gallons does it pump per hour?

   1,837 gal/hr

   Per day? 44,100 gal/day

8. Convert the following common fractions to four-place decimal fractions and the decimal fractions to common fractions of lowest terms:

   a. 9/16 = 0.5625

   b. 3/32 = 0.0938

c. $^{31}/_{64}$ = _____ 0.4844

d. 2.65625 = _____ 2 $^{21}/_{32}$

e. 0.0625 = _____ $^{1}/_{16}$.

9. A trucker has to deliver a load in four days. His destination is 1,832 miles away. He travels 450.6, 520.8, and 250.3 miles in the first three days. How far does he have to travel the fourth day to meet his delivery date?

_____ 610.3 miles

10. A pump delivers 0.635 gallons of water at each stroke, and it operates at 28 strokes per minute. If 42 gallons equals 1 barrel, how many barrels will it pump in 6 hours?

_____ 152.4 bbl

11. Calculate the weight of the following shipment of material:

a. 12 stiffeners, each of which is 10'0" long and weighs 2.22 pounds per foot.

_____ 266.4 lb

b. 10 stiffeners, each of which is 8'0" long and weighs 2.22 pounds per foot.

_____ 177.6 lb

12. The electric motor on a centrifugal pump uses electricity at the rate of 1.65 kilowatts. The cost of electricity for the pump is 5.33 cents per kilowatt-hour. What is the cost of electricity to operate the pump for 24 hours?

_____ $ 2.11

13. Discounting waste, how many pins 2.75" long can be cut from a ½" bar that is 14¾' long?

_____ 64 pins

14. The cylinders of a gasoline engine are 4¾" in diameter, and the manufacturer recommends that pistons be fitted with a clearance of 0.0035". What diameter should the pistons for this engine be? (Express the answer in inches and carry it out to four decimal places.)

_____ 4.7430 inches

15. A lease is sold for $380,000, one-half of which is paid in cash. One-fourth of the production is to be applied to the remainder of the note. Oil sells at $29.00 per barrel, and the lease has an allowable of 64 barrels per day. (Round off answers to the nearest whole numbers.)

a. How much is still owed on the note? _____ $ 190,000

b. How many barrels must be sold to equal the balance owed? _____ 6,552 bbl

c. How many barrels must be produced on the lease to pay off the note? _____ 26,207 bbl

d. What amount of the daily revenue from the lease will be applied to the note? _____ $ 464

e. How many days will it take to pay off the note? _____ 409 or 410 days

16. A drilling crew adds 2¼ pounds of CMC per barrel of mud in order to treat it. How many barrels can be treated with twelve 100-pound sacks of CMC?

_____ 533 ⅓ bbl

17. Solve the following problems:

  a. $3^3 - 4^2 =$ _____ 11

  b. $\sqrt{36} \times 58 =$ _____ 348

  c. $92^4 =$ _____ 71,639,296

  d. $\sqrt{42.27} - 2 =$ _____ 6.346

  e. $10^6 \div 10^2 =$ _____ 10,000

18. *API Specification (Spec) 5L for Line Pipe* shows pipe dimensions for standard threaded line pipe. It lists the following outside and inside diameters for such pipe. What are the wall thicknesses of each size of pipe?

	Nominal Pipe Size (Inches)	Outside Diameter	Inside Diameter	Wall Thickness
a.	½	0.840	0.622	0.109 in
b.	1	1.315	1.049	0.133 in
c.	2	2.375	2.067	0.154 in
d.	2½	2.875	2.469	0.203 in
e.	3	3.500	3.068	0.216 in

19. The recoverable butane content of a certain gas is 0.75 of a gallon per 1,000 cubic feet (0.75 gal/Mcf), the throughput of the plant is 15,000,000 cubic feet per day (15 MMcf/day), and the wholesale price of the butane is 60.9 cents per gallon.

  a. How many gallons of butane will be recovered in 24 hours? _____ 11,250 gal

  b. How many gallons will be recovered in three weeks? _____ 236,250 gal

  c. How many gallons will be recovered in thirty days? _____ 337,500 gal

  d. What is the value of the daily production? _____ $ 6,851.25

  e. What is the value of the annual production? _____ $ 2,500,706.25

20. The following problems are designed for practice in mental arithmetic. See if you can figure out the answers without writing anything down.

  a. If the daily production from a well is 150 barrels of fluid (oil and water) of which 47.9 barrels is oil, the amount of water produced in 30 days is (select one): 1,437; 2,053; 3,063; or 7,185 barrels.

  _____ 3,063

  b. Twenty-five pounds of red brass at 34 cents per pound costs: $40.00; $8.50; $85.00; $15.50.

  _____ $ 8.50

  c. The decimal equivalent of 11/4 is: 2.65; 2.55; 2.75; 2.25.

  _____ 2.75

  d. The time and one-half rate for a person whose normal hourly pay is $6.18 is: $13.09; $12.27; $8.36; or $9.27 per hour.

  _____ $ 9.27

  e. If 0.064 barrels of liquid are removed from each thousand cubic feet (Mcf) of gas at a gasoline plant, then the plant removes 640; 6,400; 640,000; or 64,000 barrels from 100 MMcf.

  _____ 6,400

2
THE CALCULATOR

OBJECTIVES

Upon completing chapter two, the student will be able to—

1. Choose a calculator that matches calculation needs with calculator design.
2. Explain the differences and similarities of the three types of notation commonly used in calculators.
3. Describe the functions of commonly used keys on a calculator.
4. Solve addition, subtraction, multiplication, and division problems on a calculator.
5. Solve square root and percentage problems using a calculator.
6. Perform chain calculations using the memory function of a calculator.

INTRODUCTION

Desktop and handheld, or pocket, calculators represent a major technological advancement. However, calculators are so common that few people realize just how technologically advanced they are. What is more, simple handheld calculators are so inexpensive that companies often give them away for promotional purposes. Further, those who have access to the World Wide Web can download various kinds of calculators to use on their computer screens.

The earliest calculators offered four basic functions: addition, subtraction, multiplication, and division. Later, manufacturers added percentage and square-root functions. Today's sophisticated models not only include trigonometric, logarithmic, and graphing functions, but also include preprogrammed scientific operations, which rapidly calculate problems that would require many steps on a four-function calculator.

A calculator eliminates the tedious aspects of carrying out calculations. For example, you may recall that working the square-root problem in Chapter 1 was a long, drawn-out process. But, to obtain the root on a calculator with a square-root function, merely enter the number you wish to find the square root of and press the square-root function key. The calculator displays the square root in a fraction of the time required to work it out on paper.

Calculators are valuable tools for solving mathematical problems related to many industrial operations. For example, in oilwell drilling operations, high-pressure fluids from a formation—a kick—may enter the wellbore. Crewmembers must quickly recognize and control a kick to prevent the well from blowing out. Supervisory personnel on rigs can use calculators to quickly determine actions needed to control the well. For example, they can calculate the new mud weight required to contain the pressure, determine pump circulating pressures, and find

39

the number of strokes of the mud pump required to get the mud from the surface to the bit, all of which are vital to controlling the well.

Pipeline construction workers may use calculators to figure pipe buoyancy, while refinery employees may use them to determine heat transfer and material balance. Electricians may use a calculator to determine total load amperes, voltage drops in wiring, or power consumption. In virtually every industry, personnel use calculators for estimating the cost of materials and labor for a job.

CHOOSING A CALCULATOR

Calculators are available in a wide variety of capabilities and prices. The least expensive models not only perform addition, subtraction, multiplication, and division, but also percentages and square roots. Most also have a memory. The user can store a set of numbers in the calculator's memory and retrieve it later without having to reenter the numbers.

For example, suppose you need to convert several measurements in feet to metres. To convert feet to metres, you multiply the number of feet by 0.3048. By entering 0.3048 into the calculator's memory, you may recall this conversion factor by pressing a single key rather than entering 0.3048 each time you wish to make the conversion. Also, when solving complicated equations, you can store part of the solved equation in the calculator's memory. Then, you can retrieve the partial solution later when it is required to solve the entire problem. Later, this chapter covers both these uses of a calculator's memory.

More expensive models offer trigonometric and logarithmic functions, among other things. Also, advanced models are equipped with more than one level of memory. That is, you can store parts of a calculation in more than one place. Although this chapter does not cover it because programming is beyond the scope of this manual, some calculators are programmable: you can load special mathematical operations into the calculator, which allow it to perform advanced functions.

Calculator Features

When choosing a calculator, it is important to match your calculation needs with the calculator's capabilities. If you anticipate solving involved calculations, then consider buying a sophisticated model. On the other hand, if most calculations are little more than solving addition, subtraction, multiplication, and division problems, then a simple, inexpensive model is adequate.

Most of today's handheld calculators are battery operated or have a power cell that operates the calculator when it is struck by light. Manufacturers often call such calculators solar powered, but sunlight is not needed to operate them. Ordinary indoor or outdoor light is adequate. If batteries operate the calculator, it may come with an AC adapter-charger, which not only powers the calculator, but also charges the batteries when the calculator is plugged in. Desktop calculators may plug into a normal electrical outlet.

Keyboards have the usual operational symbols of $+$, $-$, $\times$, $\div$, and $=$. Advanced models contain other symbols, such as %, $\sqrt{}$, x^2, y^x, log, and sin. (This chapter discusses these keys shortly.) The keyboard also has a period key (.), which is a decimal point. The number of symbols and characters vary with the complexity and price of the calculator. The manufacturer may print the numbers and symbols directly on the keys, on the case near the keys, or on both. In this chapter, reference to a key is by its symbol regardless of its printed location. Figure 2.1 shows typical calculator keyboards.

Small calculators display numbers and other entries in a window usually located at the top of the calculator. Light emitting diodes (LEDs) or, more often in today's models, liquid crystal displays (LCDs) show the numbers and symbols in the window. (Some desktop calculators also print out characters on a paper tape.) The display should be easy to read—that is, the size, color, intensity, and visibility of the symbols and numbers should be readable in sunlight as well as in artificial light.

Figure 2.1   Typical calculator keyboards

The calculator you use with this text should be able to show eight or more digits in the display and should have—

1. *number keys*, which include the ten digits of

$$\boxed{0}\ \boxed{1}\ \boxed{2}\ \boxed{3}\ \boxed{4}\ \boxed{5}\ \boxed{6}\ \boxed{7}\ \boxed{8}\ \boxed{9}$$

2. *operation keys* for the four basic arithmetic operations, which are

$$\boxed{+}\ \boxed{-}\ \boxed{\times}\ \boxed{\div}$$

3. a *decimal* $\boxed{\cdot}$ *key*

4. an *equals* $\boxed{=}$ *key* or, with calculators that use reverse Polish notation, an ENTER $\boxed{\text{ENT}}$ or EXECUTE $\boxed{\text{EXE}}$ key. (Reverse Polish notation is discussed shortly.)

Most calculators also offer square root $\boxed{\sqrt{}}$ and percent $\boxed{\%}$ keys, which are useful not only for solving many common problems, but also for the exercises in this text. If you choose a scientific calculator, it should have at least the following keys:

$$\boxed{y^x}\ \boxed{x^2}\ \boxed{(}\ \boxed{)}\ \boxed{x^y}\ \boxed{\text{SIN}}\ \boxed{\text{COS}}\ \boxed{\text{TAN}}\ \boxed{+/-}\ \text{or}\ \boxed{\text{CHS}}\ \boxed{\text{LOG}}\ \boxed{\text{LN}}\ \text{or}\ \boxed{\text{lN}}\ \boxed{1/x}$$

You will learn about these keys shortly.

Calculator Notation

A major consideration in selecting a calculator is the notation, or logic, the calculator uses to perform its work. Notation is a system of characters, symbols, and expressions used to represent mathematical operations. Notation is sometimes called logic because logic, in this sense of the term, refers to the way in which the calculator operates to perform its functions.

Generally, calculators use three types of notation, or logic: arithmetic, algebraic, and reverse Polish (pronounced poe-lish). Reverse Polish notation is so called because it evolved from work the Polish mathematician Jan Lukasiewicz did in the 1920s. His notation scheme became known as Polish notation. Then, an Australian computer scientist, Charles Hamblin, refined Lukasiewicz's work in the late 1950s and, because Hamblin reversed the order in which Polish notation was written, he called it reverse Polish notation and abbreviated it as RPN. Polish and reverse Polish notation is particularly useful in calculators and computers because of the manner in which these devices perform their mathematical operations.

Arithmetic Notation

The arithmetic notation calculator has a simple keyboard and is easy to use. Arithmetic calculators are among the least expensive and are often given away as promotional gifts. An arithmetic calculator does addition, subtraction, multiplication, and division in the same order in which the keys are pressed. (This book shows the pressed keys in boxes; the final answers are not boxed.) So, pressing the keys in the order of 4 + 3 × 5 = results in the answer 35 appearing on the display and is shown as

$$\boxed{4}\ \boxed{+}\ \boxed{3}\ \boxed{\times}\ \boxed{5}\ \boxed{=}\ 35.$$

Figure 2.2 shows the steps in graphic form.

First, enter 4 by pressing the 4 key on the calculator. The number 4 appears in the display. Then, press the plus (+) key; the number 4 remains in the display until you press the 3 key, at which time 3 shows on the display. Next, press the times key (×) and 7 is displayed. Finally, press 5 and the equals (=) key to display the answer of 35. As previously mentioned, besides the four basic operations, arithmetic calculators may also have percentage, square root, and memory functions.

ENTRY	DISPLAY
4	4
+	4
3	3
×	7
5	5
=	35

Figure 2.2   Entries and displays on an arithmetic logic calculator

Algebraic Notation

A calculator with algebraic notation operates the same as an arithmetic notation calculator except that it performs operations in a specific order. An algebraic calculator does powers first, which are usually entered in the calculator by pressing a key labeled x^y. Then, such calculators perform multiplication (¥), division (÷), and finally addition (+) and subtraction (−). As you will learn in Chapter 5, algebra problems are also computed in this order. For example, to compute

$$2 + 4 \times 3^2$$

press the keys in the order of:

$$\boxed{2}\ \boxed{+}\ \boxed{4}\ \boxed{\times}\ \boxed{3}\ \boxed{x^2}\ \boxed{=}\ 38.$$

When the $\boxed{=}$ key is pressed, the calculator does its work, but it does not perform the operations in the order entered. Instead, the calculator first computes 3 to the second power, producing 9. Next, it multiplies the 9 times 4 and then adds the resulting 36 to the 2 to get the answer, 38. Figure 2.3 shows the entries and displays in graphic form.

ENTRY	DISPLAY
2	2
+	2
4	4
×	4
3	3
x^2	9
=	38

Figure 2.3   Entries and displays on an algebraic notation calculator

Calculators with algebraic notation may be inexpensive, medium-priced, or expensive, depending on their capabilities. In addition to the four basic operations, percentage, and square root, many medium-priced and more expensive models can square a displayed number, calculate the inverse of the displayed number, perform operations within parentheses first so that the contents are treated as one unit, and change the sign of a displayed number.

Some algebraic calculators have a memory system that can store several pieces of information rather than just one. To store several entries, the operator usually presses location keys after the memory storage key. Information can be stored and recalled at any time. Calculators with several advanced keys and memory are often called scientific calculators.

Reverse Polish Notation

An enter, or execute, key characterizes a reverse Polish calculator—that is, it has a key labeled [ENTER] [ENT] [EXECUTE] or [EXE]. Although using an RPN calculator takes practice, once it is mastered, it is versatile, dependable, and reduces the number of key strokes required for many calculations.

RPN calculators work with a memory stack, which can be thought of as a ladder with several rungs. The display represents the bottom rung of the ladder, and each rung above the display is a memory level capable of holding one piece of information. When the [ENTER] key is pressed, the displayed number climbs one rung up the ladder and also pushes whatever else is on the ladder up one rung. With RPN calculators, the numbers that are keyed in first are worked on last. When a number is entered, the memory stack stores it for later use. For example, to solve the problem

$$5 + (2 \times 4) = ?$$

press the following calculator keys:

[5] [ENTER] [2] [ENTER] [4] [×] [+] 13.

Figure 2.4 shows the entries, displays, and the content that is in the memory of a reverse Polish (RPN) calculator for the preceding problem.

RPN-Algebraic Combination

Some calculators combine RPN and algebraic notation. Many of these combination calculators also feature a large display screen on which users can display graphs. Thus, they are usually called graphing calculators. With a graphing calculator, to solve the problem

$$5 + (2 \times 4) = ?$$

press the keys in the following order:

[5] [+] [(] [2] [×] [4] [)] [ENTER].

When enter is pressed, the answer of 13 is displayed. Figure 2.5 shows the sequence. Notice that the display not only shows the number pressed, but also graphically shows the problem as the numbers and symbols are put into the calculator. This graphic display allows users to check the entries and ensure that they are correct. If an entry is wrong, cursor keys on the calculator allow users to enter the display, remove the wrong entry, and key in the correction.

If the problem is stated

$$(5 + 2) \times 4 = ?$$

the calculator sequence is

[(] [5] [+] [2] [)] [×] [4] [ENTER] 28.

ENTRY	DISPLAY	MEMORY
5	5	
ENTER	5	5
2	2	5
ENTER	2	2 5
4	4	2 5
×	8	5
+	13	

Figure 2.4 Entries, displays, and memory in an RPN calculator

ENTRY	DISPLAY
5	5
+	5 +
(	5 + (
2	5 + (2
×	5 + (2 ×
4	5 + (2 × 4
)	5 + (2 × 4)
ENTER	5 + (2 × 4)
=	13

Figure 2.5 Entries and displays in a graphing calculator

Notation Differences

To illustrate the differences in the arithmetic, algebraic, and reverse Polish notation, review the following problem and consider the calculator's interpretation of the keystrokes when entered in this order:

$$2 + 3 \times 4 =$$

The arithmetic logic calculator interprets the keystrokes as

$$(2 + 3) \times 4 = 20.$$

The algebraic logic calculator performs the multiplication first and works the problem as

$$2 + (3 \times 4) = 14.$$

The RPN calculator does not work with this sequence of keystrokes. For the arithmetic interpretation, the problem is entered as

$$\boxed{2}\ \boxed{\text{ENTER}}\ \boxed{3}\ \boxed{+}\ \boxed{4}\ \boxed{\times}$$

and for the algebraic interpretation on an RPN calculator, the problem is entered as

$$\boxed{2}\ \boxed{\text{ENTER}}\ \boxed{3}\ \boxed{\text{ENTER}}\ \boxed{4}\ \boxed{\times}\ \boxed{+}.$$

Computer programmers were among the first to recognize the advantages of reverse Polish notation. The system scans an expression from left to right and therefore carries out the calculation immediately as it is encountered. This action means that less storage space is required on a calculator's memory chip. Also, fewer keystrokes are needed. Unlike algebraic logic, where the operation signs are sandwiched between the numbers, in reverse Polish they are generally entered last and are not stored. Not having to store the operation signs means less storage space is needed.

Many companies manufacturing small, nonscientific calculators use arithmetic logic with single-memory storage. Hewlett-Packard was the first manufacturer to use reverse Polish notation in scientific calculators and continues to produce them, as do Texas Instruments, Casio, and others. Texas Instruments, Casio, Sharp, and other manufacturers also make graphing calculators as well as algebraic calculators.

Writing Problems

The way a calculator interprets a problem brings up an important point in mathematics: the problem must be written properly to obtain the desired results. Consider the problem $2 + 3 \times 4$. Parentheses must be used before the problem can be correctly solved—that is, the correct result cannot be obtained unless the parentheses are placed in the correct position. The correct answer is 20 if the problem is stated as $(2 + 3) \times 4$; however, the correct answer is 14 if the problem is stated as $2 + (3 \times 4)$.

Note, too, that the way you enter the problem depends on the type of calculator in use. For example, if the problem is stated as $(2 + 3) \times 4$ and you are using an algebraic calculator, you must enter the problem as

$$\boxed{(}\ \boxed{2}\ \boxed{+}\ \boxed{3}\ \boxed{)}\ \boxed{\times}\ \boxed{4}\ \boxed{=}\ 20.$$

With an arithmetic calculator you enter

$$\boxed{2}\boxed{+}\boxed{3}\boxed{\times}\boxed{4}\boxed{=} 20.$$

On the other hand, if the problem is stated as 2 + (3 × 4) and you are using an algebraic calculator, you must enter

$$\boxed{2}\boxed{+}\boxed{(}\boxed{3}\boxed{\times}\boxed{4}\boxed{)}\boxed{=} 14.$$

However, with an arithmetic calculator, enter the problem (2 + 3) × 4 as

$$\boxed{2}\boxed{+}\boxed{3}\boxed{\times}\boxed{4}\boxed{=} 14.$$

From the preceding examples, you can see that it is vital to be familiar with your calculator and know how to properly enter problems into it. With practice, proper calculator operation becomes easy.

UNDERSTANDING CALCULATOR FUNCTIONS

In using a calculator, two important considerations are (1) how to enter the numbers into the calculator and (2) how to tell the calculator which operation to perform. Successful operation of any calculator depends on your reading and following the instructions provided by the manufacturer. Knowing exactly how the calculator functions saves time and frustration and prevents errors.

Basic Operations

Calculators require that you press keys to instruct the calculator to perform. This section describes several keys and shows their symbols in the margin. Be aware, however, that the keys on your calculator may be different. Therefore, use the following discussion to get a basic understanding of the keys, but when using your own calculator, refer to its instructions.

Power Switch. Some calculators have a power switch and some do not. Light-powered models may require that you press an all clear or some other key to turn on the calculator. Light-powered models usually turn off automatically when not in use for a short period. Battery-operated or plug-in models, on the other hand, usually have a power switch. In any case, once you turn on the calculator, it clears the display and erases whatever is stored in the memory unless the calculator has a continuous memory. Calculator memory is discussed shortly. As you will learn, some calculators store what has been put in their memory even after the calculator has been turned off.

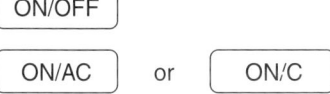

Numbers 0–9. Pressing a number key displays the number and enters it in the calculator.

Clear All. Pressing the clear-all key removes, or erases, whatever was in the calculator, so a new problem can be started. As mentioned earlier, on some models, this key also turns on the calculator.

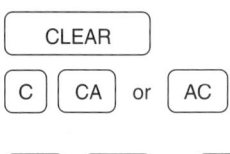

Clear Entry or Clear Display. Calculators with this key allow you to clear only the last entry into the calculator. A ce or cd key is handy when entering a large number of figures and you enter a wrong figure after you have already entered several. By pressing the clear entry key, you merely remove the last entry and not those that preceded it. This function may be combined with the clear-all function for a dual-duty key: touched once, it clears the last entry; touched again, it clears all.

ENT or ENTER

Enter. On RPN calculators, pressing this key either enters the displayed number into the calculator's memory or yields the answer to the problem. (Sometimes, it is labeled execute or exe.)

.

Period. Pressing the period on the keyboard inserts a decimal point.

+

Addition. Pressing the plus key instructs the calculator to add the next entered (or previously entered) quantity to the displayed number.

−

Subtraction. Pressing the minus key instructs the calculator to subtract the next (or previously entered) quantity to the displayed number.

×

Multiplication. Pressing the times key instructs the calculator to multiply the displayed number by the next entered quantity.

÷

Division. Pressing the division key instructs the calculator to divide the displayed number by the next entered quantity.

=

Equals. Pressing the equals key completes the previously entered operation and displays a result. Arithmetic and algebraic calculators use this key. RPN calculators use an entry key.

%

Percent. This key shifts the decimal point of the number shown in the display two places to the left because (as you will learn later in this manual) percentage is a way of stating a portion of a quantity expressed in hundredths (0.01). In other words, finding 25% of a quantity is the same as multiplying the quantity by 0.25.

$\sqrt{x}$ or $\sqrt{}$

Square root. Pressing the square-root key finds the square root of the displayed number.

Memory Functions

Most calculators come with a memory. In less expensive models, the memory has only one level, so the calculator can retain only one piece of information. More expensive calculators store many pieces of information when location keys are pressed after the memory key. Information can be stored and recalled at any time. This section discusses memory functions to give you an idea of their capability. Just as the basic keys on your calculator may be different, so may the memory keys on your calculator be different from the keys shown here. Therefore, always refer to the instructions that accompany the calculator.

M + or STO

Store in Memory. This key stores a displayed quantity in a specific place, or level, in the memory.

M + or 2nd SUM

Memory Addition. You can add or subtract a number to a number already in the memory by pressing the add memory key. However, on some calculators, pressing this key stores a number or operation in the memory. On calculators with second and third function keys (see the section on scientific functions), you may have to press two keys to add to the memory.

M −

Memory Subtraction. Pressing this key subtracts the displayed number from a number already stored in the memory. Or, on some calculators, this key clears (erases) any numbers stored in the memory.

Memory Multiplication. Pressing this key multiplies a number already stored in the memory by the number shown in the display, leaving the result in the memory.

$\boxed{\text{M}\times}$ or $\boxed{\text{PROD}}$

Memory Division. This key divides the number already stored in the memory by the number shown in the display, leaving the result in the memory.

$\boxed{\text{M}\div}$

Recall from Memory. Pressing this key recalls memory contents and displays, but does not change, memory contents.

$\boxed{\text{RM}}$ or $\boxed{\text{RCL}}$

Clear Memory. This key (some calculators require pressing more than one key) clears numbers stored in memory.

$\boxed{\text{CM}}$ $\boxed{\text{MC}}$ or $\boxed{\text{ON/C}}$ $\boxed{\text{STO}}$

Add to Memory. This key adds a displayed number (negative or positive) to a number in the memory in a specific level. (x and x represent the numbers added.)

$\boxed{\text{SUM}}$ and $\boxed{\text{X}}$ $\boxed{\text{X}}$

Multiply By Number in Memory. This key multiplies a displayed number by a number stored in the memory in a specific level.

$\boxed{\text{PRD}}$ and $\boxed{\text{X}}$ $\boxed{\text{X}}$

Exchange. This key swaps the displayed number with that in the memory.

$\boxed{\text{EXE}}$ or $\boxed{\text{X} = \text{M}}$

Clear Memory Level. This key clears data stored in a specific level as indicated by its two-digit location number.

$\boxed{\text{CM}}$ and $\boxed{\text{X}}$ $\boxed{\text{X}}$

Scientific Functions

The scientific, or slide-rule, calculator provides many functions, including trigonometric and logarithmic operations. Following are functions found on many scientific calculators. Remember, as always, that calculators are different. So, your calculator may or may not have the keys shown. Or, it may have additional keys that are not illustrated here. What is more, you may have to press a second or third function key to make it work as desired. Therefore, you must use and refer to the instructions that come with the calculator and become familiar with the calculator's operation and keys.

Second Function Key. This key allows you to gain access to a key's second function. Put another way, on some calculators, one key performs two functions. To change a key's function, press the second function key first. Pressing it makes a key assume a different (a second) function. For example, a square-root sign ($\sqrt{}$) may be on the key itself and a cube-root sign ($^3\sqrt{}$) may be printed on the case just above it. (Or, the second function may be on the key itself, but in letters smaller than the key's first function.) Pressing the $\boxed{\sqrt{}}$ key alone gives the square root of the entered number. But, pressing the second function key and then the $\boxed{\sqrt{}}$ key makes the key function as a cube root ($^3\sqrt{}$) key. Usually, a key's second function is printed on the key or on the case of the calculator in the same color as the second function key's color. For example, if the second function key is orange, a key's second function is also printed in orange.

$\boxed{\text{2ndF}}$ $\boxed{\text{2nd}}$ or $\boxed{\text{F}}$

Third Function Key. As with the second function key, you press this key first to make a key function as desired. For example, the key's first function may be marked LN, its second function e^x, and its third function $^3\sqrt{}x$. A key's third function is usually printed on the calculator's case and the color of the third function key and the label usually match.

$\boxed{\text{3rdF}}$ $\boxed{\text{3rd}}$ or $\boxed{\text{F3}}$

Square. After entering a number, pressing this key squares the number. For example, press $\boxed{2}$, $\boxed{x^2}$, and $\boxed{=}$. The answer is 4 because 2^2 is 4.

$\boxed{x^2}$

[1/x] **Reciprocal.** Pressing this key calculates and displays the reciprocal of the displayed number. A *reciprocal* is a number related to another number by the fact that when the two numbers are multiplied together, the product is 1. For example, the reciprocal of 1/16 is 16/1 (or 16) because 1/16 × 16/1 = 16/16, or 1. In the same way, the reciprocal of 4 is 1/4, because 4/1 × 1/4 = 1. One use of the reciprocal key occurs when converting English, or conventional, units of measure to metric units and metric units to English. Suppose, for example, that you know that to convert feet to metres, you multiply the number of feet by 0.3048. However, also suppose that you need to convert metres to feet but that you do not know the conversion factor. If you know the conversion factor of 0.3048, and your calculator has a reciprocal key, it is easy to determine the factor required to convert metres to feet. Simply enter .3048 and press the reciprocal key to obtain 3.2808. This number is the conversion factor needed to change feet to metres. For example, 9 feet × 0.3048 = 2.7432 metres. Because the reciprocal of 0.3048 is 3.2808, multiply 2.7432 by 3.2808 to obtain 8.9999, or 9 feet. The reciprocal key also has other uses as you will see in Chapter 7, "Trigonometry."

[x ↔ y] **x–y interchange.** After entering a number or obtaining a result, this key changes the order within the calculator so that the number entered first or a previous result is interchanged with that entered second or a displayed result. The ability to interchange entries is useful in solving complex equations.

[( ] [ )] or [[ ] [ ]] **Parentheses or Brackets.** When you press the left parenthesis or bracket, enter an operation, and press the right parenthesis or bracket, the calculator performs the operation within the parentheses or brackets first so that the contents are treated as one unit.

[+/–] or [CHS] **Change Sign.** After entering a number, pressing this key changes the sign of the displayed number from positive to negative or vice versa.

[π] **Pi.** On calculators with a pi sign key (π), pressing this key enters the constant, 3.14159265359, to the number of significant figures allowed by the calculator—for example, 3.1416 is pi carried to five significant figures, which is sometimes abbreviated as 5 s.f. (Chapter 6 covers the use of pi in calculations.) If your work regularly requires determining the area of circles and spheres, your calculator should have the pi sign. Be aware that not all of them do.

[EE] [EXP] or [y^x] **Enter Exponent.** Calculators with this key allow a special exponential notation, which is handy for dealing with large numbers. It is easier to express large numbers as powers of 10. For example, 10,000,000, which is ten million, can be expressed as 10^7. Notice that ten million, or 10^7, is 1 followed by 7 zeros. So, raising 10 to the 7th power is the same as writing 1 followed by 7 zeros. Expressing large numbers by raising a base value to a power makes them easier to write and to deal with. For example, the average distance from the sun to the earth is 93 million (93,000,000) miles. This figure may also be expressed as 9.3×10^7—that is, 93 followed by 6 zeros. On a calculator with an [EE] or [EXE] key, enter [9] [.] [3] [EE] [7] = 93,000,000. Some calculators allow you to enter an exponent with an y^x key, which is discussed below.

[SIN] [COS] [TAN] **Trigonometric Functions.** These keys allow you to enter and find trigonometric functions (chapter 7 covers trigonometric functions).

[SIN⁻¹] [COS⁻¹] [TAN⁻¹] **Inverse Trigonometric Functions.** These keys calculate the inverse trigonometric functions. (Chapter 7 also covers inverse trigonometric functions). Usually, these keys are accessed after pressing the second function key.

Logarithm. Entering a number and then pressing this key calculates and displays the logarithm to base 10 of the displayed number. For example, to find the log of 532, enter ⑤ ③ ② [LOG]. Pressing [LOG] displays the answer of 2.7259. (Chapter 1 described logarithms.)

[LOG]

Antilogarithm. Entering a number and pressing this key (usually after pressing the calculator's second function key) calculates and displays the antilogarithm of the displayed number. For example, enter ② . ⑦ ② ⑤ ⑨. Then, press [2nd] to access the antilogarithm key; finally, press 10^x. The calculator displays 531.9857, which rounds to 532. (Chapter 1 described antilogarithms.)

[10^x]

Natural Logarithm. Entering a number and pressing this key calculates and displays the natural logarithm of the displayed number. For example, enter ② ⓪, press [IN] (or [LN]). The calculator displays 2.9957, the natural logarithm of 20. (Chapter 1 described natural logarithms.)

[IN] or [LN]

Natural Antilogarithm. Entering a number and pressing this key calculates and displays the natural antilogarithm of the displayed number. (Usually, you press the second function key first to access this key.) Just as an antilogarithm is the reverse of a logarithm to the base 10, a natural antilogarithm is the reverse of a natural logarithm. To find the natural antilogarithm of a number, enter the number and press the exponential function key (e^x). For example, enter ② . ⑨ ⑨ ⑤ ⑦, press [2nd], and press [e^x]. The answer is 19.9993 (rounds to 20), the natural antilogarithm of 2.9957.

[e^x]

Power. After entering a number, pressing this key raises the entered number (y) to a power determined by the second number entered (x). (Chapter 1 discussed raising a number to powers). For example, press ② [y^x] ④ and [=]. The answer is 16 because two to the fourth power (2^4) is 16. Also, you can carry out numbers multiplied by 10 to a given power. For example, to carry out 9.3×10^7, press ⑨ . ③ [×] [10] [y^x] ⑦ [=] 93,000,000. The first number entered must be positive; the second one, however, may be negative.

[y^x]

xth Root. After entering the desired number, and, usually the second function key, pressing this key calculates the desired (x) root of the first number entered (y) provided the number is not negative. For example, press [16] [2nd] [$x\sqrt{y}$] ④ and [=]. The answer is 2 because the fourth root of 16 is 2.

[$x\sqrt{y}$]

Roll Stack. Pressing one or the other of these keys allows the user of an RPN calculator to review the contents of the working registers or to reposition numbers.

[R↓] and [R↑]

Fraction Key. A calculator with this key allows you to perform operations with proper and improper fractions. For example, to multiply ¾ × 3, press ③ [$a^b/_c$] ④ [×] ③ [=]. The answer is 2¼. Some calculators display fractions with a ⌐ symbol between the nominator and denominator. So, the calculator displays ¾ as 3⌐4, and 2¼ as 2_1⌐4.

[$a^b/_c$]

OPERATING A CALCULATOR

When properly operated, a calculator conveniently and accurately solves problems. Most calculators are easy to operate if the instructions furnished by the manufacturer are followed. Many operations may be obvious; however, some valuable features may be available but not immediately evident by simply looking at the keyboard. So, as mentioned many times, the first step in operating any calculator is to read the manufacturer's instructions.

Most booklets or instruction sheets provide example calculations. These examples can help you understand functions. Working the examples in the instructions is often a good way to learn the function of the many keys on the calculator.

As pointed out earlier, a calculator functions according to its notation, or operating system. The simplest is the arithmetic operating system. More involved calculations require the algebraic or RPN operating systems.

Solving Problems with an Arithmetic Calculator

The arithmetic calculator (fig. 2.6) adds, subtracts, multiplies, and divides two numbers in logical order with the operation key pressed between the entry of the numbers. For example, to add two numbers, press the keys in this sequence:

6 + 3 = 9.

The display shows 9 as soon as the equal key = is pressed. Likewise, to subtract, multiply, or divide, press the keys in the order of the operation:

6 − 3 = 3.
6 × 3 = 18.
6 ÷ 3 = 2.

Figure 2.6   Arithmetic calculator

In chain calculations (a calculation that is followed by another calculation that is followed by another calculation, and so on, the sequence of operations is the same as entered. For example, in the chain calculation

6 − 4 × 5 = 10.

the calculator subtracts 4 from 6 first, then multiplies the resulting 2 by 5 to give the answer 10. An arithmetic calculator treats a chain calculation where multiplication or division precedes addition or subtraction in the sequence the operation is entered. For example, in the problem

4 × 5 + 9 = 29.

the calculator multiplies 4 by 5 to give 20, then adds 9 to 20 to give the answer 29.

When using an arithmetic calculator to calculate a chain of operations that requires algebraic notation, it is easier to make the calculations if the arithmetic calculator has a memory. All but the least expensive calculators have a memory. However, even with models that feature memory, it may be only one level, so the calculator can remember only one piece of information. In this case, the memory solves one part of a problem, holds it, solves another part of the problem, retrieves the results of the first problem, and then performs a final operation using the results of both parts.

For example, when using an arithmetic calculator to work the problem

$$(12 + 5) \times (23 - 4)$$

follow this entry sequence:

12 + 5 = M+ 23 − 4 × MR = 323.

Figure 2.7 shows the entries and displays in graphic form.

ENTRY	DISPLAY
12	12
+	12
5	5
=	17
M +	17
23	23
−	23
4	4
×	19
MR	17
=	323

Figure 2.7   Entries and displays on an arithmetic calculator using memory key

In this sequence, the calculator stores the sum of the first two numbers in its memory, performs the subtraction in the next set, then multiplies this difference by the sum in its memory to get the answer 323. Note that the equal key $\boxed{=}$ is not pressed to complete the subtraction; the times key $\boxed{\times}$ activates the operation. Also, most calculators give an indication when a number other than zero is in its memory. Usually, an M appears in the display. Be sure that the calculator's memory is cleared (erased) before you begin using it. Most calculators store numbers in their memory until you take steps to remove it. For example, turning off the calculator may not clear the memory. So, to clear a calculator's memory, check the instructions.

Memory is also useful when it is necessary to make several calculations when one of the numbers does not change. For example, suppose you wish to convert several measurements in feet to metres. Because the conversion factor is 0.3048, enter this number into the calculator's memory. Then, simply recall this stored factor and multiply it by the number of feet. You do not have to enter 0.3048 each time because the calculator's memory stores the number.

Example Problem: Convert 3.5, 8.25, 10, and 22.75 feet to metres.

Solution: Make sure the calculator's memory is clear and then enter the feet-to-metres conversion factor into the calculator's memory. Next, multiply the quantity of feet to be converted to metres.

The sequence is:

$\boxed{.3048}$ $\boxed{\text{M+}}$ $\boxed{\times}$ $\boxed{3.5}$ $\boxed{=}$ 1.0668

$\boxed{\text{RM}}$ $\boxed{\times}$ $\boxed{8.25}$ $\boxed{=}$ 2.5146

$\boxed{\text{RM}}$ $\boxed{\times}$ $\boxed{10}$ $\boxed{=}$ 3.0480

$\boxed{\text{RM}}$ $\boxed{\times}$ $\boxed{22.75}$ $\boxed{=}$ 6.9342

Pressing $\boxed{\text{M+}}$ stores .3048 in memory. (On some calculators, this key is labeled $\boxed{\text{STO}}$.) After making the first conversion, press $\boxed{\text{RM}}$ to recall .3048 in the calculator's memory. (On some calculators, the memory-recall key is labeled $\boxed{\text{RCL}}$.) Next, press $\boxed{\times}$ and enter the second number. Finally, press $\boxed{=}$ to obtain the answer. Repeat the sequence for the remaining conversions.

A calculator's percent key $\boxed{\%}$ is useful for dividing numbers by 100 and for calculating markup and markdown percentages. Dividing a number by 100 merely moves the decimal point of that number two places to the left. Most calculators require the percentage number to be entered last. For example, to find 60% of 200, the keys are pressed in the following order:

$\boxed{2}$ $\boxed{0}$ $\boxed{0}$ $\boxed{\times}$ $\boxed{6}$ $\boxed{0}$ $\boxed{\%}$ 120.

In effect, the calculator multiplies 200 by 0.6 to give 120.

Example Problem: A business borrows $5,000 at a simple interest rate of 7.75%. How much interest will it pay at 7.75%?

Solution: Enter the figures as follows:

$\boxed{5}$ $\boxed{0}$ $\boxed{0}$ $\boxed{0}$ $\boxed{\times}$ $\boxed{7}$ $\boxed{.}$ $\boxed{7}$ $\boxed{5}$ $\boxed{\%}$ 387.5.

The business will pay $387.50 in simple interest.

For a markup or add-on problem, the $\boxed{+}$ key is used instead of the $\boxed{\times}$ key.

Example Problem: You have bought a carload of tires for $34.50 each. How much would you have to sell each for to make a 35% profit?

Solution: To work this problem without a ⬚%⬚ key, multiply $34.50 times 0.35 and add the product to $34.50 to get $46.58. On a calculator with a ⬚%⬚ key, enter:

$$\boxed{3}\boxed{4}\boxed{.}\boxed{5}\boxed{0}\boxed{+}\boxed{3}\boxed{5}\boxed{\%}\ \text{46.575 or \$46.58.}$$

In any case, you would have to sell each tire for $46.58 to make a 35% profit.

For a markdown, or discount, problem, the ⬚−⬚ key is used instead of the ⬚+⬚ key.

Example Problem: You normally sell an item for $398.99, but you discount the item by 16%. What is the new selling price?

Solution:

$$\boxed{3}\boxed{9}\boxed{8}\boxed{.}\boxed{9}\boxed{9}\boxed{-}\boxed{1}\boxed{6}\boxed{\%}\ \text{335.15.}$$

The new selling price is $335.15.

Using the percent key greatly simplifies an operation.

Example Problem: Find the tax (8.5%) and the tip (15%) on an $8.93 luncheon; $2.00 of the luncheon is for a glass of wine that is not taxable, but you want to pay the tip on the total. How much money will you have to pay?

Solution: First, find the cost of the lunch without the wine:

$$8.93 - 2.00 = 6.93.$$

Then, find the tax on 6.93 and add it to the total:

$$6.93 \times 0.085 = 0.5891 + 6.93 = 7.52.$$

Next, add 7.52 to 2.00 and add 15% to all for the total amount:

$$7.52 + 2.00 = 9.52$$
$$9.52 \times 0.15 = 1.43$$
$$1.43 + 9.52 = 10.95.$$

A calculator with a percent key can handle all these calculations by entering the following:

$$\boxed{8}\boxed{.}\boxed{9}\boxed{3}\boxed{-}\boxed{2}\boxed{.}\boxed{0}\boxed{0}\boxed{=}\boxed{6.93}\boxed{+}\boxed{8.5}\boxed{\%}\boxed{7.519}$$
$$\boxed{+}\boxed{2}\boxed{.}\boxed{0}\boxed{0}\boxed{=}\boxed{9.519}\boxed{+}\boxed{1}\boxed{5}\boxed{\%}\ \text{10.946.}$$

So, the total cost of the luncheon is $10.95.

Many arithmetic logic calculators can also find the square root of a number. Simply enter the number into the display, then press the key with the radical sign ($\sqrt{}$). The calculator displays the answer. For example, to find the square root of 84, press the keys in the following order:

$$\boxed{8}\boxed{4}\boxed{\sqrt{}}\ \text{9.16515139.}$$

Note that you do not press the ⬚=⬚ key to obtain the result.

Using an Algebraic Calculator

In an algebraic calculator (fig. 2.8), two-number operations are performed in the same sequence as in the arithmetic calculator—that is, numbers and operations are entered as they appear. Chain calculations are also entered as they appear; however, the sequence of operations actually taking place in the calculator follows algebraic logic. So, the calculator computes powers first, multiplication and division second, and addition and subtraction last.

For example, the problem

$$2 + 4 \times 3^2 =$$

is entered as:

$$\boxed{2}\;\boxed{+}\;\boxed{4}\;\boxed{\times}\;\boxed{3}\;\boxed{x^2}\;\boxed{2}\;\boxed{=}\;.$$

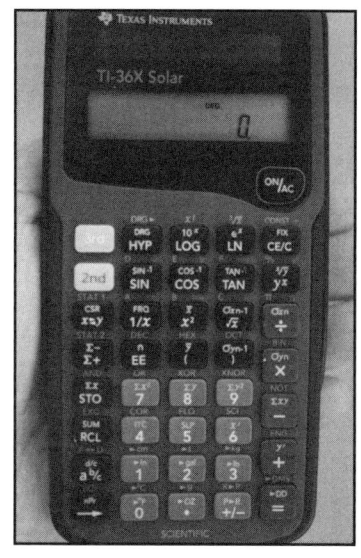

Figure 2.8   Algebraic logic scientific calculator

When the $\boxed{=}$ key is pressed, the calculator begins to work, but it does not perform the operations in the order given. Instead, it first calculates 3 to the 2nd power to produce 9; next, it multiplies the 9 by 4 to give 36; finally, it adds 36 to the 2 to give the answer, 38.

Most algebraic calculators have parenthesis or bracket keys, which allow you to properly group the expressions. For example, the problem

$$(2 + 3) \times (4 + 5) = 45$$

if entered without parentheses in the order shown, is interpreted by the calculator as

$$2 + (3 \times 4) + 5 = 19.$$

Unfortunately, 19 is the wrong answer. So, you must also enter the parentheses as well as the numbers and operational signs in their proper order for the calculator to come up with the right answer of 45.

An algebraic calculator may have a single-level memory system or one that allows storage of more than one piece of information by pressing location keys after the memory key. To use this type of memory, calculators may have the following keys:

$$\boxed{STO}\;,\quad \boxed{RCL}\;,\quad \boxed{SUM}\;,\;\text{and}\;\boxed{PRD}\;.$$

If the calculator's memory has many levels (storage places), you must tell the calculator at what level to store the information. If a two-digit number designates the level, you must enter two digits (01, 12, etc.). To bring back a number stored in the memory, press $\boxed{RCL}$ and the keys for the two-digit location where the number is stored. To add a displayed number to a number in the memory, press the $\boxed{SUM}$ key and the two-digit memory location keys. Likewise, to multiply a displayed number by a number in the memory, press the $\boxed{PRD}$ key and the two digits. With single-level memory, the same keys are used with the location keys, or some calculators have simply the $\boxed{STO}$ and $\boxed{RCL}$ keys.

Also keep in mind that to access some of these keys, you may have to press the $\boxed{2nd}$ or $\boxed{3rd}$ function keys. For example, on some calculators, the $\boxed{RCL}$ and $\boxed{SUM}$ key may be shared. In this case, to use the $\boxed{RCL}$ key, you simply press it; however, to use the $\boxed{SUM}$ key, you press $\boxed{2nd}$ first and then the $\boxed{SUM}$ key. Pressing $\boxed{2nd}$ gives the calculator access to the $\boxed{SUM}$ key rather than the $\boxed{RCL}$ key.

Some problems require both parentheses and memory functions to reach the correct answer. In this type of problem, it is especially important to clear the memory and the display before starting calculations. To clear memory in your calculator, refer to the instruction booklet. Different calculators use different keys to clear memory; so, how you clear your calculator's memory depends on the specific calculator.

Example Problem:

$$\frac{32.64 + 18}{5.32} \times \frac{44.9 + 16.33}{81} = \quad ?$$

Solution: Treat the problem in two parts. First, write the first part for the calculator as:

$$(32.64 + 18) \div 5.32.$$

So, enter the figures in the following order after you have cleared the display and memory:

$$\boxed{(}\,\boxed{3}\,\boxed{2}\,\boxed{.}\,\boxed{6}\,\boxed{4}\,\boxed{+}\,\boxed{1}\,\boxed{8}\,\boxed{)}\,\boxed{\div}\,\boxed{5}\,\boxed{.}\,\boxed{3}\,\boxed{2}\,\boxed{=}\ \text{STO}.$$

Pressing $\boxed{\text{STO}}$ stores the answer to the first part of the problem in memory. Now, do the second part of the problem in the same way. Write the second part as $(44.9 + 16.33) \div 81$ and enter

$$\boxed{(}\,\boxed{4}\,\boxed{4}\,\boxed{.}\,\boxed{9}\,\boxed{+}\,\boxed{1}\,\boxed{6}\,\boxed{.}\,\boxed{3}\,\boxed{3}\,\boxed{)}\,\boxed{\div}\,\boxed{8}\,\boxed{1}\,\boxed{=}\ .$$

To multiply the two answers, press:

$$\boxed{\times}\ \boxed{\text{RCL}}\ \boxed{=}\ .$$

The answer is 7.19550543, or rounded off to two places, 7.20. (Fig. 2.9 shows all the entries and displays.)

When solving complicated problems, keep in mind that a calculator with algebraic notation performs operations in a specific order. Since some calculators do not have parentheses keys, it is necessary to solve parts of an involved problem in a specific order—not necessarily in the order of appearance. The following problem is solved without parentheses.

Example Problem:

$$8 + \frac{30 + \sqrt{12 - (4 \times 2)}}{7 + (2 \times 3)^2} = ?$$

Solution: Solve the denominator part of the problem first and store it in the memory for later use. First, multiply 2 by 3 and square the answer. If 2 is not multiplied by 3 first, the calculator squares the 3 and gets a wrong answer. Once the problem in parentheses is solved, add 7 to it and then store the answer, 43, in the memory. To solve for the denominator, press the keys in this order:

$$\boxed{2}\,\boxed{\times}\,\boxed{3}\,\boxed{=}\,\boxed{x^2}\,\boxed{+}\,\boxed{\text{STO}}\ \text{1.}$$

If the calculator has parentheses, you can solve the denominator as:

$$\boxed{7}\,\boxed{+}\,\boxed{(}\,\boxed{2}\,\boxed{\times}\,\boxed{3}\,\boxed{)}\,\boxed{x^2}\,\boxed{=}\,\boxed{\text{STO}}\ \text{1.}$$

Now solve the numerator part of the problem. The parentheses are not a problem here because the calculator automatically does multiplication first. Solve the part of the problem under the radical sign, then add the answer to 30:

$$\boxed{1}\,\boxed{2}\,\boxed{-}\,\boxed{4}\,\boxed{\times}\,\boxed{2}\,\boxed{=}\,\boxed{\sqrt{}}\,\boxed{+}\,\boxed{30}\,\boxed{=}\ .$$

ENTRY	DISPLAY
ON/C	0
(	0
32.64	32.64
+	32.64
18	18
)	50.64
÷	50.64
5.32	5.32
=	9.518796992
STO[1]	9.518796992 M
(	0
44.9	44.9
+	44.9
16.33	16.33
)	61.23
÷	61.23
81	81
=	0.755925925
×	0.755925925
RCL[1]	9.518797992
=	7.19550543

Figure 2.9   Entries and displays on an algebraic logic calculator

The answer, 32, is not stored in the memory because this numerator is going to be divided by the denominator, 43, which is recalled from the memory:

$$\boxed{\div}\ \boxed{RCL}\ 1\ \boxed{=}$$

Finally, add 8 for the answer:

$$\boxed{+}\ \boxed{8}\ \boxed{=}\ 8.744186047,\text{ or rounded off to two places, }8.74.$$

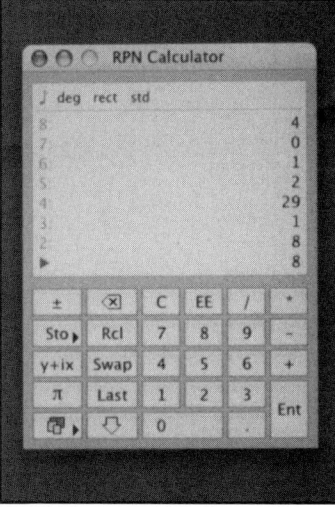

Figure 2.10   An RPN calculator

Using an RPN Calculator

To solve a problem with an RPN calculator (fig. 2.10), you must enter the problem in a specific order. After the calculator stores parts of the problem in its stack, press the operation keys. Operations can be grouped into two categories. An operation with only one variable (one number to be dealt with, such as x, x^2, $1/x$) not only acts on the display contents, but also the answer shows in the display. An operation with two or more variables (addition, subtraction, multiplication, and division where more than one number is involved) acts on the display contents and on the contents of the next higher memory level in the stack. The answer shows in the display, but all stack contents move down one level. The numbers that are keyed in first are worked on last, and operation keys are entered in reverse order of their use. Numbers are automatically stored in the memory and worked on in reverse order. After each number to be put into the memory stack is pressed, the $\boxed{ENTER}$ key must be pressed to move the number into the stack.

Example Problem:

$$1 + \frac{2 + \sqrt{33 - 4}}{1 + (7 \times 3)^2} = ?$$

Solution: Enter the parts of the problem in the calculator as follows:

$$\boxed{1}\ \boxed{ENTER}\ \boxed{2}\ \boxed{ENTER}\ \boxed{3}\ \boxed{3}\ \boxed{ENTER}\ \boxed{4}\ \boxed{-}\ \boxed{\sqrt{}}\ \boxed{+}\ \boxed{ENTER}$$

$$\boxed{1}\ \boxed{ENTER}\ \boxed{7}\ \boxed{ENTER}\ \boxed{3}\ \boxed{\times}\ \boxed{x^2}\ \boxed{+}\ \boxed{\div}\ \boxed{+}\ 1.1067.$$

To understand what happens in an RPN calculator, follow the display and memory of the problem as each entry is made (fig. 2.11). The number 1 was entered into the memory stack first and left there to be worked on later. Next, the numerator of the fraction was solved and stored for later use. The numerator was solved by displaying and entering the numbers from left to right. The last number, 4, was displayed but not entered so that it could be subtracted from 33 to get an answer of 29. Working backwards, the square root of 29 was found and added to the 2. The answer to the numerator, 7.38, was entered in the memory stack.

Next, the denominator of the fraction was solved in the same way, entering the numbers from left to right and then working backwards with the operations between these numbers. Because the 3 and 7 are in parentheses, they are one unit, and they had to be multiplied together first to get an answer of 21. Then the square of 21 was found and added to the 1. The result was 442, the answer to the denominator. This answer was not entered into the memory stack, as the next step was to divide the numerator by the denominator (7.38 ÷ 442). Finally, the result of the division was added to the whole number 1 to obtain the solution, 1.0167.

ENTRY	DISPLAY	MEMORY
1	1	
ENTER	1	1
2	2	1
ENTER	2	1 2
33	33	1 2
ENTER	33	1 2 33
4	4	1 2 33
−	29	1 2
√	5.38	1 2
+	7.38	1
ENTER	7.38	7.38
1	1	1 7.38
ENTER	1	1 7.38 1
7	7	1 7.38 1
ENTER	7	1 7.38 1 7
3	3	7 1 7.38 1
×	21	1 7.38 1
x^2	441	1 7.38 1
+	442	7.38 1
÷	0.0167	1
+	1.0167	

Figure 2.11   Stack memory of an RPN calculator

Practice Problems

The following problems can be solved by any calculator. Experiment with your calculator, read the instructions, and discover how it works. Then solve these problems, rounding off the answer to two decimal places.

1. $32.4 + 68.3 + 17.1 =$ _____

2. $981.503 + 576 + 834.334 - 1,000 =$ _____

3. $5,621 - 5.86 - 7.90 - 3 - 569 =$ _____

4. $53.85 \times 28 + 40.75 =$ _____

5. $(91 - 52) \times 48 =$ _____

6. $8.9 \times 0.87 \times 0.065 =$ _____

7. $(575 \times 286) \div 300 =$ _____

8. $2,500 \times (25 \div 50)^2 =$ _____

9. $350 + \sqrt{3,368 - 4} =$ _____

10. 46% of $3,500,152 =$ _____

11. $72 - [(530 - 468) \div (95 \times 0.052)] =$ _____

12. How far can a car be expected to travel without a refill if it gets 22 miles per gallon and its tank holds 14 gallons?

    _____

13. If 9.8 pounds of peaches cost \$12.65, what is the selling price per pound?

    _____

14. A bag of nails regularly sells for \$3.50. It is now advertised at 20% off. If you bought 5 bags at the sale price and paid a 5% sales tax, what is your total bill?

    _____

15. $\dfrac{14.9 \times (13.5 - 12.2)}{35.5 - 13.5} \times 1,450 =$ _____

SELF-TEST

2. The Calculator

Multiply each correct answer by five to arrive at your percentage of competency.

Solve the following problems using a calculator. Show all places.

1. Find the sum of 8, 9, 7, 5, 4, 6, 2, 1, and 3.

   _____

2. How much is 30.1 plus 3.01 plus 0.301 plus 0.0301?

   _____

3. Add 4.3, 6.75, 8.67, 12.0, and 0.16.

   _____

4. 10 less 1.67 =

   _____

5. $6 + 5 - 14 + 4.5 - 0.5 =$

   _____

6. $734 - 30.6 + 90.3 - 428.1 =$

   _____

7. $(5{,}200 \times 3) - 1{,}300 =$

   _____

8. $\$1.00 + 36¢ - \$0.21 - 15¢ + \$1.00 =$

   _____

9. $0.052 \times 11.2 \times 4{,}000 =$

   _____

10. $2{,}329.6 \div 4{,}000 \div 0.52 =$

    _____

11. $13.2 + \dfrac{290}{0.052 \times 5{,}500} =$

    _____

12. $(12.3 - 9.5) \times 3{,}000 \times 0.052 =$

    _____

13. $(3{,}090 \div 1.32) - (9.5 - 9.3) + 3{,}000 \times 0.052 =$

    _____

14. $6.10^3 \div (4.38 + 3.25)^2 =$

_____

15. $0.624 + \sqrt{3,481} - (600 - 490) \div 240 =$

_____

16. $1,000 + 4.5^2 \times (0.8 \times 680) =$

_____

17. $72 + \sqrt{530 - 468} - (95 \times 0.052) =$

_____

18. $12.45 + 15\% =$

_____

19. 6% of $\$1,195 =$

_____

20. $3,454$ less $33.3\% =$

_____

3
Number Relations

OBJECTIVES

Upon completion of chapter three, the student will be able to—

1. Discuss the relation of percent to the whole, and calculate a given percent of a given number.
2. Change percent to hundredths and hundredths to percent.
3. Solve for base, rate, or percentage in percent problems.
4. Read, write, and determine ratios of one quantity to another.
5. Solve problems involving direct proportion.
6. Solve problems involving inverse proportion, including pulley and gear ratio problems.
7. Find the average, or mean, of a set of statistics.
8. Find the median and mode in a body of data.
9. Use reference tables for extracting information.
10. Interpolate additional numerical data from information given in a table.
11. Name the parts of a table and construct a table from given data.
12. Extract approximate statistics from a graph, noting trends.
13. Determine the best method for depicting numerical information.
14. Plot and draw a bar graph, a line graph, and a circle graph.

INTRODUCTION

Number relationships, charts, graphs, and tables can help in making calculations and decisions. Charts, graphs, and the like can show trends, illustrate deductions, and help make decisions based on numerical facts. Common relationships are percentage, ratio, proportion, average, mean, median, and mode. Graphs, tables, and charts are often used to depict relationships.

PERCENTAGE

Percent expresses a proportion of an amount in hundredths. Percent means for or out of each hundred. The symbol for percent is % and it expresses quantity in relation to a whole. It is only used specifically and always with a number. For example:

$$26 \text{ percent} = {}^{26}\!/_{100} = 26\% = 0.26.$$
$$14\tfrac{1}{2} \text{ percent} = {}^{145}\!/_{1,000} = 14.5\% = 0.145.$$

To change from percent to hundredths, shift the decimal point two places to the left and drop the percent symbol. For example, to change 32% to hundredths, drop the % symbol and move the decimal point two places to the left, which yields 0.32. In other words, 32% = 0.32. To change a decimal fraction to percent, move the decimal point two places to the right and add the percent symbol; for example, 0.64 becomes 64%.

Example Problem: Find 14% of 430.

Solution:

14% = 0.14
0.14 × 430 = 60.2.

Thus, 14% of 430 is 60.2.

If you have a calculator with a % key, all you do is enter:

430 × 14% = 60.2.

Example Problem: Find 4½% of 85.

Solution:

0.045 × 85 = 3.825,

or, using the calculator,

85 × 4.5% = 3.825.

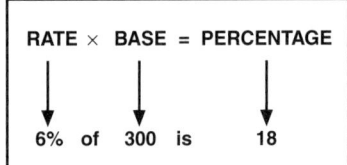

Figure 3.1   *Percent relations*

Base, Rate, and Percentage

Base, rate, and percentage are terms used in percentage problems. *Base* is the quantity of which a percentage is desired. *Rate* is a desired percentage of the base. And, *percentage* is the product of the rate times the base. For example, figure 3.1 shows that 6% of 300 is 18. In this case, 6% is the rate, 300 is the base, and 18 is the percentage. The relationship of base, rate, and percentage can be expressed as

percentage = rate × base.

Three types of percentage problems involve finding one of these elements when the other two are known. First, when the base and rate are known, percentage is found by multiplying the base times the rate.

Example Problem: How much is 8% of $625?

Solution: In this example, 8% is the rate and $625 is the base. So, rate times base is

$625 × 0.08 = $50.00.
Eight percent of $625 is $50.00.

Example Problem: If a woman earns $120 and saves 12½% of it, what percentage of her earnings did she save?

Solution:

12½% = 12.5% = 0.125
$120 × 0.125 = $15.00.
The percentage saved is $15.00.

The second percentage problem involves solving for rate when the base and percentage are known. The formula for finding rate when base and percentage are known is

$$rate = percentage \div base.$$

Example Problem: What percent of $500 is $125?

Solution: Here, the percentage is $125 and the base is $500. So, the percentage divided by the base is

$125 \div 500 = 0.25$, or 25%

or $\dfrac{125}{500} = 0.25$, or 25%.

Example Problem: An oil tank contains 130 barrels, of which 2.6 barrels are sediment and water (S&W). What percent of S&W does it contain?

Solution:

$\dfrac{2.6}{130} = 0.02$, or 2% S&W.

The third percentage problem involves knowing the rate and percentage but not the base. The formula for finding base is

$$base = percentage \div rate.$$

Example Problem: An oilwell produces 95% water, and the allowable oil production for a specified period is 325 barrels. What is the total fluid production needed to produce this much oil?

Solution: Oil comprises 5% (rate) of the fluid, so

$100\% - 95\% = 5\%$.

The allowable, 325 barrels, is the percentage; and the total fluid production is the base. Thus,

$base = \dfrac{325}{0.05} = 6{,}500$ barrels.

Example Problem: 43.75 is 5% of what number?

Solution:

$\dfrac{43.75}{0.05} = 875.$

Practice Problems

Round off all answers to two places.

1. An item costs $52 plus 5% sales tax. What is its total price?

_____

2. How much is 8½% of 200?

_____

3. Six percent of a person's earnings goes into a retirement fund. If this person earns $1,940 per month, what amount goes toward the person's retirement each month?

_____

4. An alloy contains 83.3% tin, 5.6% copper, and 11.1% antimony. How much of each metal is contained in 500 pounds of this alloy?

   a. Tin _____

   b. Copper _____

   c. Antimony _____

5. In a shipment of pressure gauges, 15% were damaged, leaving 170 that could be used. How many gauges were in the shipment?

_____

6. A supply house gives a discount of ⅓ off from the prices listed in its catalog. What will be the net price of four 3-inch gate valves that list at $92 each?

_____

7. A carpenter orders 240 two-by-fours but sends 36 of them back because they are defective. What percent did he return?

_____

8. A job in a machine shop required 27 hours to complete. Of this, 14½ hours were spent on the lathe, 2¼ hours on the drill press, 6¾ hours on the shaper, and 3½ hours in welding. What percent of the time did each process require?

   a. Lathe _____

   b. Drill press _____

   c. Shaper _____

   d. Welding _____

9. A certain oil yields 39% gasoline. How many barrels of this crude does it take to produce 500 barrels of gasoline?

_____

10. The following problems provide practice in mental arithmetic. Try to figure out the answers in your head and then write them down.

   a. Find 25% of 88. (Hint: Multiplying by 0.25 is the same as dividing by 4.) _____

   b. If a person earning $400 per week receives a 25% raise and then a 20% cut in salary, what is the person's new salary? _____

   c. If 42 gallons of oil are removed from a full 100-barrel tank, what percent of oil is left in the tank? _____

   d. If ⅛ of a lease has been surveyed, what percent remains to be surveyed? _____

   e. What percent of 110 is 55? _____

RATIO AND PROPORTION

Ratio

Ratio is a proportional relationship between two numbers or quantities. Ratio describes how two different things relate numerically to one another. For example, the statement, "the ratio of trucks to tank cars is two to one" means that the number of trucks is twice as great as the number of tank cars. So, if the ratio statement is about a company that owns tank cars and trucks, then the company may own 50 tank cars and 100 trucks.

The ratio of one quantity to another is obtained by dividing the first quantity by the second. Thus, ratio can also be defined as the relative size of two quantities expressed as the quotient of one divided by the other. For example, the ratio of $6 to $2 is three to one because $\$6/\$2$ equals $\frac{3}{1}$.

A colon placed between two quantities expresses the ratio of one quantity to the other. For example, the ratio of $6 to $2 can be written as $6:$2 and spoken as "six dollars is to two dollars." This ratio may also be written as the fraction $\$6/\$2$.

Example Problem: What is the ratio of 12 to 4?

Solution: Applying the definition for ratio, divide 12 by 4, which is written as $^{12}/_4$ = 3. In effect, dividing 12 by 4 reduces the ratio to 3 to 1. That is, the ratio of 12 to 4 is the same as the ratio of 3 to 1.

Proportion

Proportion is a relation of equality between two ratios. The ratio of 2 to 4 is the same as the ratio of 6 to 12. That is, 2 to 4 is proportional to 6 to 12 or 2 is to 4 as 6 is to 12 (2:4 = 6:12). Proportions are mostly used when one ratio is known and only part of another ratio is known. Since ratios in a proportion are equal in value, the missing part of a ratio can be found. As with ratios, proportions can be expressed in two ways: a colon inserted between the two numbers or as a fraction:

$$2:4 = 6:12$$

or

$$^{12}/_4 \; = \; ^6/_{12}.$$

Written in either form, this proportion is read as "two is to four as six is to twelve."

The first and fourth terms of a proportion (in this case, 2 and 12) are the *extremes*, and the second and third terms (4 and 6) are the *means*. A key consideration is that in any proportion, the product of the extremes is equal to the product of the means. For example, in the proportion 2:4 = 6:12,

$$2 \times 12 \text{ (the extremes)} = 24$$
$$4 \times 6 \text{ (the means)} = 24$$
$$24 = 24.$$

Direct Proportion

Proportions can be either direct or indirect (also called inverse). In direct proportion, each term increases or diminishes as the term on which it depends

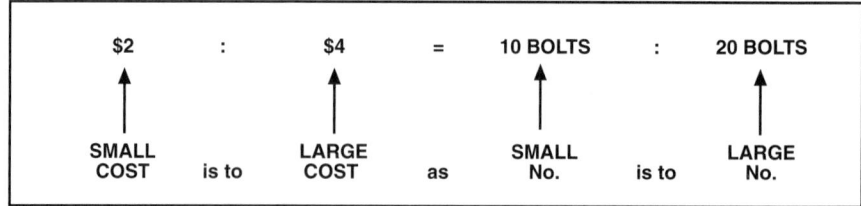

Figure 3.2   Direct proportion

increases or diminishes. For example, figure 3.2 shows that a small cost is to a large cost as a small number of objects (bolts in this case) is to a large number of objects. Put another way, fewer bolts cost less than more bolts and more bolts cost more than fewer bolts. This statement is a direct proportion because cost increases as the number of bolts increases and vice versa.

To solve direct proportion problems, three terms must be known. If three terms are known, then the fourth term can be determined.

Example Problem: If 12 hacksaws cost $155.88, how much do 100 hacksaws cost?

Solution: The three known quantities in this problem are a small number of hacksaws, a small cost, and a large number of hacksaws. Because the number to express the cost of the large number of hacksaws is unknown, call the unknown quantity *x* and set up the proportion: 12:100 = $155.88:*x*. Note that this proportion reads 12 is to 100 as $155.88 is to *x*.

Mathematically, the proportion can be written as:

$$\frac{12}{100} = \frac{155.88}{x}.$$

To solve the proportion, merely cross-multiply the ends and the means. (Cross-multiplying means to multiply the numerator on one side of the fraction by the denominator on the other side. In this problem, the result of the cross-multiplication is:

$$12x = 15{,}588.$$

As you will learn in chapter 5, "Principles of Algebra," to solve $12x = 15{,}588$, divide 15,588 by 12, which equals 1,299. Thus, $x = \$1{,}299$ the cost of 100 hacksaws.

Inverse Proportion

In inverse, or indirect, proportion, each term increases as the term on which it depends decreases; or, each term decreases as the term on which it depends increases. For example, figure 3.3 shows that a slow speed is to a fast speed as a

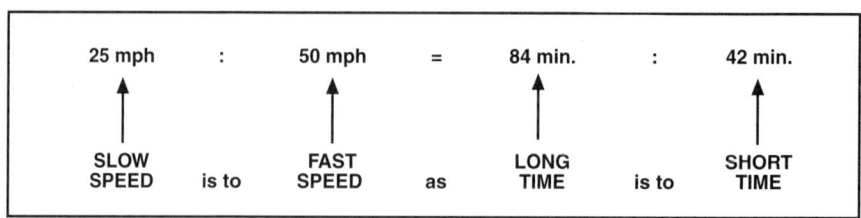

Figure 3.3   Inverse proportion

long time is to a short time. That is, the slower the speed is, the more time it takes and the faster the speed is, the less time it takes. The ratios vary in the opposite order—as one quantity goes up, the other goes down and vice versa. The speed of a vehicle is inversely proportional to the time needed to travel a certain distance. For example, a car moving at 25 miles per hour can cover 35 miles in 84 minutes, while at 50 miles per hour it can go the same distance in 42 minutes.

Another example of inverse proportion involves belt- or chain-driven pulleys. The pulley directly connected to a motor or other power source is the driver pulley. The other pulley is the driven pulley. As the driver pulley rotates, the belt or chain drives the other, driven pulley. The bigger the pulley is, the slower it rotates. For example, the 12-inch pulley (pulley B) in figure 3.4 rotates slower than the smaller, 8-inch pulley (pulley A). Increasing the size of pulley A decreases its speed; likewise, decreasing the size of pulley B increases its speed.

Example Problem: If the speed of pulley B in figure 3.4 is 250 revolutions per minute (rpm), and it is 12 inches in diameter, how many rpm does pulley A turn if it is 8 inches in diameter?

Solution: Since the speeds of the two pulleys are inversely proportional to their diameters, write the proportion:

$$S_A : S_B = d_B : d_A \text{ or } \frac{S_A}{S_B} = \frac{d_B}{d_A}$$

where

S_A = speed of pulley A
S_B = speed of pulley B
d_B = diameter of pulley B
d_A = diameter of pulley A.

Then substitute the known values and let x stand for the speed of pulley A:

$$x{:}250 = 12{:}8, \text{ or } \frac{x}{250} = \frac{12}{8}$$
$$8x = 250 \times 12 = 3{,}000$$
$$x = 3{,}000 \div 8 = 375.$$

The speed of pulley A is 375 rpm.

A simple rule to remember in calculating speeds and diameters of pulleys is that the speed of the driver pulley multiplied by its diameter is equal to the speed of the driven pulley multiplied by its diameter. The unknown or missing quantity is designated by a symbol such as x, and its value is found as previously shown.

Gears are similar to pulleys (fig. 3.5). Gears, like pulleys, rotate. However, instead of belts or chains, gears have teeth, or cogs, that intermesh. As the driver gear turns, the intermeshing teeth cause the driven gear to rotate. A rule for determining speeds and diameters of gears is that the speed of the driver multiplied by its number of teeth is equal to the speed of the driven gear multiplied by its number of teeth.

Example Problem: When gear A in figure 3.5 makes 5 revolutions, how many revolutions does gear B make?

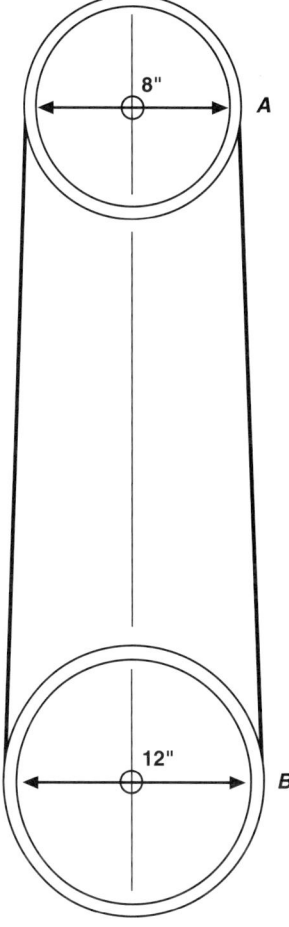

Figure 3.4   Pulley arrangement

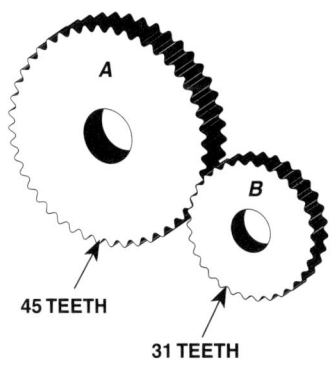

Figure 3.5   Gear arrangement

Solution: Using the gear ratio rule, set up the proportion:

$$S_A \times T_A = S_B \times T_B$$

where

S_A = speed of gear A
T_A = number of teeth in gear A
S_B = speed of gear B
T_B = number of teeth in gear B.

Substituting the known values and letting x equal the unknown:

$$5 \cdot 45 = x \cdot 31$$
$$31x = 225$$
$$x = 7.258.$$

Gear B turns 7.258 revolutions each time gear A makes 5 revolutions.

Practice Problems

1. What is the ratio of 125 to 5?

   _____

2. Jones' salary is $250 and Smith's is $150. What is the ratio of Jones' salary to Smith's?

   _____

3. Soft pine wood weighs 35 pounds per cubic foot; steel weighs 490 pounds per cubic foot. Find the ratio of these densities.

   _____

4. In treating drilling mud, 60 sacks of barite are added to 100 barrels of mud to increase its weight 1 pound per gallon. How many sacks are required to raise the weight of 850 barrels of mud from 9.4 to 11 pounds per gallon?

   _____

5. If 3 inches on a blueprint represents 12 feet, how many feet does a line that is 10 inches long represent?

   _____

6. A job must be completed in 6 days. Records show that 20 workers completed a similar job in 15 days. How many workers must be put on the job to insure its completion in the given time?

   _____

7. A car traveling at 48 miles per hour takes 50 minutes to go from lease A to lease B. At what speed must it go to make the trip in 30 minutes?

   _____

8. The motor shown in the sketch below drives the small, 4-inch pulley at 1,740 rpm. How many revolutions per minute does the large, 16-inch pulley make?

_____

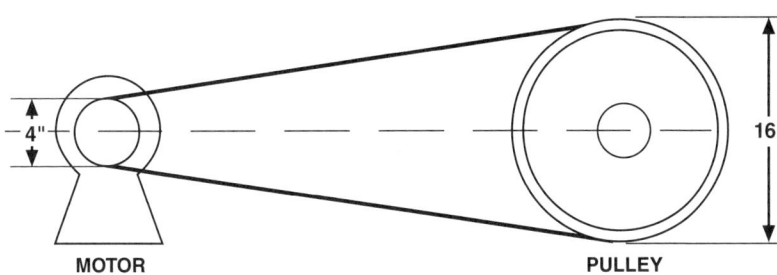

9. The motor shown in the sketch below runs at 1,800 rpm and has a 4-inch pulley. The countershaft pulleys are 8 inches and 18 inches in diameter, and the lineshaft is 16 inches in diameter. How many revolutions per minute does the lineshaft make?

_____

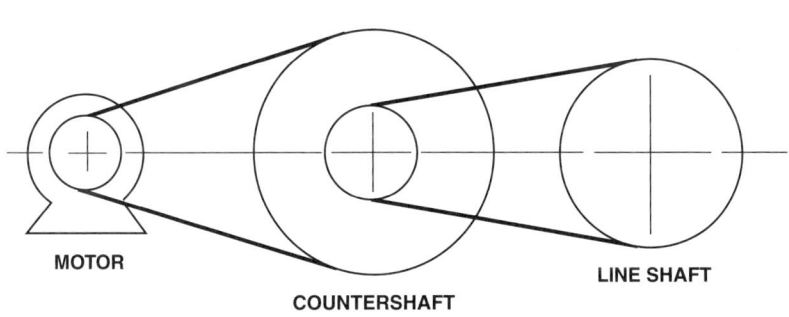

10. If a 48-tooth gear meshing with a 64-tooth gear has a speed of 165 rpm, what is the speed of the 64-tooth gear?

_____

MEAN (AVERAGE), MEDIAN, AND MODE

Four other terms in number relationships are average, mean, median, and mode. Average and mean are identical statistically; however, the term mean is usually employed when discussing median and mode.

Average, or Mean

The *average*, or *mean*, of something is a number that typifies a set of numbers or represents the midpoint between two extremes. Knowing average quantities, such as the average specific gravity of a fluid, the average cost of an item, and the average temperature of a liquid can help analyze problems and costs. Two kinds of averages are straight average and weighted average.

Straight Averages

To find a *straight average*, total the quantities and divide by the number of quantities involved.

Example Problem: Find the average cost of a particular handheld calculator if four telephone bid requests resulted in quotes of $15.00, $16.50, $12.99, and $11.65.

Solution: Total the bid prices:

$$15.00 + 16.50 + 12.99 + 11.65 = 56.14.$$

Then, divide this total by the number of bids:

$$56.14 \div 4 = 14.035.$$

The average cost of this handheld calculator is $14.04.

Weighted Averages

In many cases, a *weighted average* best expresses the average value. For example, a person buys 3 screws at 50¢ each and 17 screws at 80¢ each. To determine the average cost of all the screws, do not use a straight average. A straight average comes out to 65¢ because:

$$(50 + 80) \div 2 = 65.$$

However, more screws were bought at the higher price, so the 80¢ price should be weighted more in deciding an average cost. To determine the true average cost, find the total cost first:

$$(0.50 \times 3) + (0.80 \times 17) = \$15.10,$$

and then divide by the total number of screws:

$$\$15.10 \div 20 = 75.5¢.$$

The average cost per screw is 75.5¢.

An equation for finding a weighted average is

$$A = \frac{(N_a \times V_a) + (N_b \times V_b) + (N_c \times V_c) + \ldots}{a + b + c = \ldots}$$

where

A = weighted average
N = number of items
V = value of each item
a, b, c = items.

The ellipsis (...) in the equation simply means that it continues on in the same manner for the required number of values and items in the calculation.

Example Problem: A service station owner totaled her sales for the day as follows: 1,644 gallons of regular at $1.349/gal; 15,206 gallons of regular unleaded at $1.449/gal; 7,310 gallons of super unleaded at $1.549/gal. What are the average earnings per gallon for the day?

Solution: Using the equation for finding a weighted average, substitute the figures:

$$A = \frac{(1{,}644 \times 1.349) + (15{,}206 \times 1.449) + (7{,}310 \times 1.549)}{1{,}644 + 15{,}206 + 7{,}310}$$

$$= (2{,}217.756 + 22{,}033.494 + 11{,}323.190) \div 24{,}160$$

$$= 35{,}574.44 \div 24{,}160 = 1.4725.$$

The average earnings per gallon for that day was $1.473.

Median

A *median* in a body of data is the middle value. That is, the items of data greater than the median are equal to the items of data less than the median. For example, in the following temperature readings,

$$50°, 53°, 57°, 61°, 64°, 70°, \text{ and } 73°,$$

the median is 61° because three values (50°, 53°, and 57°) are less than 61°, and three values (64°, 70°, and 73°) are greater than 61°.

Where an even number of values exists, the median must fall between two values. For example, in the series of numbers 3, 7, 9, 11, 13, and 16, the median lies between 9 and 11. In this case, the median is 10 because it falls between 9 and 11.

Example Problem: Find the median in the following set of speeds:

$$25, 30, 45, 55, 58, 60, 65, \text{ and } 70 \text{ mph.}$$

Solution: The median of these eight numbers lies between the fourth and fifth numbers—55 and 58 mph. To find the median, add the two numbers and divide the sum by 2:

$$55 + 58 = 113 \div 2 = 56.5.$$

The median speed is 56.5 mph.

Mode

In a set of data, the *mode* is the value that occurs most frequently. Technically speaking, mode (which is also called the norm) is the value that has the highest frequency within a statistical range. For example, in measurements such as

$$4", 5", 5", 7", 6", 3", 5", 3", \text{ and } 5",$$

the mode is 5" because that value occurs more often than any other value in the data.

Example Problem: Find the mode in the following set of numbers:

$$30, 45, 65, 10, 45, 30, 50, 55, 65, 45.$$

Solution: The mode is 45 because it occurs three times, which is more than any other number.

Practice Problems

1. In the offshore waters of Louisiana, 118 rotary rigs were drilling in 2000. In 2001, 119 were drilling, and in 2002, 92 rigs were drilling. What is the average number of rotary rigs operating during the three years?

2. A company needed three reports duplicated. The cost of duplicating the reports was quoted as follows:

   Report A—500 copies at 35¢ per copy;

   Report B—250 copies at 50¢ per copy;

   Report C—100 copies at 70¢ per copy.

   What is the weighted average cost per copy?

3. Give the mean of the following numbers:

   9, 14, 52, 66, 99.9, 34.5, 7½, 47.8, and 37.5.

4. Find the median of the following monthly wages paid by one company to its supervisors:

   $1,550; $2,350; $1,245; $2,050; $1,860.

5. What is the mode in the following set of measurements:

   15", 1', 30.48 cm, 18", 1½', 0.610 m, and 12"?

GAS PIPELINE PRESSURE READINGS

Time	Pressure (psig)
1 P.M.	675
2 P.M.	670
3 P.M.	665
4 P.M.	655
5 P.M.	650

Figure 3.6   Simple table

Figure 3.7   Simple graph

TABLES AND GRAPHS

Representing numerical information so that it can be easily grasped is important to anyone using numbers on the job. A random list of numbers may hide important trends, whereas a table or a graph based on these numbers might show these trends clearly. For example, compare the following ways to present data. First, read the sentence, "At 1:00 P.M., the pressure reading on a gas pipeline was 675 psig; at 2:00 P.M., 670 psig; at 3:00 P.M., 665 psig; at 4:00 P.M., 655 psig; and at 5:00 P.M., 650 psig." Now, look at the same information in tabular form (fig. 3.6) and on a graph (fig. 3.7). Carefully reading the sentence and the table reveals that the pressure is falling steadily, but the graph shows the trend at a glance.

Using Reference Tables

Large amounts of information can be presented in a relatively small amount of space when presented in a table. Moreover, a table can show the data in a way that makes it easy to understand.

Reference information can be conveniently presented in tabular form. Also, knowing how to read and use tables can be important when solving mathematical problems. For example, figure 3.8 is a table that gives the squares, cubes, square roots, and cube roots for numbers from 1 to 50. You can use the table to find any of these properties for a given number, or for finding the number if one of the properties is known.

Example Problem: Find the cube of 47, using the table in figure 3.8.

Solution: Find the number 47 in the left-hand column labeled No. Then, move horizontally across the table to the cube column. The answer—103,823—is found where the two columns intersect.

You can also use the table to find an unknown number. For example, you can find the value for *a* in the equation

$$a^3 = 39,304.$$

Locate the number 39,304 in the cube column, then read the answer, which is 34, across from it in the No. column. Thus, 34^3, is 39,304.

Interpolation

Sometimes, even a comprehensive table does not show the specific information needed for working problems. Fortunately, when you need a value that falls between the figures given in a table, you can interpolate to come up with the value. To *interpolate* means to insert or estimate values between two known values. For example, let's say you need the square root of 800. Looking at figure 3.8, note that the table does not give this information. However, it does show that the square root of 784 is 28 and the square root of 841 is 29. So, the square root of 800 is between 28 and 29. To interpolate this information, first find the difference in the two numbers given:

$$841 - 784 = 57.$$

Then, find the difference in the number desired and the smaller known number:

$$800 - 784 = 16.$$

The number 16 shows that the square root of 800 is ¹⁶⁄₅₇, or 0.28, more than the square root of 784, which is 28. So,

$$28 + 0.28 = 28.28.$$

Therefore, 28.28 is the square root of 800 (or $28.28^2 = 800$). If you have a calculator with the square root function, use it to check the answer. Depending on how many places your calculator carries the solution to, the answer should be something like 28.28427125, which is very close to 28.28. (Actually, $28.28^2 = 799.7584$, which, for most purposes, is close enough.)

Constructing Tables

Most tables are constructed to present numerical or statistical information in the form of percentages, amounts of money, number of occurrences, and the like. In any table, one set of information represents the primary focus of the table, and a second set of data represents varying conditions or other related variables.

SQUARES, CUBES, SQUARE ROOTS, AND CUBE ROOTS OF NUMBERS

No.	Square	Cube	Sq. Root	Cube Root
1	1	1	1.00	1.00
2	4	8	1.41	1.26
3	9	27	1.73	1.44
4	16	64	2.00	1.59
5	25	125	2.24	1.71
6	36	216	2.45	1.82
7	49	343	2.65	1.91
8	64	512	2.83	2.00
9	81	729	3.00	2.08
10	100	1,000	3.16	2.15
11	121	1,331	3.32	2.22
12	144	1,728	3.46	2.29
13	169	2,197	3.61	2.35
14	196	2,744	3.74	2.41
15	225	3,375	3.87	2.47
16	256	4,096	4.00	2.52
17	289	4,913	4.12	2.57
18	324	5,832	4.24	2.62
19	361	6,859	4.36	2.67
20	400	8,000	4.47	2.71
21	441	9,261	4.58	2.76
22	484	10,648	4.69	2.80
23	529	12,167	4.80	2.84
24	576	13,824	4.90	2.88
25	625	15,625	5.00	2.92
26	676	17,576	5.10	2.96
27	729	19,683	5.20	3.00
28	784	21,952	5.29	3.04
29	841	24,389	5.38	3.07
30	900	27,000	5.48	3.11
31	961	29,791	5.57	3.14
32	1,024	32,768	5.66	3.17
33	1,089	35,937	5.74	3.21
34	1,156	39,304	5.83	3.24
35	1,225	42,875	5.92	3.27
36	1,296	46,656	6.00	3.30
37	1,369	50,653	6.08	3.33
38	1,444	54,872	6.16	3.36
39	1,521	59,319	6.24	3.39
40	1,600	64,000	6.32	3.42
41	1,681	68,921	6.40	3.45
42	1,764	74,088	6.48	3.48
43	1,849	79,507	6.56	3.50
44	1,936	85,184	6.63	3.53
45	2,025	91,125	6.71	3.56
46	2,116	97,336	6.78	3.58
47	2,209	103,823	6.86	3.61
48	2,304	110,592	6.93	3.63
49	2,401	117,649	7.00	3.66
50	2,500	125,000	7.07	3.68

Figure 3.8   Reference table

A simple table is constructed with the primary set of variables (known as *boxheads*) across the top of the table and the second set of variables (known as the *stub*) down the left side of the table (fig. 3.9). The *body* of the table presents the desired statistics. The *table's title* is important because it affects the amount of information that the boxhead must present. Boxheads or stubs also need units of measure to qualify them, such as percent, dollars, inches, metres, degrees Fahrenheit, and so forth.

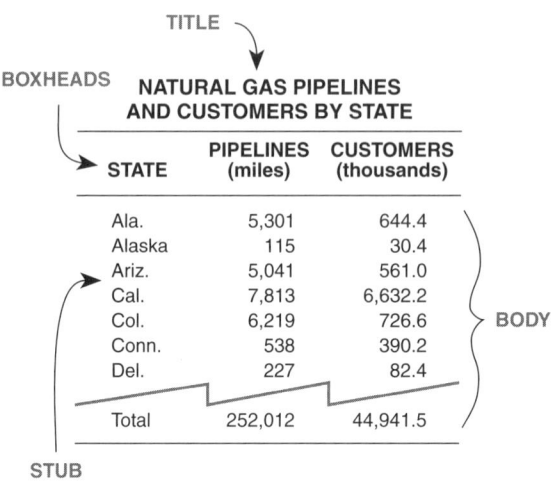

Figure 3.9   Parts of a table

Example Problem: According to projections published by the *Pipeline & Gas Journal* in 2000, the number of miles of main gas distribution utility piping to be installed between 2000 and 2004 were as follows: 2000—14,000; 2001—14,200; 2002—14,210; 2003—14,200; and 2004—15,100. The number of miles of service lines during the same period was projected to be the following: 2000—15,000; 2001—15,600; 2002—15,500; 2003—15,900; and 2004—15,900. Prepare a table that presents these statistics as well as the total number of miles of piping installed for each year.

Solution: Pick out the primary focal point of the table and use these points as boxheads: Year, Main, Service, and Total. For these heads to be meaningful, a unit of measurement (miles) should be added to Main, Service, and Total. Then, the years become the stub, and the statistics become the body as shown in figure 3.10.

PROJECTED GAS UTILITY PIPING TO BE INSTALLED

YEAR	MAIN (mi)	SERVICE (mi)	TOTAL (mi)
2000	14,000	15,000	29,000
2001	14,200	15,600	29,800
2002	14,210	15,500	29,710
2003	14,200	15,900	30,000
2004	15,100	15,900	31,000

Source: "Gas Distribution Utility Piping Market Statistics," *Pipeline & Gas Journal*, 2000.

Figure 3.10   Table data

Plotting Graphs

A *graph* is a diagram that indicates relationships between two or more variables. Types of graphs include bar graphs, line graphs, and circle graphs, which are also called pie charts. An axis is one of usually two straight lines on a line graph—that is, a line graph usually has two axes. Reference points called coordinates are placed on the axes (fig. 3.11). Axes are commonly drawn perpendicular to each other. The horizontal axis is the *abscissa*, or X axis; the vertical axis is the *ordinate*, or Y axis.

The location of any point on a line graph is given as its perpendicular distance from the two axes. For example, point *A* in figure 3.11 is located 5 units from the X axis and 4 units from the Y axis, or 4 across and 5 up. These dimensions are often referred to as coordinates; in this case, the coordinates of point *A* are 4, 5.

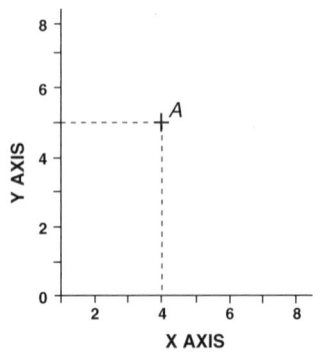

Figure 3.11   Pair of reference axes

Arrange Data in Tabular Form

Suppose that you wish to plot a graph for the number of rigs that were running at XYZ Drilling Company for the months January through June. Before a graph is plotted, arrange the data in tabular form. Arrange the data to be plotted on the X axis in one column—the months in this case—and the data for the Y axis—the number of rigs in this case—in a parallel column with corresponding items opposite each other (fig. 3.12).

Choose Scale and Plot Points

The scale chosen for a graph depends on the amount of space available and on the type of data to be shown. Scales are normally chosen so that the maximum value along the Y axis uses about 80 to 90 percent of the available space, and the X axis completely fills the available space. Standard graph paper is often used as a base and it determines the size and scale of the graph. Using the data arranged in tabular form and the scale, the point can be plotted (fig. 3.13).

The next step is to decide the type of graph to show the plotted information.

XYZ DRILLING COMPANY RIG ACTIVITY

X AXIS	Y AXIS
Month	No. Rigs
January	13
February	11
March	8
April	12
May	14
June	19

Figure 3.12   Information arranged in tabular form

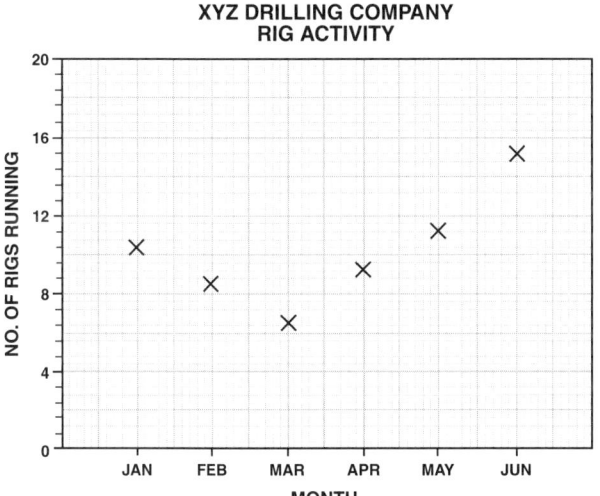

Figure 3.13   Points plotted for a graph

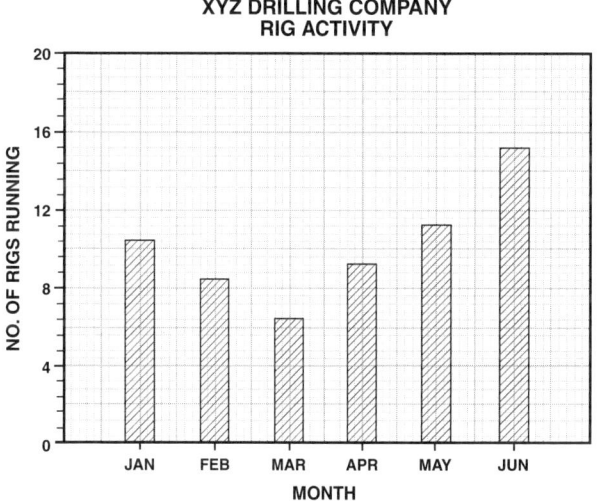

Figure 3.14   Bar graph showing tabular data given in figure 3.12

Types of Graphs

As previously mentioned, three types of graphs are the bar graph, the line graph, and the circle graph. Line graphs may have broken lines or curved lines, depending on the data used. Each type of graph offers information at a glance, such as a trend, a comparison, or a curve.

Bar Graphs

The *bar graph* is a good way to present data that represents a series of observations made at periodic intervals. For example, the bar graph in figure 3.14 depicts the tabular information given in figure 3.12 using the points plotted in figure 3.13. The graph shows the relationship between the number of rigs running in January, February, March, and so on. A bar graph is a good way to

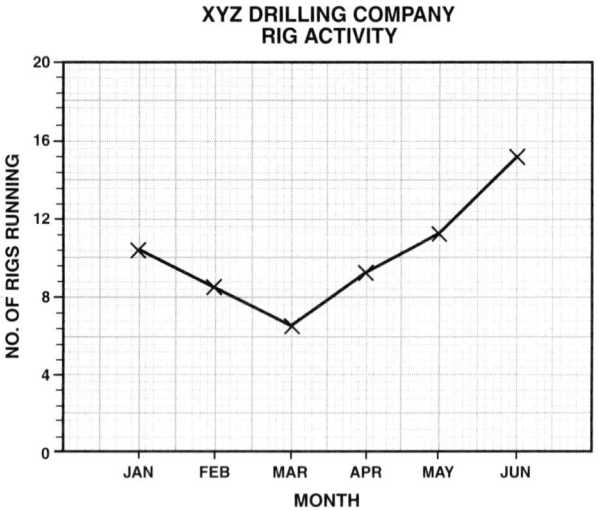

Figure 3.15   Line graph showing tabular data given in figure 3.12

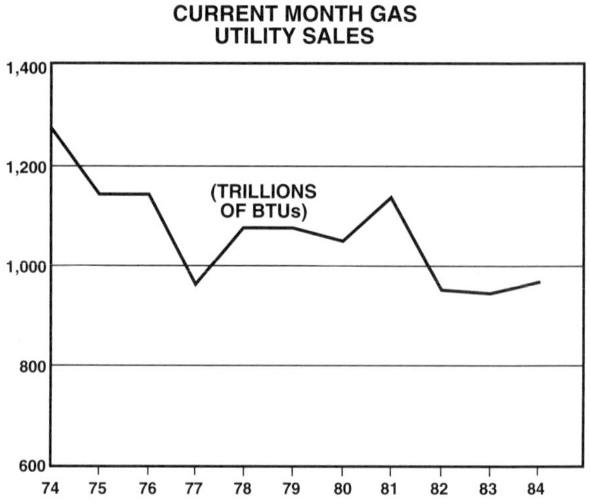

Figure 3.16   Broken-line graph

present unrelated facts—that is, facts where one fact is not dependent on the other.

Line Graphs

The *line graph* can also be used to depict unrelated facts (fig. 3.15); however, it is more commonly used to plot two related variables, such as temperature and time or pressure and time. If the information represents irregular variations, it is plotted by a broken line joining the known points (fig. 3.16; also see fig. 3.15). If the information represents variables depending directly on each other, a curved line may be used to join the plotted data (fig. 3.17).

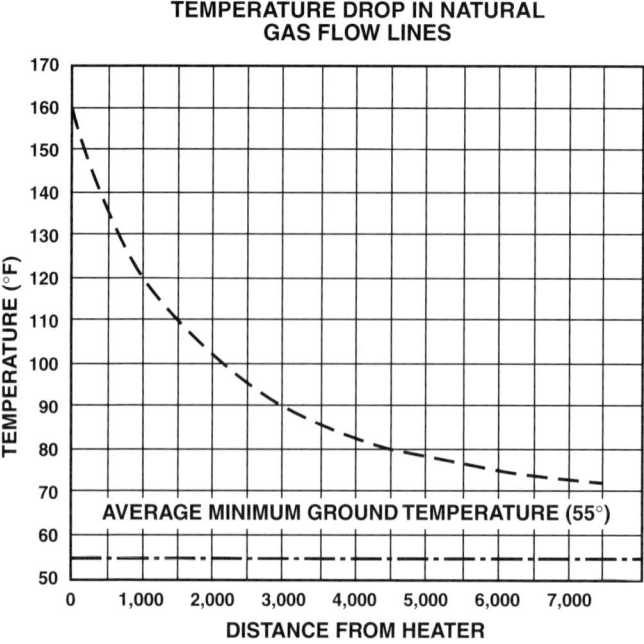

Figure 3.17   Curved-line graph

Curved-line graphs are also used to show relationships in general—without specific statistics. For example, figure 3.18 is a graph that plots the vapor pressure for a pure substance at increasing pressures and temperatures.

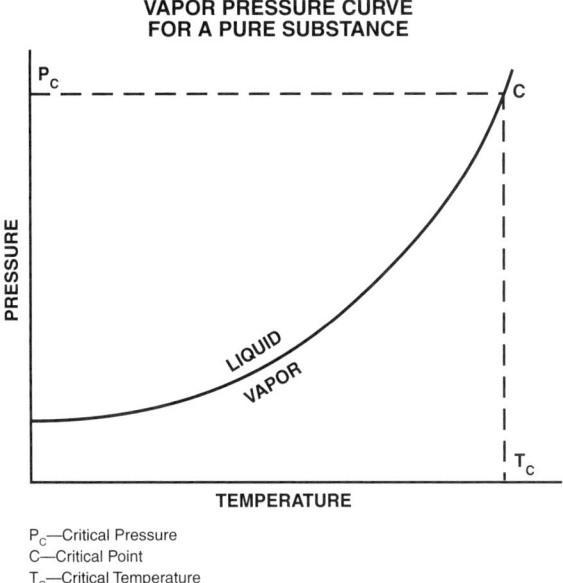

P$_C$—Critical Pressure
C—Critical Point
T$_C$—Critical Temperature

Figure 3.18   Curved graph showing a general relationship

Circle Graphs

A *circle graph*, also called a *pie chart*, is often used to compare the various parts of a whole to each other and to the whole. With 360° equaling the whole, the circle is divided so that each 1% of the whole is represented by 3.6°. To construct a circle graph, it is necessary to find the size of the angle for each part. Find angle size by using the formula

$$\frac{\text{part}}{\text{whole}} \times 360° = \text{angle}.$$

When each angle is determined, it is plotted with a protractor and lines are drawn inside a circle.

Example Problem: A well produces 180 barrels of fluid a day. Ninety barrels are oil, 45 barrels are free water, 36 barrels are emulsion, and 9 barrels are sediment. Make a circle graph, or pie chart, to illustrate this distribution.

Solution: Calculate the angles of each "slice of pie" using the formula:

part ÷ whole × 360° = angle.
90 ÷ 180 × 360° = 180° (oil)
45 ÷ 180 × 360° = 90° (free water)
36 ÷ 180 × 360° = 72° (emulsion)
9 ÷ 180 × 360° = 18° (sediment)

Then plot the graph as shown in figure 3.19.

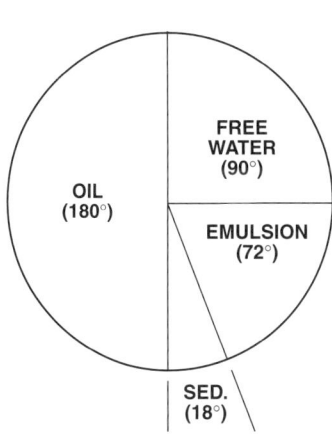

Figure 3.19   Circle graph or pie chart

Practice Problems

1. A worker earned $286.50 one week, of which $60 was spent on rent, $70.80 on food, $7 on laundry, $32.50 on gasoline, $63.40 on incidentals, and the rest was saved. Plot a circle graph showing how the worker's weekly salary was distributed. (Round off degrees to whole numbers before plotting.)

2. Using the graph of OPEC production and demand shown below as a reference, give approximate answers to the following questions.

    a. In what month and year did the excess capacity cease, creating a supply shortage?

        _____

    b. From 1981 to 1982, did the demand for OPEC oil increase or decrease?

        _____

   c. How much was the excess capacity of OPEC oil production at the
       beginning of 1976?

       _____

   d. In 1984, what was the average daily OPEC capacity to produce oil?

       _____

   e. In what year did OPEC production capacity begin to gradually level
       out?

       _____

3. The pressure of a certain volume of gas at 50°F is 1,000 psig. At 60°F the
    pressure increases to 1,020 psig; at 70°F to 1,040 psig; at 80°F to 1,060
    psig; and at 100°F to 1,100 psig. Plot a smooth curve from this data on a
    line graph.

4. The total miles of gas pipe installed by type between 1995 and 2000 are to
    be put into a table. Statistics for plastic pipe installed are 1995—15,985;
    1996—11,640; 1997—15,991; 1998—14,576; 1999—18,826; and 2000—
    18,912. Similar statistics for steel pipe are 1995—12,663; 1996—9,718;
    1997—6,157; 1998—7,212; 1999—8,145; and 2000—8,334. Construct a
    table giving this information.

AVERAGE DAILY WORLD CRUDE PRODUCTION
(in millions of bbl)

YEAR	OFFSHORE	TOTAL
1976	8.5	58.5
1977	10.0	56.5
1978	11.0	60.0
1979	12.0	62.5
1980	13.5	59.0

5. Total world crude production varied from year to year between 1976 and 1980, but offshore crude production showed a steady growth. Using the statistics given in the table to the left, construct a bar graph that shows total production and offshore production. (Note: Shade a part of each bar to show the offshore production.)

SELF-TEST

3. Number Relations

*Multiply each question/problem answered correctly by five
to arrive at your percentage of competency.*

1. What is the ratio of 25¢ to $1.00?

_____ 1:4 _____

2. If ten men can dig 125 feet of ditch in one day, how many feet can twenty-five men dig?

_____ 312·5 ft _____

3. One well produces 75 barrels per day, while a nearby well makes 225 barrels per day. What is the ratio of the production of the first well to the production of the nearby well?

_____ 1:3 _____

4. A pickup runs 42 miles on 3 gallons of gas. How many gallons are needed for a 200-mile trip?

_____ 14·28 gal _____

5. If 16 hammers cost $256, how much do 36 hammers cost?

_____ $576.00 _____

6. A worker can finish 9 bolts in 12 minutes. How many hours will it take the worker to finish 165 bolts?

_____ 3·66 hrs _____

7. A 6-inch drive pulley makes 800 revolutions per minute. What is the speed of a 10-inch pulley that is driven by the 6-inch pulley?

_____ 480 rpm _____

8. An 18-inch pulley on an engine is driving an 11-inch band wheel. The speed for the 18-inch pulley is 180 rpm. What is the speed of the band wheel?

_____ 294·54 rpm _____

9. A pinion gear with 16 teeth, running at 1,200 rpm, drives a gear that has 96 teeth. How fast does the gear run?

_____ 200 rpm _____

10. The U.S. wholesale price of kerosene over a 4-month period was $42.20, $42.97, $39.97, and $37.71 per barrel. What was the average price per barrel for this period?

_____ $40·71/bbl _____

11. A derrick and a 10-foot pole nearby both cast a shadow as shown below. The derrick shadow is 64 feet, while the pole shadow is 5 feet. How high is the derrick?

128 ft

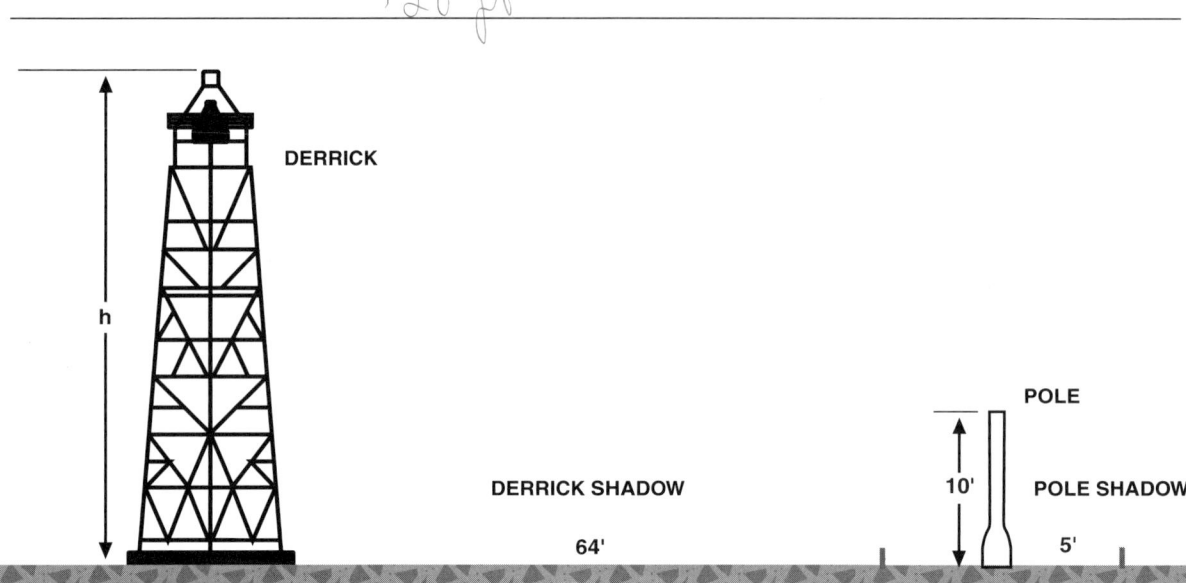

12. Maximum daily temperatures at a plant during one month were (in °F) 46, 47, 51, 50, 45, 48, 61, 59, 55, 54, 47, 52, 55, 46, 39, 43, 52, 56, 49, 51, 53, 56, 63, 51, 54, 59, 58, 49, and 52. Find the mean, median, and mode for this set of temperature data.

Mean _____

Median _____

Mode _____

13. A pipeline buys 1 million cubic feet (MMcf) of natural gas at $2.67 per thousand cubic feet (Mcf), 48 MMcf at $2.71 per Mcf, and 36 MMcf at $2.56 Mcf. What is the average cost per Mcf?

_____

14. The Smith family used a total of 7,284 kilowatt-hours (kWh) of electricity for the months designated in the table. (The table also shows cost per kWh for each month and the monthly cost per kWh.) Calculate the average price paid per kWh by the Smiths during the six-month period.

_____

MONTH	NO. kWh	CENTS/kWh
July	1.360	7.50
August	1,424	7.52
September	1,292	7.55
October	1,150	7.50
November	960	7.25
December	1, 098	6.97

15. Five quarts of treating solution are added to every 100 barrels of production from a well. How many quarts of treating solution are required for 780 barrels of production?

_____

16. Plot a line graph showing the fluctuation of active rotary drilling rigs in the year 1982, using the following information.

MONTH	NO. RIGS
January	4,436
February	4,160
March	3,816
April	3,460
May	3,128
June	2,980
July	2,746
August	2,620
September	2,483
October	2,402
November	2,500
December	2,696

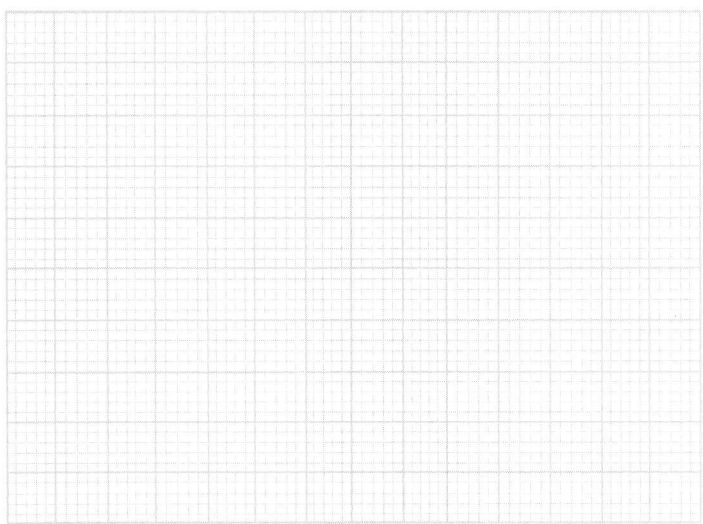

17. Construct a table showing the U.S. wholesale product prices. The prices for gasoline, kerosene, fuel oil, and residual fuel for the last three months in 1982 were as follows: gasoline—$39.68, $38.87, and $36.88 per barrel; kerosene—$42.20, $42.97, and $36.88 per barrel; fuel oil—$40.06, $40.79, and $37.84 per barrel; and residual fuel—$24.78, $26.06, and $24.57 per barrel.

18. Between 1973 and 1982, there was a decline in daily consumption of residual oil in the United States, Japan, and Western Europe, while consumption increased in the rest of the free world. Show this comparison in a bar graph, using the following figures:

MILLION BARRELS/DAY

	1973	1982
U.S.	2.8	1.7
Japan	2.6	1.4
Western Europe	4.5	2.9
Rest of Free World	3.2	3.6
Total Free World	13.1	9.6

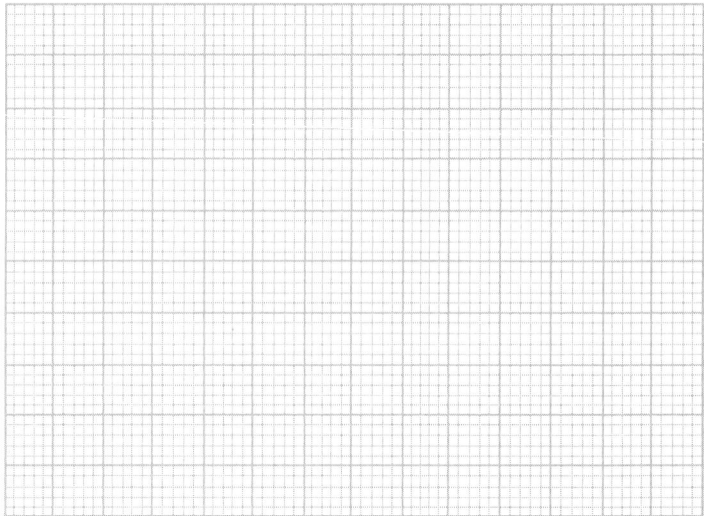

19. Steam-cracking naphtha yielded the following petrochemicals in the percentages shown: ethylene, 23%; propylene, 16%; butene, 9%; butadiene, 5%; fuel gas, 15%; steam-cracked naphtha, 26%; and fuel oil, 6%. Construct a circle graph showing these percentages. Be sure to label each section of the graph. (Round off degrees to whole numbers before plotting.)

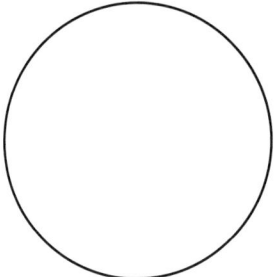

20. The amount of coal produced in the United States between 1970 and 1980 was as follows:

YEAR	MILLION SHORT TONS
1970	612.7
1971	560.9
1972	602.5
1973	598.6
1974	610.0
1975	654.6
1976	684.9
1977	697.2
1978	670.2
1979	781.1
1980	829.7

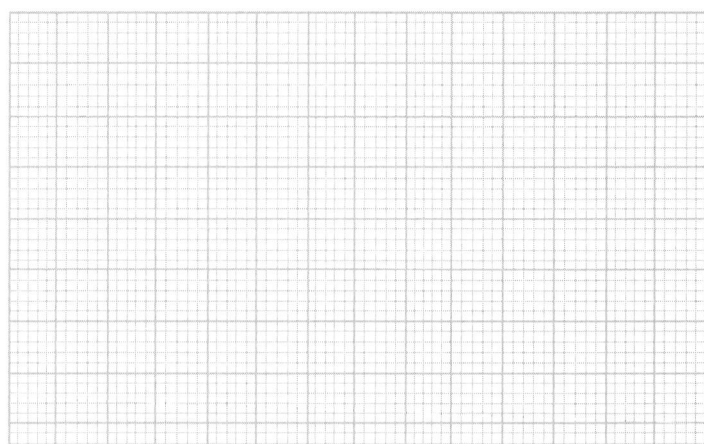

Present this information graphically.

4

Some Physical Quantities and Their Measurement

OBJECTIVES

Upon completion of chapter four, the student will be able to—

1. Describe the difference between fundamental quantities and derived quantities.
2. Convert measurements from U.S. conventional units to SI metric units and from SI metric units to conventional units.
3. Convert units of measurement to equivalent units in order to solve problems.
4. Use and read proper symbols and abbreviations for common measurement units.
5. Measure distance using a ruler or scale.
6. Solve problems concerning length, area, and volume measurements.
7. Convert temperatures from one scale of measurement to another.
8. Work in all increments of time measurement.
9. Solve problems involving measurements of weight, mass, force, work, power, pressure, density, and specific gravity.
10. Solve problems dealing with the measurement units of electricity including volts, amperes, and ohms.
11. Read simple electrical circuits and solve problems involving voltage, current, and resistance.
12. Solve for kilowatt-hours and other electrical power measurements.

INTRODUCTION

Quantity has many definitions. As used here, a *physical quantity* is something that has dimensions and can be measured, such as length, mass (weight), and time. These three physical quantities are *fundamental quantities*. A fundamental quantity cannot normally be divided into other quantities. Additional fundamental quantities are electric current, luminous intensity, temperature, and the amount of a substance. Thus, seven fundamental quantities exist: (1) length, (2) mass, (3) time, (4) electricity, (5) luminous intensity, (6) temperature, and (7) the amount of a substance.

Scientists derived several nonfundamental quantities from the seven fundamental quantities. Consequently, they called nonfundamental quantities derived quantities. For example, velocity is a derived physical quantity because it is composed of distance (length) and time. So, when we speak of velocity, we speak of it in terms of miles per hour, feet per second, and so forth. Another example of a derived quantity is pressure. It is a measure of a force on a given area—for example, pounds per square inch.

TABLE 4.1
Conventional and Metric (SI) Units of Measurement
for Some Physical Quantities

Quantity	Conventional		Metric (SI)	
	Unit	Symbol	Unit	Symbol
Fundamental Quantities				
Length	foot	ft	metre	m
Mass	pound	lb	kilogram	kg
Time	second	sec	second	s
Temperature	degree Fahrenheit	°F	degree Celsius	°C
	degree Rankine	°R	Kelvin	K
Electrical current	ampere	A	ampere	A
Amount of substance	mole	mol	mole	mol
Luminous intensity	candela	cd	candela	cd
Derived Quantities				
Acceleration	feet per second per second	ft/sec^2	metre per second per second	m/s^2
Area	square feet	ft^2	square metre	m^2
Density	pound per cubic foot	lb/ft^3	kilogram per cubic metre	kg/m^3
Electrical potential	volt	V	volt	V
Electrical resistance	ohm	Ω	ohm	Ω
Electrical capacitance	farad	F	farad	F
Energy	kilowatt-hour	kWh	kilowatt-hour	kWh
Force	pound-force	lb_f	newton	N
Frequency	hertz	Hz	hertz	Hz
Power	foot-pound per second	ft-lb/sec	joule per second	J/s
	horsepower	hp	watt	W
	watt	W	watt	W
Pressure	pound-force per square inch	psi	pascal	Pa
Velocity	foot per second	ft/sec	metre per second	m/s
Volume	cubic foot	ft^3	cubic metre	m^3
	gallon	gal		
	barrel	bbl		
Work	foot-pound	ft-lb	joule	J
			newton-metre	N•m

Another term for a fundamental quantity is *dimension*. Dimensions include distance (length), time, and mass (weight). So, we can say, for example, that the dimension of a room is 12 feet wide by 14 feet long by 8 feet high. We can also say that it takes two hours to complete a journey, and that a car weighs 3,250 pounds.

A *unit* is a standard measure of a quantity. For example, a unit of length is the foot. In the measurement system commonly used in the United States, which is the English, or conventional, system, each quantity usually has more than one unit. Thus, the unit of measure for length is not only the foot, but also the inch, yard, and mile, to name but a few. Similarly, we measure mass, or weight, in tons, pounds, drams, ounces, and other units. We measure time in years, months, weeks, days, hours, minutes, and seconds. Table 4.1 shows the base units for fundamental quantities, as well as derived quantities.

Several years ago, scientists from all over the world held a special Conference on Weights and Measures. One result of the conference was a new system of measurement, which some parts of the world adopted—notably Canada and the United Kingdom. This system of measurement is based on the metric system. Abbreviated in virtually all languages as SI (from the French, le Système International d'Unités), it has only one unit for each fundamental quantity: *metre* for length, *kilogram* for mass, *second* for time, and so on. (One exception is temperature measurement; for normal applications, degree Celsius is used, but for scientific uses, degree Kelvin is preferred.) Measurements larger or smaller than the base units are expressed in multiples or submultiples of the fundamental unit, such as kilometres (1,000 metres) and millimetres (0.001 metre).

Because the SI system employs the British spelling of many of the terms, this book follows those spelling rules. The unit of length, for example, is *metre*, not *meter*. (Note, however, that the unit of weight is *gram*, not *gramme*.)

The SI metric system enjoys a distinct advantage over the conventional, or English, system (which virtually only the United States uses) because it uses powers of ten and prefix names and symbols. Using powers of ten makes it easy to convert from one unit to another: simply multiply or divide by ten. For those in the U.S., think of how easy it is to convert from dollars to cents in the U.S. monetary system. For example, if you have 100 pennies, you also have 1 dollar. Similarly, 10 dimes make 1 dollar and 10 ten-dollar bills add up to $100. The same holds true for the metric system: 100 centimetres is a metre, 1,000 metres is a kilometre, a millimetre is $\frac{1}{1,000}$ (0.001) of a metre, a decimetre is $\frac{1}{10}$ (0.1) of a metre, and so on.

In the conventional system of weights and measures, we use many units to express the value of a fundamental quantity. Typically, these units are clumsily related to other units that measure the same physical quantity. For example, inches are divided into quarters, eighths, and sixteenths; a foot has 12 inches; a yard has 3 feet, or 36 inches; and a mile has 5,280 feet, or 1,760 yards.

On the other hand, a single unit in the SI metric system represents each fundamental quantity. Measurements much larger or much smaller than the base unit are expressed by multiplying that unit by 10^n, where n is an exponent that can be a positive (+) or a negative (–) whole number. For example, a kilometre is 1,000 metres and can be expressed as 10^3 metres. A millimetre is $\frac{1}{1,000}$ (0.001) metre, or 10^{-3} metre. A megametre is 10^6 or 1 million (1,000,000) metres. The SI system encourages the use of $\pm$ values of n equal to ±3, ±6, ±9, ±12, ±15, and ±18. These increments allow accurate expression of magnitude and, at the same time, reduce the number of prefixes and symbols required. Table 4.2 lists the names and symbols of the prefixes and their numerical values. In an effort to keep the system simple, SI discourages the use of deca, deci, hecto, and centi,

TABLE 4.2
SI Metric Prefixes

Numerical	Name	Symbol
*10^1	deca	da
*10^2	hecto	h
10^3	kilo	k
10^6	mega	M
10^9	giga	G
10^{12}	tera	T
10^{15}	peta	P
10^{18}	exa	E
*10^{-1}	deci	d
*10^{-2}	centi	c
10^{-3}	milli	m
10^{-6}	micro	μ
10^{-9}	nano	n
10^{-12}	pico	p
10^{-15}	femto	f
10^{-18}	atto	a

SI discourages using these values and names.

their symbols, and their numerical values. However, because people in countries that already use the metric system have employed these prefixes for some time, they continue to be used.

LENGTH, AREA, AND VOLUME

Length is a common measurement. It is one of the fundamental quantities and is the base unit for area and volume. Area is the product of two lengths, and volume is the product of three lengths.

Length

Length can be measured in many different units. Two common ones in the conventional system are the *foot* and the *inch*. For measuring small values, the inch is expressed in mils, which is short for thousandths of an inch. In close-tolerance work—for example, measuring the clearance between a bearing cap and a shaft—machinists and mechanics use values of ten-thousandths of an inch. Generally, the conventional system uses the foot to measure intermediate lengths or distances. However, units of length overlap considerably. For example, people in the pipeline industry commonly refer to a pipeline that is 3 or 3½ feet in diameter as a 36- or a 42-inch pipeline.

Other common units of length in the conventional system are *yards* and *miles*. Yards are well known on U.S. football fields and also measure lengths of material, such as fabric or carpet. Miles are commonly used to measure distances of travel or distances between points on the earth's surface.

Where the distances are enormous, such as in interstellar space, scientists often use special units of length. For example, astronomers often use light years to express the distance between stars and galaxies. A *light year* represents the distance light travels in a year's time. Since light travels at 186,000 miles per second, a light year is about 5.87 trillion miles.

A disadvantage of the conventional measurement system is that more than one unit is frequently used to express a measurement. For example, a length of pipe may be 18 feet, 7 inches long. Two units, feet and inches, are used to state the length. The same piece of pipe in SI metric is 5 metres, 66 centimetres long, which is stated and written as 5.66 metres. This same disadvantage is encountered in weight measurement where typically pounds and ounces are used to express weight.

As previously mentioned, in the SI system, a simple relation exists among metres, centimetres, and millimetres. Each is easily converted to the others by using multiples or submultiples of 10 as a multiplier. For example, 1 metre equals 10^2 (or 100) centimetres, and 10^3 (or 1,000) millimetres. Conversely, a millimetre equals 10^{-3} (or $\frac{1}{1,000}$) metre, and a centimetre equals 10^{-2} (or $\frac{1}{100}$) metre. In decimal fractions, 10^{-3} is 0.001 and 10^{-2} is 0.01. Notice that when 10 is raised to a power, the exponent's value represents the number of zeros that follow 1. For example, 10^5 is 1 followed by 5 zeros, or 100,000. Similarly, 10^{-5} is 1 over 1 followed by 5 zeros or $\frac{1}{100,000}$. Expressed as a decimal fraction, 10^{-5} is 0.00001. The negative exponent indicates the number of decimal places to the right of the decimal point—in the case of 10^{-5}, it is five.

Example Problem: Express 0.360 metre in centimetres and millimetres.

Solution: A metre has 10^2 centimetres or 10^3 millimetres. Multiplying by 10^2 moves the decimal point two places to the right to obtain centimetres:

$$0.360 \times 10^2 = 36.0 \text{ centimetres.}$$

Multiplying by 10^3 moves the decimal point three places to the right:

$$0.360 \times 10^3 = 360 \text{ millimetres.}$$

Example Problem: Express 10^{-6} metres as a decimal fraction.

Solution: 10^{-6} is 0.000001 metres, which is 1 millionth of a metre (a micrometre, or a micron).

Example Problem: Express 57 centimetres as a decimal fraction of a metre.

Solution: Since each metre contains 100 centimetres, 57 centimetres is $^{57}/_{100}$, or 0.57 metre.

Symbols and Abbreviations

When dealing with length measurement, symbols and abbreviations are convenient to use. A length of 10 feet, 2 inches can be written as 10'2", or 10 ft, 2 in. Yard and mile are abbreviated as yd and mi, respectively. Metres, kilometres, centimetres, and millimetres, are symbolized by m, km, cm, and mm. Metric symbols are never followed by a period. A period after yd, ft, and mi is optional. However, the abbreviation for inch, in., should have a period to avoid confusion with the word in.

Measuring Devices

To measure length, rules, tape measures, surveying chains, and calipers can be used. On one extreme is a surveying chain, which is a long tape measure, 100 feet or more in length. On the other extreme, a caliper measures short lengths with great accuracy. For example, a micrometer caliper (fig. 4.1) measures in mils (thousandths of an inch), or it may be calibrated in micrometres (in millionths of a metre).

Small parts of an inch are usually expressed in fractional form: ½, ¼, ⅛, ¹⁄₁₆, ¹⁄₃₂, and ¹⁄₆₄. Sizes smaller than ¹⁄₆₄ are typically measured in mils (0.001 in.) or, for even smaller values, in ten-thousandths of an inch (0.0001 in.). Many rules and tapes are divided into inches and fractional parts as small as ¹⁄₁₆ or ¹⁄₃₂ of an inch (fig. 4.2).

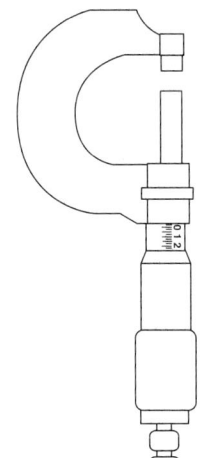

Figure 4.1   Micrometer caliper

Example Problem: Using figure 4.2, give the distance between *A* and *E* and between *G* and *H*.

Solution:

$$A \text{ to } E = 2^3/_{16} \text{ in.}$$
$$G \text{ to } H = {}^7/_{16} \text{ in.}$$

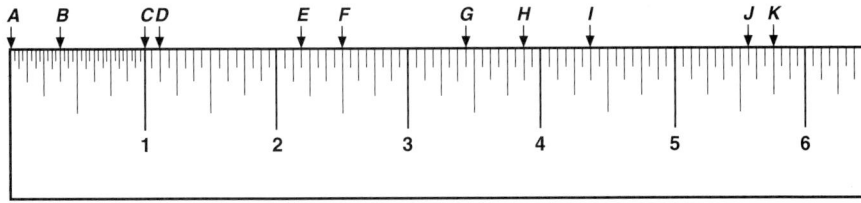

Figure 4.2   Section of a conventional ruler, divided into sixteenths

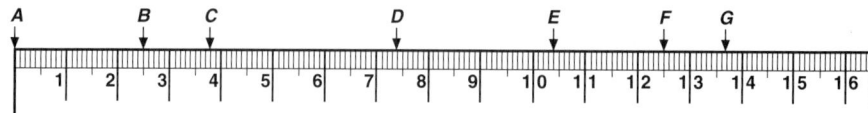

Figure 4.3   Section of a metric ruler, divided into tenths

Each centimetre on an SI metric ruler is divided into 10 millimetres. Figure 4.3 shows the true sizes of these small metric units.

Example Problem: Using figure 4.3, give the distance between *A* and *B*.

Solution: Each gradation is 1 millimetre, which is ⅒ of a centimetre, or 0.1 cm. Contained in the distance between *A* and *B* are 2 cm and 5 mm. Written as a decimal,

$$A \text{ to } B = 2.5 \text{ cm.}$$

Besides converting from metres to centimetres or another metric length measurement, it is often necessary to convert from metric units to conventional units and vice versa. To convert from conventional to metric units, the metric equivalent is used as a multiplication factor. For example, the metric equivalent of 1 mile is 1.609 kilometres (table 4.3). Thus, to convert 4 miles to kilometres, multiply:

$$4 \times 1.609 = 6.436 \text{ km.}$$

TABLE 4.3
Units of Length Measurement

Conventional Unit	Conventional Equivalents	Metric Equivalents
inch (in.)	1,000 mils 0.08333 ft	2.54 cm 25.4 mm
foot (ft)	12 in.	0.3048 m
yard (yd)	3 ft 36 in.	0.9144 m
rod (rd)	5.50 yd 16.5 ft	5.029 m
statute mile (mi)	1,760 yd 5,280 ft	1,609 m 1.609 km

Metric Unit	Metric Equivalents	Conventional Equivalents
metre (m)	100 cm 1,000 mm	39.37 in. 3.28 ft
kilometre (km)	1,000 m	3,280 ft 1,093 yd 0.62 mi
centimetre (cm)	10 mm 0.01 m	0.394 in.
millimetre (mm)	0.001 m 0.1 cm	0.0394 in.

Example: Convert 6 inches to millimetres; then change millimetres to centimetres.

Solution: From table 4.3, find the millimetre equivalent to 1 inch: 25.4. Then multiply:

$$6 \times 25.4 = 152.4 \text{ mm}$$
$$152.4 \div 10 = 15.24 \text{ cm.}$$

To convert from metric to conventional, the same principle applies. Find the conventional unit equivalent and multiply this factor by the metric measurement.

Example Problem: A famous World War II cannon had a bore of 88 millimetres. What is the diameter of this gun's bore in inches?

Solution: Since 1 millimetre equals 0.0394 inch, multiply:

$$88 \times 0.0394 = 3.4672.$$

The diameter of the bore is 3.4672 inches.

Area and Volume

Area and volume are closely related to length. As stated earlier, area is the product of two lengths, and volume is the product of three lengths. Sometimes, however, the results of these multiplications are further multiplied by a constant, which, as you may recall, is a quantity that retains a fixed value throughout a particular set of calculations. For example, in calculating the area of circles and spheres, a constant is required, as you will see shortly.

Figure 4.4 shows examples of area and volume in their simplest forms: a square and a cube. Side lengths (labeled *a*) for both the square and the cube are equal. The area of the square is simply

$$a \times a = a^2$$

and the volume of the cube is

$$a \times a \times a = a^3.$$

If *a* is equal to 3 ft, the square has an area of

$$3 \times 3 = 9 \text{ ft}^2$$

and the cube has a volume of

$$3 \times 3 \times 3 = 27 \text{ ft}^3.$$

Note that you treat the units of length (feet) as mathematical entities to square and cube them. In the example, each side of the square is 3 feet and each side of the cube is 3 feet. Thus, the units of feet are the same as the number 3.

Of course, figures and objects do not always have equal sides. The rectangle in figure 4.5 has two sides equal to *a* and two sides equal to *b*. The rectangular solid in figure 4.5 has three sides with length *a*, three with length *b*, and three with length *c*. Although the sides are of different lengths, the areas and volumes are calculated the same:

$$\text{area} = a \text{ ft} \times b \text{ ft} = ab \text{ ft}^2.$$
$$\text{volume} = a \text{ ft} \times b \text{ ft} \times c \text{ ft} = abc \text{ ft}^3.$$

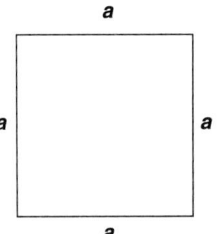

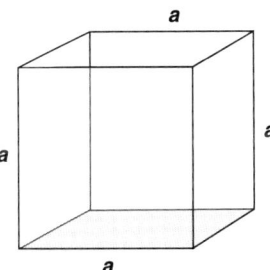

Figure 4.4   Equal-sided square and cube

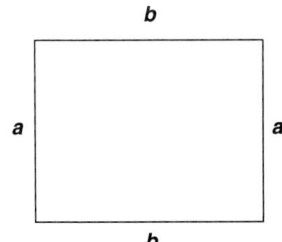

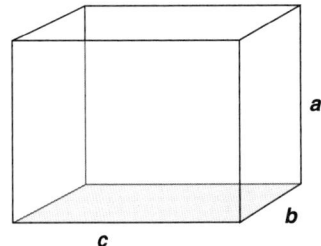

Figure 4.5   A rectangle and a rectangular solid

All plane figures have area and all are determined by multiplying two values of length. (A *plane figure* is a figure that only has two dimensions.) For example, a circle has area, and calculating it involves a product of two lengths and a constant. The formula for the area of a circle is πr^2, where r is the radius of the circle, and π (called pi) is a constant equal to 3.1416. Pi (π) represents the ratio of the circumference to the diameter of a circle. The double use of the radius r provides the product of two lengths —that is,

$$r \times r = r^2.$$

Example Problem: Determine the area, in square inches, of a circle whose radius is 6 inches.

Solution: Because the equation for determining the area of a circle is πr^2, first square 6, then multiply it by π:

$$6^2 = 36 \times 3.1416 = 113.1 \text{ square inches (in.}^2).$$

A formula to determine the area of a triangle is $\frac{1}{2}b \times a$, or $\frac{1}{2}(ab)$ where b is the length of the base and a is the altitude (height) of the triangle (fig. 4.6).

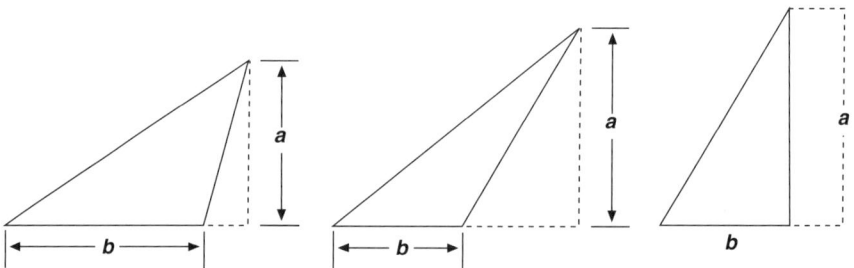

Figure 4.6   Several forms of triangles

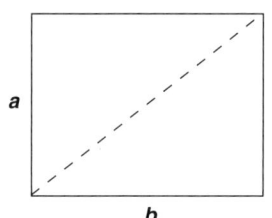

Figure 4.7   The area of a triangle

Be sure to measure the base and altitude along lines that are perpendicular to one another as figure 4.6 shows with dotted lines. The formula for determining the area of a triangle is based on the area of a rectangle. Figure 4.7 shows that a diagonal line that halves the rectangle divides it into two equal triangles. The area of the rectangle is $a \times b$; so, half of a rectangle (which is a triangle) is $\frac{1}{2}(a \times b)$, or $\frac{1}{2}(ab)$.

Example Problem: Determine the area, in square inches, of a triangle with a base of 1 foot and an altitude (height) of 18 inches.

Solution: First, convert feet to inches; in this case, 1 foot = 12 inches. Then, because the equation for determining the area of a triangle is $\frac{1}{2}(ab)$, multiply a times b; then, multiply the product by one half (0.5):

$$18 \times 12 = 216 \times 0.5 = 108 \text{ square inches (in.}^2).$$

Areas and volumes have an unlimited number of shapes and sizes. Both are characterized by having dimensions of length as principal components. The only other component is a constant, such as $\frac{1}{2}$ (for the area of a triangle), π (for the area of a circle), or $\frac{4}{3}\pi$ (for the volume of a sphere). Chapter 6, "Practical Geometry," further explains formulas for calculating areas and volumes.

Basic Conventional Units

In the English, or conventional, system of measurement, square inch, square foot, and square yard are the basic values of *area*, and the cubic inch, cubic foot, and cubic yard are the basic values of *volume*. Some units of area and volume have special names in the conventional system, but all are based on these few units. For example, a cord, which is a measure of a quantity of firewood, is a volume 4 by 4 by 8 feet.

Square feet commonly express the floor area in buildings. Fluid ounces, quarts, gallons, and barrels are important units of *liquid measure*. Natural gas and liquids such as drilling mud are often measured in cubic feet. Cubic yards often designate quantities of material such as sand, gravel, and concrete.

A fluid ounce is one of many ambiguous units in the conventional system. The ounce, either troy or avoirdupois, is a measure of weight or mass, while the fluid ounce is a unit of volume. Actually, 1 fluid ounce of water (1.8047 in.^{3}) weighs 1 ounce, avoirdupois, but 1 fluid ounce of alcohol, which is less dense than water, weighs somewhat less. Incidentally, the weight terms avoirdupois and troy are French in origin. Avoirdupois comes from the old French phrase *avenir de peis*, which translates as goods of weight. Avoirdupois weight is based on a 16-ounce pound and is the most common weight unit used in the conventional system. Troy weight, on the other hand, is based on a 12-ounce pound. Named after the city of Troyes, France, where it originated, troy weights are employed in weighing precious metals and gemstones. For example, ½ troy pound of gold weighs 6 troy ounces.

Metric Units

The traditional metric system has many units for area and volume—square centimetres and square metres for area; and cubic centimetres, litres, and cubic metres for volume. SI metric recognizes only the square metre (m^2) for area and the cubic metre (m^3) for volume. However, the cubic centimetre (formerly cc, now cm^3, or 10^{-6} m^3) and litre are so entrenched in laboratory and commercial use, respectively, that many people still employ them. A litre is 1,000 cubic centimetres, or 1.0571 U.S. quarts. Nowadays, many technicians also use a millilitre (mL), which is almost the same as a cubic centimetre, to measure volume.

Land Measurement Units

The acre, square mile, section, and township are important units of land measurement in the U.S. In countries using the metric system, land is measured in square metres or in hectares. A hectare contains 10,000 square metres.

Standard U.S. linear land measures are the mile, rod, and foot. In the conventional system, the standard measure for land area is the section or acre. A section is 1 square mile; an acre is 43,560 square feet or 160 square rods. A square acre measures 208.71 feet on each side. Half-sections and quarter-sections are common divisions of land and correspond to 320 acres and 160 acres, respectively. Often, people refer to the size of land in acres.

Board Measurement Units

An interesting and practical measurement peculiar to English-speaking countries is for lumber. Lumber is often measured in *board feet* (bd ft). A variation of volume measurement, a board foot contains 144 cubic inches of wood. A board that is 1 inch thick, 12 inches wide, and 12 inches long contains 1 board foot. Lumber is usually described as having inches of thickness, inches of width, and feet of length. For example, a piece of wood that is approximately 2 inches thick by 4 inches wide and 12 feet long is usually written as 2" × 4"–12'. In the same way, a 16-foot long piece of lumber that is approximately 1 inch by 12 inches is written as 1" × 12"–16'.

Although the actual thickness and width of a length of lumber is usually in fractions of an inch, fractional values are not considered when calculating the amount of board feet in lumber. For instance, a two by four (2" × 4") is really closer to 1⅝" or 1¾" by 3⅝" or 3¾", but board-foot calculations stick with 2" × 4".

Fractional lumber dimensions occur because rough-sawn lumber—that which comes directly from the sawmill—is smoothed (planed) to make it more usable for construction. In their rough-sawn state, boards are quite close to whole-number dimensions. However, planing the rough-sawn boards removes enough material to give them fractional dimensions. For example, 2 × 4s are normally planed to an actual thickness and width of about 1⅝" × 3⅝".

In any case, the equation for board feet is

$$\text{bd ft} = \frac{\text{thickness (in.)} \times \text{width (in.)} \times \text{length (ft)}}{12.}$$

Example Problem: Find the board feet in a 2" × 4"–12'.

Solution:

$$\text{bd ft} = \frac{2" \times 4" \times 12'}{12}$$

$$= \frac{8 \times 12}{12}$$

$$= \frac{96}{12}$$

$$\text{bd ft} = 8.$$

A 2" × 4"–12' contains 8 board feet. Notice that you do not have to convert the length measurement of feet into inches.

Conversions

Converting from one unit to another is an important step when solving area and volume problems. In some cases, it is necessary to change one conventional unit to another, such as square inches to square feet or cubic feet to barrels. Moreover, it is often necessary to change from conventional units to SI metric units and vice versa. Tables 4.4 and 4.5 show several conversion factors. To convert from one conventional unit to another, find the conventional unit of the measurement in the left-hand column of the appropriate table. Then, find its equivalent conventional value in the units desired (middle column) and multiply this figure by the number of original conventional units.

Example Problem: How many gallons does a container with a capacity of 532 cubic inches hold?

TABLE 4.4
Units of Area Measurement

Conventional Unit	Conventional Equivalents	Metric Equivalents
Square inch (in.²)	0.007 ft² 0.00077 yd²	645.16 mm² 6.451 cm²
Square foot (ft²)	144 in.² 0.111 yd²	0.093 m²
Square yard (yd²)	1,296 in.² 9 ft²	0.836 m²
Acre	43,560 ft² 4,840 yd²	4,046.45 m² 0.405 hectare
Square rod (rd²)	30.25 yd² 0.0006 acre	25.293 m²
Square mile (mi²)	640 acres 102,400 rd²	2.59 km²
Section	1 mi²	2.59 km²
Township	36 mi²	93.24 km²

TABLE 4.5
Units of Volume Measurement

Conventional Unit	Conventional Equivalents	Metric Equivalents
Cubic inch (in.³)	0.00058 ft³ 0.0043 gal	16,387 mm³ 16.387 cm³
Cubic foot (ft³)	1,728 in.³ 7.48 gal	0.028 m³
Fluid ounce (oz)	1.8047 in.³	29.573 cm³
Quart	57.75 in.³ 32 fluid oz	0.00378 m³ 0.946 litre
U.S. gallon	231 in.³ 128 fluid oz 0.1337 ft³	0.00378 m³ 3.784 litres
Barrel (petroleum)	5.615 ft³ 42 U.S. gal	0.1589 m³ 158.928 litres

Solution: Using table 4.5, locate cubic inch in the left-hand column and its equivalent in gallons (0.0043) in the middle column. Multiply the equivalent by 532 cubic inches:

$$0.0043 \times 532 = 2.2876 \text{ gal.}$$

The container holds 2.2876 gallons.

To convert conventional units to metric units, multiply the metric equivalent (right-hand column) by the number of conventional units desired.

Example Problem: How many square metres does 8 acres contain?

Solution: Find acre in the left-hand column of table 4.4, then its metric equivalent (4,046.45 m²) in the right-hand column. Multiply:

$$4{,}046.45 \times 8 = 32{,}371.6 \text{ m}^2.$$

Eight acres contains 32,371.6 square metres.

To convert metric units to conventional units, two ways are available. One is to divide the metric units by the metric equivalent.

Example Problem: How many U.S. gallons in 2 cubic metres?

Solution: Table 4.5 shows that 1 gallon has 0.00378 cubic metres; so, divide:

$$2 \div 0.00378 = 529.1 \text{ gal.}$$

Two cubic metres contain 529.1 gallons.

If you prefer to multiply rather than divide (sometimes, multiplying is easier than dividing especially when making complex calculations), and you have a calculator with a reciprocal key (1/x), you can find the reciprocal of 0.00378 ($\frac{1}{0.00378}$ or 1 ÷ 0.00378), which is 264.55026. (One cubic metre contains 264.55026 gallons.) In this case, multiply 264.55026 by 2 cubic metres:

$$2 \times 264.55026 = 529.1 \text{ gallons.}$$

Practice Problems

Refer to tables 4.3, 4.4, and 4.5 to work these problems.

1. Using figure 4.2, give the dimensions of the following:

    a. *A* to *B* _____

    b. *A* to *K* _____

    c. *C* to *F* _____

    d. *F* to *J* _____

    e. *D* to *I* _____

2. Referring to figure 4.3, give the dimensions of the following in centimetres:

    a. *A* to *D* _____

    b. *A* to *C* _____

    c. *D* to *F* _____

    d. *F* to *G* _____

    e. *B* to *E* _____

3. A football field in the United States is 100 yards between goal lines. How many metres is this distance?

_____

4. How many centimetres are in 12 inches?

   _____

5. Two towns are 32 miles apart. How many kilometres separate the towns?

   _____

6. A man is 6 feet, 4 inches tall. Express his height in metres.

   _____

7. Express 240 centimetres in the following units. Show all places.

   a. Metres _____

   b. Millimetres _____

   c. Kilometres _____

   d. Micrometres _____

   e. Megametres _____

8. How many square yards are in 1,000 square metres?

   _____

9. How long does it take a pump that pumps 50 gallons per minute to fill a 1,000-barrel tank?

   _____

10. How many cubic inches are contained in 47 gallons?

    _____

11. How many cubic inches are in a board foot?

    _____

12. Calculate the board feet in a 1" × 6"—8'.

    _____

13. A woman is building a patio to be 8 feet wide by 10 feet long. The concrete forms are 6 inches deep and will be filled to the top. How many cubic yards of concrete will she have to buy?

    _____

14. How many square miles are in an acre?

    _____

15. A realtor is dividing a section of land into 2-acre home sites. How many home sites will there be in all?

    _____

16. How many cubic yards are in a cubic metre?

    _____

17. How many litres does it take to make a U.S. barrel of petroleum?

_____

18. What cubic-foot capacity is needed to hold 53 gallons?

_____

19. A lease containing 530 hectares is how many acres?

_____

20. A right-of-way that is 100 yards wide measures 13,492 rods long. How many square miles are contained in the right-of-way?

_____

TABLE 4.6
Common Units of Time

Unit	Equivalent Values
Century	100 years
Decade	10 years
Year	12 months 365+ days* 8,760 + 6 hours*
Week	7 days 168 hours 10,080 minutes
Day	24 hours 1,440 minutes
Hour	60 minutes 3,600 seconds
Minute	60 seconds
Second	0.0167 minute

*A year has 365 days plus 6 hours. The 6 hours are added up and applied every 4 years to form leap year, which has 366 days.

TIME AND TEMPERATURE

Measuring Time

We measure time in many units, but the most common are years, months, weeks, days, hours, minutes, and seconds (table 4.6). A year is based on the time required for the earth to orbit the sun; so, it is also called a solar year. Slightly more than 365 days make up a year. In reality, it takes earth 365¼ days to make one orbit around the sun. Thus, a year is 365 days plus about 6 hours, or one-fourth of a 24-hour day. To keep calendars on track, every 4 years we add up the 6 hours and create an extra 24-hour day. We call this 366-day year a *leap year*. The 12 months making up a year have either 30 or 31 days, except for February, which has 28 days (excluding a leap year when it has 29). To keep track of how many days are in a particular month, remember the doggerel:

> *Thirty days hath September,*
> *April, June, and November.*
> *All the rest have thirty-one,*
> *Save for February,*
> *Which has 28 alone.*

Of course, you also have to remember that in a leap year, February has 29 days.

A day is 24 hours long. To keep track of time, conventional clocks are divided into two 12-hour periods—A.M. (midnight to noon) and P.M. (noon to midnight). The abbreviation A.M. is for the Latin *ante meridiem*, which means before noon. The abbreviation P.M. is for the Latin *post meridiem*, which means after noon. So, to state the correct time with conventional clocks, the time must be followed by A.M. or P.M.—for example, 10:07 A.M. or 4:10 P.M. As for the question: "Is 12 o'clock that follows 11:00 A.M., 12:00 A.M. or 12:00 P.M.?" the answer is that it is neither. Instead, it is 12 noon, just as 12 o'clock that follows 11:00 P.M. is 12 midnight.

To avoid confusion, the military has long used a 24-hour clock. Thus, they designate midnight as 0000, 1:00 A.M. as 0100 hours (spoken as oh-one-hundred hours), 2:00 A.M. as 0200, and so on. Noon is 1200 (spoken as twelve-hundred hours), 6 P.M. is 1800, and midnight is 2400 or 0000. Countries that use the metric system often use a 24-hour clock. In such cases, the time is written or spoken as, for example, 0935 (oh-nine-thirty-five) or 2115 (twenty-one-fifteen). With a 24-hour clock A.M. and P.M. are not required.

We divide each hour into 60 minutes and 3,600 seconds. The second is the fundamental unit of time in the SI system. The standard second is the *ephemeris* second, an astronomical term having to do with the time it takes the earth to make one orbit around the sun. So, at one time, scientists defined a second in terms of the solar year, but, today, they define it in terms of the frequency of radiation from cesium atoms. Cesium is a soft, silvery white metal that is liquid at room temperature. Its radiation frequency is 9,192,631,770 hertz (cycles per second). Because scientific equipment can measure this frequency accurately, scientists set clocks by it, which makes the clocks very accurate. Indeed, a cesium clock loses only 1 second of accuracy in 1 to 4 million years.

Seconds, minutes, hours, and days are the important units for describing the time rate at which something happens—for example, how many revolutions per minute (rpm) a motor makes as it rotates, how much fuel an aircraft consumes per hour of flight, and how many miles per hour a car travels.

Scientists measure very short intervals of time in milliseconds (ms or 10^{-3} s), microseconds (μs, or 10^{-6} s), and nanoseconds (ns, 10^{-9} s). A millisecond is one thousandth of a second, a microsecond is one millionth of a second, and a nanosecond is one billionth of a second.

Example Problem: An electric motor turns at 1,800 revolutions per minute (rpm). How many revolutions does it make each second?

Solution: Perhaps the best way to solve this problem is to set up a proportional equation with x representing the unknown:

$$\frac{1,800}{60} = \frac{x}{1}$$

Another way of stating the problem is 1,800 is to 60 as x is to 1. Cross multiplying the figures results in:

$$60x = 1,800 \times 1$$
$$60x = 1,800$$
$$x = 1,800 \div 60$$
$$x = 30.$$

The motor makes 30 revolutions per second.

Temperature and Measurement

The measurement and control of temperature are important aspects of everyday life and industry. Temperature, like pressure, forms a part of everyone's environment. Natural temperatures on the earth range from well below the freezing point of water to well above levels that we consider comfortable. Scientists and manufacturers often work with temperatures considerably higher and lower than normal. Theoretically, the lowest temperature is absolute zero, which is defined as the temperature at which molecular movement in a substance ceases. Up to now, laboratories have not been able to achieve absolute zero; however, they have come to within a fraction of a degree.

The conventional measurement system expresses temperature in degrees Fahrenheit (°F). Daniel Gabriel Fahrenheit, a German physicist working in the Netherlands, invented the Fahrenheit scale of temperature measurement in the early 1700s. Important points on the Fahrenheit scale are the temperature of water at its freezing point (32°F) and its boiling point (212°F).

Most of the world measures temperature in degrees Celsius (°C). This scale is named after its inventor, Anders Celsius, who was a Swedish mathematician and astronomer in the mid-1700s. In the days before SI, the original metric system used an identical scale: the centigrade scale. When the SI system was introduced, the measurement specialists needed a new name for the centigrade scale. The word *centigrade* contains an important prefix (centi) and using it would introduce an awkward anomaly into the system. The solution was to honor the inventor of the scale, Anders Celsius. On the Celsius scale, 0° is the temperature at which water freezes and 100° is the temperature at which water boils.

The *kelvin scale* is the absolute temperature scale for metric measurements. The temperature of −273.16 kelvin (K) corresponds to 0°C. The degree sign (°) is not used with kelvin, and the term kelvin is not written with an uppercase K, although an uppercase K is the symbol for degrees kelvin. Its inventor was William Thompson, who was also Lord Kelvin. Kelvin was a physicist born in Northern Ireland and devised his scale in 1848.

Another absolute temperature scale is degrees Rankine (°R). Based on the Fahrenheit scale, it serves the conventional system of measurements. It was named after William R.M. Rankine, a Scottish engineer who set it up in the mid-1800s.

The Fahrenheit and Celsius scales are in common use for everyday functions such as weather forecasting, heating, cooling, cooking, and medical care. The absolute scales are mainly used in scientific applications. For example, the various laws relating to behavior of gases require the use of absolute scales.

Figure 4.8 shows the relationship of the four temperature scales. Table 4.7 shows how these scales are related through equations.

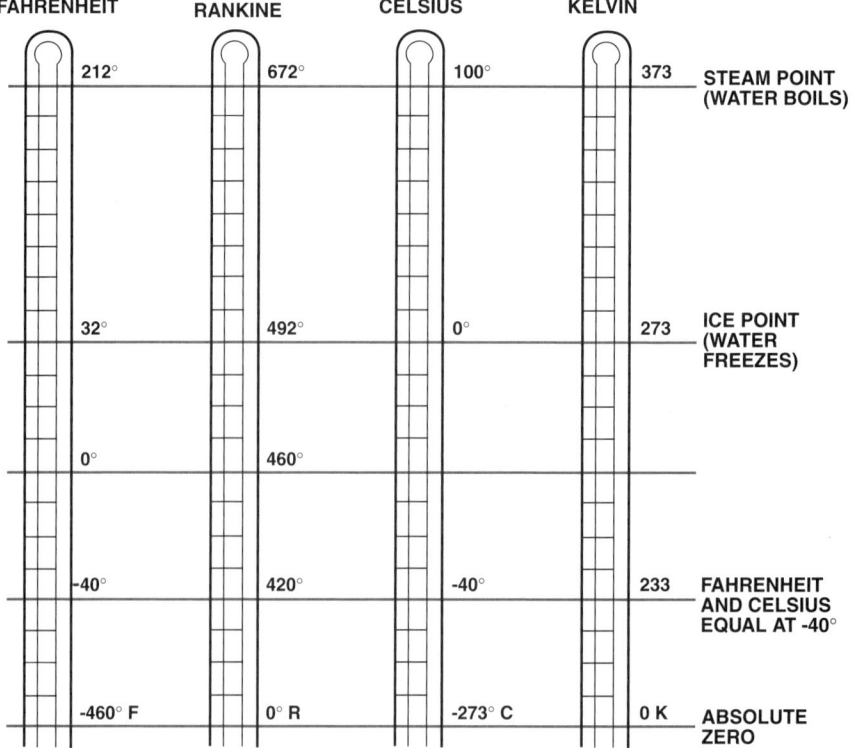

Figure 4.8   A comparison of various temperature scales

TABLE 4.7
Temperature Scales

Scale	Equivalents
°F	$\frac{9}{5}$°C + 32, or 1.8°C + 32 $\frac{9}{5}$(K – 273) + 32, or 1.8(K – 273) + 32 °R – 460
°C	$\frac{5}{9}$(°F – 32), or (°F – 32) ÷ 1.8 K – 273 $\frac{5}{9}$(°R – 492), or (°R – 492) ÷ 1.8
°R	°F + 460 $\frac{9}{5}$K, or 1.8K $\frac{9}{5}$°C + 492, or 1.8°C + 492
K	°C + 273 $\frac{5}{9}$°R, or °R ÷ 1.8 $\frac{5}{9}$(°F – 32) + 273, or (°F + 460) ÷ 1.8

Example Problem: Convert 250°F to Celsius.

Solution: Use the formula for converting Celsius to Fahrenheit.

$$
\begin{aligned}
°C &= \tfrac{5}{9}(°F - 32°) \\
&= \tfrac{5}{9}(250°F - 32°) \\
&= \tfrac{5}{9}(218) \\
&= (5 \times 218) \div 9 \\
&= 1{,}090 \div 9 \\
°C &= 121.11
\end{aligned}
$$

or

$$
\begin{aligned}
°C &= (°F - 32°) \div 1.8 \\
&= (250°F - 32°) \\
&= 218 \div 1.8 \\
°C &= 121.11.
\end{aligned}
$$

Example Problem: Convert –2.5°C to Fahrenheit.

Solution: Use the formula for converting Fahrenheit to Celsius.

$$
\begin{aligned}
°F &= \tfrac{9}{5}°C + 32 \\
&= (9 \times -2.5) \div 5 + 32 \\
&= (-22.5 \div 5) + 32 \\
&= -4.5 + 32 \\
°F &= 27.5
\end{aligned}
$$

or

$$
\begin{aligned}
°F &= 1.8°C + 32 \\
&= (1.8 \times -2.5) + 32 \\
&= -4.5 + 32 \\
°F &= 27.5
\end{aligned}
$$

Practice Problems

Refer to tables 4.6 and 4.7 to solve these problems.

1. An airliner traveled 3,975 miles from Dallas, Texas to Honolulu, Hawaii in 7½ hours.

   a. What was its average speed in miles per hour?

   _____

   b. What was its speed in miles per minute?

   _____

2. The capacity of a pump is 1,000 gallons per minute. How long will it take to fill a 500-barrel tank? (42 gal = 1 bbl)

   _____

3. For every 90 revolutions of a pumping engine, the pump rods are raised and lowered once. If there are 15 strokes per minute, what is the rpm of the engine?

   _____

4. Convert the following temperatures to Fahrenheit:
   a. 345 K __161.6°F_____
   b. 20°C __68°F_____
   c. 0°C __32°F_____
   d. 482°R __22°F_____
   e. −78°C __−108.4°F_____

5. Convert the following temperatures to Celsius:
   a. −89°F __−67.2°C_____
   b. 32°F __0°C_____
   c. 277 K __4°C_____
   d. 432°R __−33.3°C_____
   e. 10 K __−263°C_____

6. What are the Celsius temperature values for the following Kelvin values?
   a. 324 K __51°C_____
   b. 203 K __−70°C_____
   c. 293 K __20°C_____
   d. 250 K __−23°C_____
   e. 49 K __−224°C_____

7. What are the conventional absolute temperature values for the following Fahrenheit temperatures?

  a. 680°F ___ 1,140 °R _____

  b. 3°F ___ 463 °R _____

  c. 43°F ___ 503° R _____

  d. –24°F ___ 436° R _____

  e. –113°F ___ 347° R _____

8. In batching products through a pipeline, the rate of travel of the fluid through a 50-mile pipeline is 1.8 miles per hour. A new batch is started at 9:00 A.M. on a Tuesday. When will the batch first arrive at the terminal end?

  _____

9. At a loading dock for oil tankers, oil is delivered to a tanker at the rate of 10,000 barrels per hour. How many minutes will it take to load a 16,500-barrel tanker?

  _____

10. Convert 325 degrees Rankine to the metric absolute temperature equivalent.

  _____

MASS AND RELATED DERIVED QUANTITIES

Like length and time, mass is a fundamental quantity. *Mass* is the amount of material an object, or body, contains. Mass makes a body have weight in a gravitational field and creates inertia in the body. *Inertia* makes a stationary body tend to remain stationary or, if the body is moving, makes the body continue to move in a straight line until acted upon by a directional force. Thus, any force applied to start a body moving or to change the direction of a moving body has to overcome the body's inertia. Derived quantities related to mass are weight and force, density and specific gravity, work and power, and pressure.

Weight and Force

Under most circumstances, weight and mass are the same thing. And, sometimes weight and force are considered to be the same. For example, in rotary drilling operations, weight, or force, must be placed on the bit to enable it to drill. So, a common expression is weight on the bit. In SI units, weight on the bit is expressed in decanewtons, which are units of force. In conventional units, weight on the bit is expressed in pounds-force. This pound-force is, however, usually stated as simply pounds. So, in Canada, for instance, a driller may say that the force on the bit is 8,896 decanewtons, while in the U.S., a driller may say that the weight on the bit is 20,000 pounds.

In the United States, pound-mass is the unit of mass and is symbolized as lb_m to distinguish it from the pound-force, lb_f. Interestingly, the metric system's unit of mass, the kilogram, defines a pound-mass. One pound-mass (1 lb$_m$) equals 0.45359237 kilogram. The conventional pound-force is defined as the force necessary to accelerate a mass of 1 pound at 32.15 feet per second squared (32.15 ft/s^2). Table 4.8 lists the most-used units of mass, weight, and force in the conventional and SI metric systems.

TABLE 4.8
Units of Weight, Mass, and Force Measurement

Conventional Unit	Conventional Equivalents	Metric Equivalents
ounce (oz)	0.0625 lb$_m$	28.35 g
pound-mass (lb$_m$)	16 oz	0.4536 kg 453.6 g
ton (tn)	2,000 lb$_m$	0.9072 t 907.185 kg
pound-force (lb$_f$)	lb$_m$ × 32.15 ft/s^2	4.45 N

Metric Unit	Metric Equivalents	Conventional Equivalents
gram (g)	0.002 kg	0.03527 oz
kilogram (kg)	1,000 g	2.2046 lb$_m$
tonne (t)	1,000 kg	1.10231 tn
newton (N)	kg•m/s^2	0.2247 lb$_f$

Example Problem: A spring is acting against a heavy door with a force of 30 decanewtons (daN). What is the force in pounds-force (lb$_f$)?

Solution: A newton is equal to 0.2247 pounds-force (from table 4.8), and a decanewton is 10 newtons, or

$$10 \times 0.2247 = 2.247 \text{ lb}_f.$$

Then, 30 daN is

$$30 \times 2.247 = 67.41 \text{ lb}_f.$$

Example Problem: On a rotary drilling rig, the driller reports to the toolpusher that the weight on the bit is 35,800 pounds. Because the toolpusher has spent a great deal of time in Canada, he asks the driller to give him the force on the bit in decanewtons.

Solution: 1 lb$_f$ is equal to 4.45 N (from table 4.8). Since a daN is 10 N, then

$$10 \div 4.45 = 0.445 \text{ daN}.$$

Then, 0.445 daN is

$$35,800 \times 0.445 = 15,931 \text{ daN}.$$

Density and Specific Gravity

Density

Density is the weight of a substance per unit volume. A unit volume can be a cubic inch, a cubic foot, a gallon, a barrel, a cubic yard, or any of several measurement units. In conventional units, pounds and cubic feet are the most commonly paired units, although pounds per gallon is used to measure density in some instances, particularly when dealing with liquids.

In the metric system, grams per cubic centimetre (g/cm^3) is in common use, although in SI only kilograms per cubic metre (kg/m^3) is recognized as a density measure. A cubic centimetre is $(10^{-2}m)^3$, or $10^{-6} m^3$, while a gram is 10^{-3} kg; so,

$$g/cm^3 = 10^{-3} \div 10^{-6} m^3 = 10^3 kg/m^3 = 1,000 kg/m^3.$$

One gram per cubic centimetre and 1,000 kilograms per cubic metre are equal quantities as far as density is concerned and represent the density of pure water at a temperature of 4° Celsius.

The density of a material indicates the weight of a product made from the material. Weight is an important consideration when engineers design a product. For example, they favor aluminum alloy over steel as a material for aircraft because aluminum is less dense than steel for a given strength. Thus, they can design an aircraft of adequate strength but of less weight.

The weight of a large body may be calculated if the density of the material and the volume of the body are known. Table 4.9 gives the densities of several

TABLE 4.9
Densities of Some Solids, Liquids, and Gases

Material	lb/ft3	kg/m3	g/cm3 (sg)
Solids			
Gold	1,206.2	19,300	19.3
Mercury	846.0	13,500	13.5
Lead	712.5	11,400	11.4
Iron	485.0*	7,700*	7.7*
Aluminum	165.6	2,600	2.6
Wood	50.0*	800*	0.8*
Ice	56.9	900	0.9
Liquids			
Sulfuric acid	125.0	2,000	2.0
Seawater	64.3	1,030	1.03
Fresh (pure) water	62.5	1,000	1.00
Kerosene	50.0	800	0.80
Gasoline	46.8	750	0.75
Gases			
Air	0.075	1.20	0.0012
Oxygen	0.084	1.34	0.00134
Nitrogen	0.0737	1.18	0.0018
Carbon monoxide	0.0734	1.17	0.0017
Hydrogen	0.0053	0.085	0.000085

*Wood and iron vary in density. The values shown are approximate.

solids, liquids, and gases. The density of each is given in pounds per cubic foot (lb/ft³), kilograms per cubic metre (kg/m³), and grams per cubic centimetre (g/cm³).

Example Problem: The volume of a gasoline storage tank is 5,360 cubic feet. When the tank is full, what is the weight of the gasoline?

Solution: From table 4.9, find the density of gasoline in pounds per cubic foot. Then set up the proportion, 46.8 pounds is to 1 cubic foot as *x* pounds is to 5,360 cubic feet:

$$\frac{46.8}{1} = \frac{x}{5,360}$$

$$x = 46.8 \times 5,360 = 250,848.$$

The gasoline in the tank weighs 250,848 pounds.

Specific Gravity

For liquids and solids, the specific gravity (sp gr) is the ratio between the weight of a given volume of a substance and the weight of an equal volume of pure water at 39°F (water is at its densest at 39.2°F or 4°C):

$$\text{sp gr} = \frac{\text{weight of a volume of substance}}{\text{weight of an equal volume of pure water}}$$

or

$$\text{sp gr} = \frac{\text{density of substance}}{\text{density of pure water}}$$

For gases, specific gravity is the ratio between the weight of a given volume of a gas to an equal volume of air or hydrogen under prescribed conditions of temperature and pressure. (The density of a gas varies with the temperature and pressure. Thus, a specific temperature and pressure must be stated—for example, 60°F and 14.7 psi.)

$$\text{sp gr} = \frac{\text{density of gas}}{\text{density of air or hydrogen}} \text{ (at a specific pressure and temperature)}.$$

Like density, weight per volume can be expressed in lb/ft³, kg/m³, or g/cm³. When the density is expressed in grams per cubic centimetre, the specific gravity of a substance is the same as its density because the density of pure water is 1 gram per cubic centimetre, and any number divided by 1 equals the same number. Thus, the densities given under the column headed g/cm³ in table 4.9 are also the specific gravities for those substances. Incidentally, the SI system calls specific gravity "relative density."

Example Problem: What is the specific gravity of a drilling mud that weighs 78.54 lb/ft³?

Solution: From table 4.9, the density of pure water is 62.5 lb/ft³.

$$\text{sp gr} = 78.54 \div 62.5 = 1.257.$$

The same equation may be used to find the weight or density of a substance by transposing it to read:

$$\text{density of substance} = \text{density of pure water} \times \text{sp gr}$$

or

wt per vol of substance = wt per vol of pure water × sp gr.

As mentioned earlier, the specific gravities of gases are frequently expressed as a ratio between the weight of a given volume of the gas and the weight of an equal volume of air or hydrogen at a given temperature and pressure. Specific gravity of gases may also be referred to as *vapor density*.

Table 4.10 shows specific gravities based on air for several gases.

Baumé Scales

Several scales based on specific gravity have been developed for specialized use. In the early 1800s, the French chemist Antoine Baumé developed two scales for measuring liquid density, or gravity. Today, two Baumé scales are used: one for measuring the gravity of heavy liquids and another for measuring the gravity of liquids lighter than fresh water. The increments in gravity are called degrees Baumé (°Baumé). Two equations express the two scales:

1. For heavy liquids: °Baumé = 145 − 145 ÷ .
2. For light liquids: °Baumé = 140 ÷ − 130.

Note that in equation 1 if the liquid has a specific gravity of 1, the degrees Baumé is 0, but in equation 2 it is 10. In equation 2, the degrees Baumé increase as the specific gravity decreases. In equation 1, the degrees Baumé increase as the specific gravity increases.

Example Problem: A heavy syrup has a specific gravity of 1.153. What is its density in degrees Baumé?

Solution: Solve the problem by using the Baumé scale for heavy liquids:

$$°Baumé = 145 − 145 ÷ 1.153 = 19.2.$$

Example Problem: A liquid has a specific gravity of 0.7006. What is its gravity in degrees Baumé?

Solution: Solve the problem by using the Baumé scale for light liquids:

$$°Baumé = 140 ÷ 0.7006 − 130 = 69.8.$$

°API Gravity Scale

During the early 1920s, the American Petroleum Institute (API), the Bureau of Mines, and the National Bureau of Standards adopted a special gravity scale for measuring crude petroleum and some of its products. It is designated °API and bears a resemblance to the Baumé scale for light liquids:

$$°API = \frac{141.5}{} − 131.5.$$

Example Problem: A sample of crude oil has a specific gravity of 0.8692. What is its API gravity?

Solution: Use the formula for determining API gravity.

$$°API = \frac{141.5}{0.8692} − 131.5.$$

TABLE 4.10 Specific Gravity Based on Air	
Gas	Sg Gr
Air	1.00
Oxygen	1.120
Nitrogen	0.0983
Carbon monoxide	0.979
Hydrogen	0.071
Butane	2.004

$$= 162.8 - 131.5$$
$$°API = 31.3.$$

Using the Baumé and API scales expands changes in specific gravity, which are expressed in very small increments. For example, a change from 28°API to 30°API is equivalent to a change from 0.8348 to 0.8251 in specific gravity, which is a change of only 0.0097. The API scale makes the differences in crude oil densities much more evident. Notice, too, that the higher the API gravity, the lower is the specific gravity. For example, a liquid with an API gravity of 30° has a specific gravity of 0.8251, whereas a liquid with an API gravity of 28° has a specific gravity of 0.8348.

To solve for specific gravity when the °API gravity is known, the equation is:

$$= \frac{141.5}{°API + 131.5}.$$

Example Problem: What is the specific gravity of a 43°API gravity oil?

Solution:

$$= 141.5 \div (43 + 131.5)$$
$$= 0.810882$$
$$= 0.8109.$$

Pressure

One definition of pressure is that it is a force per unit area. The force can be pounds-force or newtons. The area can be square inches, square centimetres, square metres, and so on. *Pounds per square inch* (psi) is the most often used unit of pressure in the English, or conventional, system. SI stipulates that pressure be measured in *newtons per square metre* (N/m^2), which it terms a *pascal* (Pa)—that is, $1 N/m^2 = 1$ Pa. One psi equals 6,895 pascals or 6.895 kilopascals (kPa).

A measure of pressure that falls outside the definition of force per unit area is *kilograms per square centimetre* (kg/cm^2). A kilogram is the unit of mass in SI, and SI convention does not allow it to be used as a unit of force. However, kg/cm^2 is still encountered where people employ the old metric system. A kilogram-force per square centimetre is about 98 kPa, or about 14.2 psi.

Another form of pressure measurement outside the force-per-unit-area definition is the height of a liquid column, typically a column of mercury or water. In this instance, mercury or water is placed in a length of glass tubing that is bent into a U shape. Each end of the U is open and each leg is of equal height. Mercury or water is put into the tube and, under atmospheric pressure, the liquid rises in both legs of the U to an equal height. However, if pressure above atmospheric is put on one leg of the tube, the water or mercury rises in the other side. By marking the tube in inches or millimetres, the height of the rise can be read and the pressure given in inches or millimetres of mercury or water. Such equipment is called a mercury or water manometer. Manometers are often used to measure small changes in pressure, such as in the atmosphere. In this case, the manometer is often called a barometer.

As is the case with temperature, an absolute scale of pressure also exists. The air that surrounds the earth exerts a pressure of about 14.7 pounds per square inch at sea level. The pressure value of 14.7 psi is above absolute zero pressure, and it is designated as *pounds per square inch, absolute,* or psia. In SI, the value is about 101.35 kilopascals (kPa), and for mercury columns it is about 760 millimetres or 29.92 inches—that is, the pressure of the atmosphere on one side of a column of mercury in a U-tube forces the mercury to rise to a level of 760 millimetres or 29.92 inches on the other side.

A typical pressure gauge, such as a tire-pressure gauge, shows *gauge pressure* (psig). Gauge pressure means that the atmospheric pressure of 14.7 corresponds to zero on the pressure gauge. Put another way, before the gauge is attached to measure the pressure of a vessel, the gauge reads zero, although the atmosphere is placing 14.7 psi on it.

Some gauges show readings below atmospheric pressure. On such gauges, the values may be expressed in inches of mercury, vacuum. In this case, vacuum means a pressure below that of the atmosphere. For example, if a vacuum exists in a vessel, then a mercury manometer used to measure the vacuum reads lower than 29.92 inches or 760 millimetres.

In converting gauge pressure to absolute pressure, the formula is

$$psia = psig + 14.7.$$

Example Problem: A gauge attached to a vessel containing a fluid under pressure reads 25 psig. What is the absolute pressure of the fluid in this vessel?

Solution:

$$psia = 25 + 14.7 = 39.7.$$

Physical laws of gases state that the absolute pressure of a gas varies directly with its absolute temperature as

$$\frac{P_2}{P_1} = \frac{T_2}{T_1}$$

where P and T signify pressure and temperature before (P_1, T_1) and after (P_2, T_2) a change. This formula can be applied only where the volume of gas does not change.

Example Problem: Suppose the pressure on a 10 ft^3 tank of air was 30 psig at 70°F. After the sun had shone on it all morning, the temperature rose to 110°F. How much did the pressure increase?

Solution: In this case, P_1, T_1, and T_2 are known but P_2 is not. First, convert both temperatures to absolute values:

$$70° + 460° = 530°.$$
$$110° + 460° = 570°.$$

Then, convert the pressure to absolute value:

$$30\ psig + 14.7 = 44.7\ psia.$$

Then

$$P_2 \div 44.7 = 570 \div 530$$
$$P_2 = 44.7 \times 570 \div 530 = 48.07 = 48.1\ psia.$$

The pressure increase is

$$48.1 - 44.7 = 3.4\ psi.$$

Hydrostatic pressure is the force exerted by a body of fluid at rest. Hydrostatic pressure increases directly with the density and the depth of the fluid. The hydrostatic pressure of fresh water is 0.433 psi per foot of depth, or 9.792 kPa per metre of depth.

Example Problem: What is the hydrostatic pressure in psi at the bottom of a 150-foot (46-metre) column of water? In kPa?

Solution:

150 × 0.433 = 65 psi.
46 × 9.792 = 450 kPa.

WORK AND POWER

Work is the result of force that produces motion. Work is therefore expressed as a product of force and distance. In the conventional measurement system, pound is the commonly used unit of force. Foot is the common distance unit; thus, work is normally expressed in foot-pounds (ft-lb). For example, if a can of fuel weighing 35 pounds is lifted 3 feet to the bed of a truck, the amount of work done is

3 ft × 35 lb = 105 ft-lb.

Example Problem: A man pushes a hand truck with a force of 50 pounds a distance of 20 feet. How much work does he perform?

Solution:

20 ft × 50 lb = 1,000 ft-lb of work.

Power is the rate of time of doing work. Power is expressed in foot-pounds per unit of time, usually minutes but sometimes seconds or hours. Foot-pounds per minute and foot-pounds per second are common expressions for power. The standard unit for mechanical power is the horsepower (hp), adopted centuries ago. Scientists established that an average horse could do 33,000 foot-pounds of work per minute, and this unit became the basis for the standard equation

hp = ft-lb/min ÷ 33,000.

Example Problem: How much horsepower does a pump expend that lifts 5,000,000 foot-pounds of water per hour?

Solution: Since this amount is for a period of 1 hour, find the foot-pounds per minute (60 min = 1 hr):

5,000,000 ÷ 60 = 83,333.33 ft-lb/min.

Then, using the equation for finding horsepower:

83,333.33 ÷ 33,000 = 2.525 hp.

The unit of work and energy in the SI system is the newton-metre (N•m) and is called the *joule* (J). One foot-pound in the conventional system equals 1.356 joules, and 1 joule equals 0.737 foot-pound. A force of 1 newton acting through a distance of 1 metre equals 1 joule.

The SI unit of power is the joule per second (J/s) and is called the *watt* (W). One thousand watts—the *kilowatt* (kW)—is a world standard for electric power, and the kilowatt-hour (kWh) is a standard for energy use. A kilowatt-hour is equal to 3.6 megajoules (MJ) as shown by the equation

$$1 \text{ kW} = 1,000 \text{ J/s} \times 3,600 \text{ s} = 3,600,000 \text{ J} = 3.6 \text{ MJ}.$$

One horsepower is equal to 746 watts, or 0.746 kilowatts.

Example Problem: How many megajoules (MJ) are in 75 horsepower-hours?

Solution: First convert horsepower-hours to kilowatt-hours:

75 hp-hr × 0.746 kWh = 55.95 kWh.

Then convert to megajoules:

55.95 kWh × 3.6 MJ = 201.42 MJ.

Practice Problems

Refer to tables 4.8, 4.9, and 4.10 to solve these problems.

1. How many kilograms are in a conventional ton?

_____

2. The weight on the bit at a drilling rig is 120,000 lb$_f$. Express this force in decanewtons (daN) and kilonewtons (kN).

_____

3. A truck's load capacity is 6 tons. How many feet of 6-inch pipe can it haul if the pipe weighs 19 pounds per foot?

_____

4. How many pounds are in 5 kilograms?

_____

5. How much more does a cubic foot of seawater weigh than an equal volume of fresh water?

_____

6. If the specific gravity of a certain crude is 0.82140, what is its °API gravity?

_____

7. A man weighing 160 pounds climbs a 20-foot ladder. How much work has he done in the climb?

_____

8. Convert a gauge pressure of 36.9 psi to absolute pressure.

_____

9. What is the difference in the weight between a barrel of 40°API gravity oil and a 50°API gravity oil? (1 bbl = 5.6 ft^3)

_____

10. What is the pressure at the bottom of a 1,600-barrel freshwater storage tank when the height of the water is 16 feet?

_____

ELECTRICITY

Electricity is the flow of electrons through a conducting medium, such as a metal wire. The flow of electricity is much like the flow of fluid in a pipe. For one thing, pressure is needed to force the fluid to flow. Put another way, a pressure difference must exist between one point in the pipe and another point if the fluid is to flow between the two points. The same applies to electron flow—an electrical pressure difference is needed between two points in a wire or other conducting material.

In electricity, the pressure difference is called *voltage*, *potential*, *potential difference*, or *electromotive force* (emf). Two other important quantities in electrical circuitry are *current* and *resistance*. Just as voltage is similar to pressure in fluid flow, current is similar to the rate of fluid flow, and resistance is similar to friction in a fluid piping system. (When fluid flows inside a pipe, the moving fluid rubs against the sides of the pipe and friction occurs. Similarly, in electrical circuits, electron flow encounters electrical friction, which is resistance.)

The standard unit of measurement for voltage (V or E) is the *volt* (V); for current (I), the *ampere* (A); and for *resistance*, the *ohm* (Ω) (table 4.11). In terms of how we measure electricity, the science of electricity was developed after French scientists had already developed the metric system; so, electrical measurement units are the same in both the conventional and SI metric systems.

TABLE 4.11
Basic Electrical Units of Measure

Quantity	Symbol	Unit Name	Unit Symbol
Voltage	*E* or *V*	volt	V
Current	*I*	ampere	A
Resistance	*R*	ohm	Ω

Ohm's Law

In 1827, the German mathematician and physicist Georg Simon Ohm published a book that, among other things, presented a basic law of electrical behavior. Called Ohm's law, it states that the strength or intensity of an unvarying electrical current is directly proportional to the electromotive force (voltage) and inversely proportional to the resistance of the circuit. Simplified, Ohm's law shows the relationship among the three electrical quantities to be

$$I = \frac{E}{R}$$

$$E = I \times R$$

$$R = \frac{E}{I}$$

where

I = current in amperes
E = voltage in volts
R = resistance in ohms.

An easy way to remember Ohm's law is to use the Ohm's law wheel (fig. 4.9). To use the wheel, cover the unknown factor and read the resulting equation. For example, suppose the unknown factor is I. Cover I and the diagram indicates that I equals E/R. When E is the unknown factor, the equation is $E = IR$, and when R is the unknown, the equation is $R = E/I$.

Ohm's law determines (1) the amount of current that flows in a circuit, (2) the amount of resistance needed in a circuit to limit current to a given value, or (3) the voltage needed to cause a given flow of current through a particular value of resistance.

Example Problem: What voltage is needed to force 1.2 amperes through a device that has a resistance of 80 ohms?

Solution: Use the formula for finding voltage E:

E = IR
I = 1.2 amperes; R = 80 ohms
E = 1.2 × 80 = 96 volts.

Other relations exist among the three quantities, but $I = E/R$ is the basis for Ohm's law. An electrical quantity called power is also derived from this formula, and is described later.

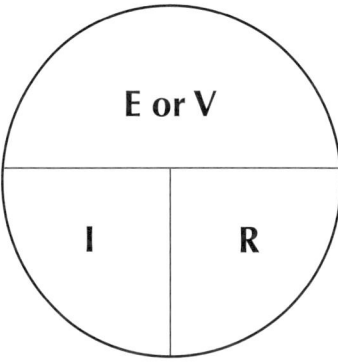

Figure 4.9   Ohm's law wheel

Circuits

Kinds of electricity include static, direct current (DC), and alternating current (AC). Both direct-current and alternating-current electricity are of great importance in everyday life, but this section only considers direct-current electricity. Direct current can be obtained from various kinds of cells and batteries, as well as from generators. (For a more detailed look at electricity and electronics, see *Basic Electricity for the Petroleum Industry*—catalog number 1.40020, ISBN 0-88698-109-3—and *Basic Electronics for the Petroleum Industry*—catalog number 1.41040, ISBN 0-88698-199-9. These manuals are available from Petroleum Extension Service, The University of Texas at Austin, www.utexas.edu/cee/petex.)

An electrical circuit provides a route for electricity to flow. A circuit can be simple or complex, depending on its function. When discussing electrical circuits, various symbols are used for electrical components. Figure 4.10 shows a few of these symbols. Electrical component symbols are a standard means of representation and are commonly used in drawing circuits.

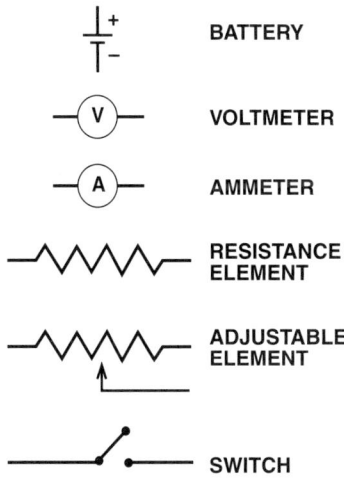

Figure 4.10   Electrical symbols

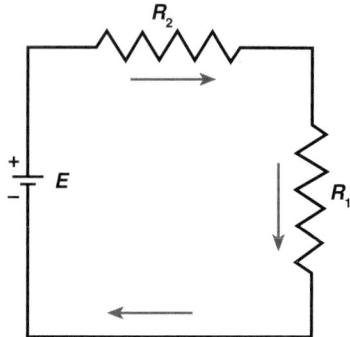

Figure 4.11   Series circuit

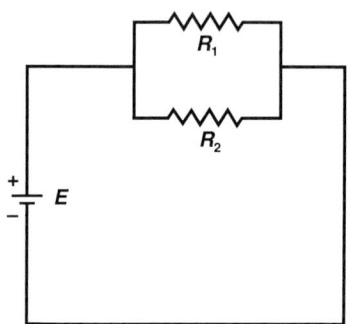

Figure 4.12   Parallel circuit

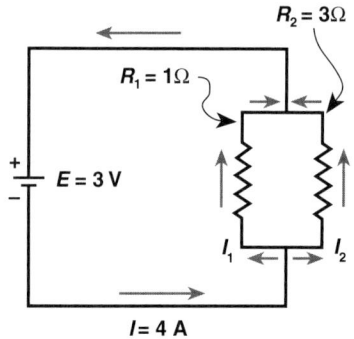

Figure 4.13   Parallel circuit with total current

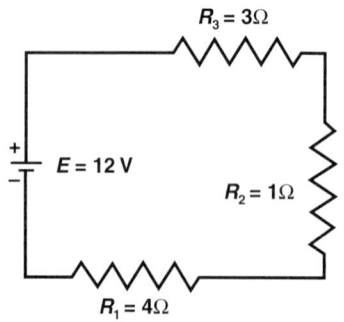

Figure 4.14   Series circuit with three resistors

Two types of circuits for direct current are series circuits (fig. 4.11) and parallel circuits (fig. 4.12). Figure 4.11 shows a simple series circuit. Electricity from the battery (E) flows through wire in which two resistors (R_1 and R_2) are installed. All of the current flows directly through each resistor. In the parallel circuit in figure 4.12, the resistors are connected in parallel, and the total current divides, with part going through one resistor (R_1) and part going through the other resistor (R_2).

In a series circuit, current flow is equal at all points. The value of the current is

$$I = \frac{E}{R_1 + R_2 + R_3 + \ldots}$$

This equation shows that all resistances in the series circuit are added together and then divided into the voltage (E) to obtain the current. The ellipsis (. . .) after the third plus sign (+) in the equation simply means that the resistances continue and would be called R_4, R_5, and so on.

In a parallel circuit, the total current equals the sum of the currents in the separate branches, and the voltage is equal in each branch to the voltage in the main part of the circuit. Figure 4.13 shows a 3-volt battery sending electricity through a circuit with two resistors, R_1 and R_2, in parallel. R_1 is a 1-ohm (Ω) resistor and R_2 is a 3-Ω resistor. A total of 4 amperes (A) of current (I) flows through the circuit. The current passing through R_1 is

$$I_1 = \frac{E}{R_1} = \frac{3}{1} = 3 \text{ amperes (A)}.$$

while at R_2, it is

$$I_1 = \frac{E}{R_2} = \frac{3}{1} = 1 \text{ A}.$$

The total current is still 4 amperes.

The effective resistance (R) of a parallel circuit with parallel resistors R_1, R_2, R_3, . . . is

$$\frac{1}{R} = \frac{1}{R_1} + \frac{1}{R_2} + \frac{1}{R_3} + \ldots$$

For the circuit shown in figure 4.13, resistance works out to

$$\frac{1}{R} = \frac{1}{1} + \frac{1}{3} = \frac{3}{3} + \frac{1}{3} = \frac{4}{3}$$

$$R^4 = 3$$
$$R = \text{¾, or } 0.75 \,.$$

Example Problem: In the circuit in figure 4.14, the initial potential is 12 volts. Calculate the voltage drop across each resistance element.

Solution: If the total potential in the circuit is 12 volts, then the total voltage drop for the circuit is 12 volts. For any resistor, the voltage drop is proportional to the value of the resistor. The total current (I) in the circuit is

$$I = 12 \div (4 + 1 + 3) = 12 \div 8 = 1.5 \text{ A}.$$

The voltage drop across each resistor is

E_1 = 1.5 A × 4 = 6 V.
E_2 = 1.5 A × 1 = 1.5V.
E_3 = 1.5 A × 3 = 4.5 V.

The sum of these voltages is equal to 12 volts, which is the initial potential in the circuit:

6.0 V + 1.5 V + 4.5 V = 12 V.

Example Problem: In the parallel circuit in figure 4.15, find the total resistance and the current in each part of the circuit.

Solution: In each branch of the circuit, the potential is 25 volts. Then at R_1, R_2, R_3, and R_4, the respective currents are

$$I_1 = \frac{25}{10} = 2.5A.$$

$$I_2 = \frac{25}{20} = 1.25A.$$

$$I_3 = \frac{25}{5} = 5.0A.$$

$$I_3 = \frac{25}{20} = 1.25A.$$

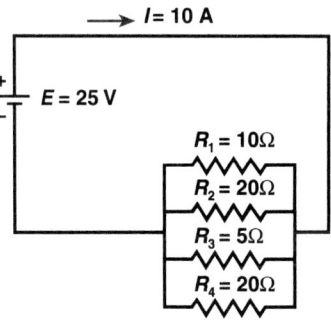

Figure 4.15   Parallel circuit with four branch resistors

The total of these currents is 10 A (2.5 + 1.25 + 5.0 + 1.25 = 10), the total current in the circuit. The total resistance, R, of the parallel arrangement of resistors in figure 4.15 is

$$\frac{1}{R} = \frac{1}{R_1} + \frac{1}{R_2} + \frac{1}{R_3} + \frac{1}{R_4}$$

$$= \frac{1}{10} + \frac{1}{20} + \frac{1}{5} + \frac{1}{20}$$

$$= \frac{2}{20} + \frac{1}{20} + \frac{4}{20} + \frac{1}{20}$$

$$= \frac{8}{20}$$

$$\frac{1}{R} = \frac{4}{10}$$

$$R = \frac{10}{4}$$

$$R = 2.5 .$$

Example Problem: Find the current in each portion of the circuit in figure 4.16 and find the total resistance of the circuit.

Solution: This problem is an example of a combination of series and parallel elements in one circuit. The current I at R_1 is 5 amperes. Thus, the voltage drop at R_1 is

$$E_1 = R_1 \times I = 100 \times 5 = 500 \text{ V}.$$

At point B,

$$I = 5 \text{ A}$$
$$E_B = 520 \text{ V} - 500 \text{ V} = 20 \text{ V}.$$

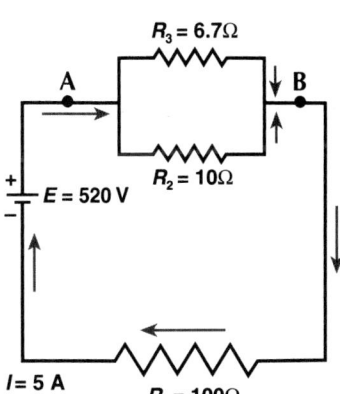

Figure 4.16   Series-parallel circuit arrangement

The current through R_2 is I_2:

$$I_2 = \frac{E_B}{R_2} = \frac{20}{10} = 2 \text{ A}.$$

The current through R_3 is I_3:

$$I_3 = \frac{E_B}{R_3} = \frac{20}{6.7} = 3 \text{ A}.$$

Effective resistance R of the parallel portion is

$$\frac{1}{R} = \frac{1}{10} = \frac{1}{6.7} = 1.4 = 4 \quad .$$

Power

Power is the rate of doing work or otherwise expending some form of physical energy. In electrical calculations, the unit of power is the watt (W). In the SI system, the watt is the unit of all physical power, both mechanical and electrical. Large amounts of power are expressed in kilowatts (kW) or megawatts (MW). Small amounts of power are typically expressed as milliwatts (mW). Power is related to the three quantities in Ohm's law—voltage, current, and resistance:

$$P = E \times I$$

where

P = power in watts
E = voltage in volts
I = current in amperes.

Since

$$I = \frac{E}{R},$$

$$P = E \times \frac{E}{R}, \text{ or } P = \frac{E^2}{R}, \text{ and}$$

$$E = I \times R.$$

Then

$$P = I \times R \times I, \text{ or } P = I^2 R.$$

The power wheel is shown in figure 4.17 and provides an easy method of determining the desired equation to use. Just cover the unit desired and the other two units are identified for the proper power equation.

Example Problem: How much current does a 100-watt, 120-volt bulb draw?

Solution: If $P = E \times I$, then

$$I = \frac{P}{E}$$

$$I = \frac{100}{120} = 0.833 \text{ A}.$$

As an energy source, electric power is sold by the kilowatt-hour (kWh). A kilowatt-hour is the amount of work done when 1,000 watts are applied for 1 hour (1,000 watts = 1 kilowatt). The number of kilowatt-hours equals the power (P) in kilowatts times the time (T) in hours:

$$P\,(\text{kW}) \times T(\text{hr}) = \text{kWh}.$$

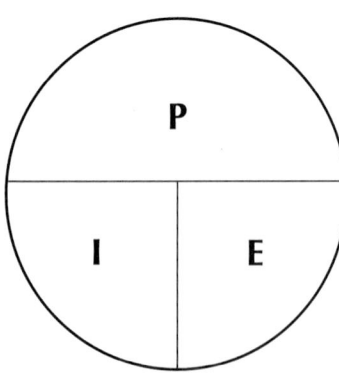

Figure 4.17   Power law wheel

Example Problem: At 6.2 cents per kilowatt-hour, how much does it cost to burn a 100-watt bulb for 100 hours?

Solution: Find the number of kilowatt-hours:

100 W = 0.1 kW

0.1 kW × 100 hr = 10 kWh.

Then, at 6.2 cents per kilowatt-hour, the cost is

6.2 × 10 = 62 cents.

Practice Problems

1. If a 12-volt battery supplies current to a circuit having 100-ohm resistance, what is the resultant current?

   _____

2. Using the circuit shown to the right, solve the problems that follow.

   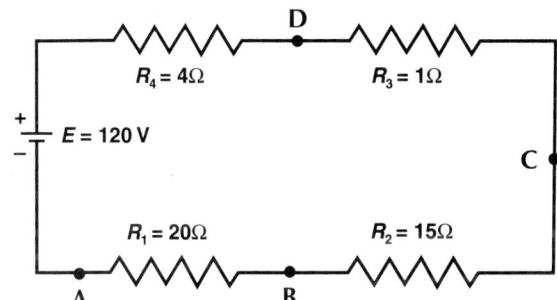

   a. What is the voltage drop across R_1?

   _____

   b. What is the current at point C?

   _____

   c. What is the voltage drop across R_4?

   _____

   d. What is the total current in the circuit?

   _____

3. An air conditioner draws 10 amps from a 115-volt circuit for 6 hours per day over a period of 25 days. How much energy in kilowatt-hours is used?

   _____

4. Using the circuit to the right, solve the following problems:

   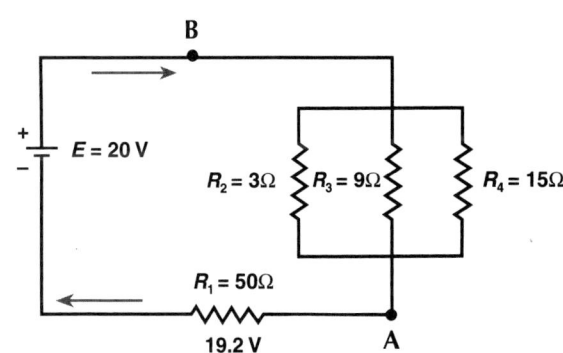

   a. What is the effective resistance of the portion of the circuit containing R_2, R_3, and R_4?

   _____

   b. What is the total current in the circuit?

   _____

   c. What is the current through R_2?

   _____

   d. What is the voltage drop across R_1?

   _____

   e. How large a resistor could be added at B to make the current in the circuit equal 0.1 ampere?

_____

5. How much power does a 20-ohm resistor absorb when it is connected across a 60-volt supply?

_____

6. How much energy in watt-hours is delivered by a 6-volt storage battery supplying an average current of 5 amps for an 8-hour period?

_____

7. What is the effective resistance of a 6-ohm and a 12-ohm resistor connected in parallel?

_____

8. In the figure below, R_2 is rated at 60 ohms.

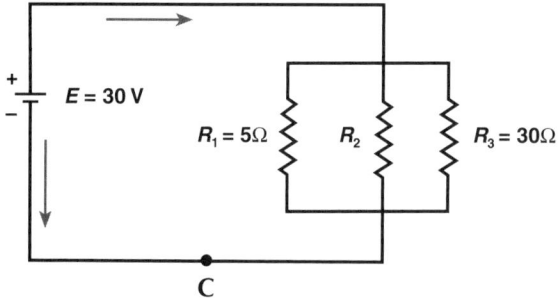

   a. What is the value of the current at point C in the circuit?

_____

   b. What is the total power consumed in the circuit?

_____

9. How much current does a 60-watt, 120-volt light bulb draw?

_____

10. What is the power of an electrical source that has a current of 10 amperes and a resistance of 5 ohms?

_____

4. Some Physical Quantities and Their Measurement

*Multiply each problem answered correctly by five to arrive
at your percentage of competency.*

Refer to all tables in the chapter to answer these questions.

1. How many miles are in a 10,000-metre race?

   _____

2. How many cubic metres are in a 500-barrel tank?

   _____

3. What is a quantity that cannot be divided into other entities called?

   _____

4. Give the appropriate abbreviation or symbol for the following units of measure:

   a. mile _____

   b. pounds per gallon _____

   c. cubic foot _____

   d. square metre _____

   e. pounds per square inch _____

   f. grams per cubic centimetre _____

   g. pound-force _____

   h. ohm _____

   i. feet per second _____

   j. revolutions per minute _____

   k. kilowatt-hour _____

5. Give the numerical value of the following metric prefixes:

   a. kilo _____

   b. deci _____

   c. milli _____

   d. micro _____

   e. mega _____

6. Convert the following measurements:

    a. 64 pounds into kilograms _____

    b. 3.25 tonnes into tons _____

    c. 80 newtons into pounds-force _____

    d. 75 kilograms into pounds-mass _____

    e. 63 grams into ounces _____

7. A 30-foot length of pipe weighs 8 pounds per foot. What is the weight of the pipe per metre?

    _____

8. How many megajoules are in 100 horsepower-hours?

    _____

9. Convert the following temperatures:

    a. 70°F to Celsius _____

    b. 324K to Celsius _____

    c. −28°C to Fahrenheit _____

    d. 203K to Rankine _____

    e. 152°F to Rankine _____

10. A cubic foot of a certain grade of iron weighs 470 pounds. What is its specific gravity?

    _____

11. The volume of a storage tank filled with fresh water is 5,360 cubic feet. What is the weight of the water in the tank?

    _____

12. How many cubic feet are in a ton of aluminum?

    _____

13. A truck winch line pulling with a force of 1,800 pounds pulls an engine 12 feet into position. How much work did the winch do?

    _____

14. A diesel engine has 375 horsepower. How many kilowatts of power does it produce?

    _____

15. A 1,200-watt heating unit is rated at 120 volts. What is its current rating?

    _____

16. If resistor R_2 in the circuit below is rated at 40 ohms, what is the voltage drop across each resistor?

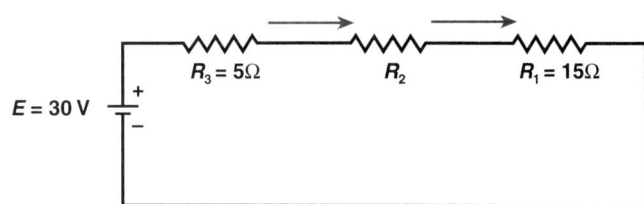

17. What is the pressure at the bottom of a 1,600-barrel water storage tank when the height of the water is 16 feet?

18. Convert the following temperatures and pressures to absolute values:

  a. 110°F _____

  b. 23°C _____

  c. 39 psig _____

  d. 16 psig _____

  e. 1,300°F _____

19. A tank of gas at 99°F had a pressure of 25 psig. When the pressure dropped to 10 psig, what was its temperature?

20. If a battery is delivering 24 volts DC at a current level of 100 amperes, what is the power level in kW?

5

Principles of Algebra

OBJECTIVES

Upon completion of chapter five, the student will be able to—

1. Use variables to represent an unknown quantity.

2. Solve an equation for an unknown quantity.

3. Convert a written problem into an equation and solve the equation.

4. Simplify an expression by removing the grouping symbols.

5. Add two or more algebraic expressions.

6. Add algebraic expressions containing a negative quantity.

7. Subtract one algebraic expression from another.

8. List the rules for multiplying algebraic expressions.

9. Multiply algebraic expressions containing exponents.

10. Divide one algebraic expression by another.

11. Transpose an equation to solve for a variable.

12. Use formulas to solve for quantities often used in industrial applications.

INTRODUCTION

In general, to calculate problems in simple arithmetic, only the numbers 0 through 9 (alone and in combination) and the signs $+$, $-$, $\times$, and $\div$ are used. Algebra goes a step further because it not only employs numbers and signs, but also letters and symbols. For example, algebra uses the letters a through z and symbols such as γ, α, and β.

The letters and symbols in algebra represent variables. A *variable* is an unspecified or unknown quantity—a quantity that varies with a problem. A variable may denote different values in different problems, but it always represents the same value in a given problem. For instance, x may stand for gallons in one problem and for feet in another, but, if it is selected to stand for gallons in a problem, it must represent gallons throughout that problem.

Algebraic expressions—combinations of symbols, letters, and signs—are used in formulas and equations. A *formula* is a symbolic expression of a general fact, rule, or principle and is often stated as an equation. An *equation* is an expression of equality between two quantities.

ALGEBRAIC EXPRESSIONS

An algebraic expression is a combination of algebraic symbols and signs. For example, $2a + 3ab - 4c$ is an algebraic expression. The letters a, b, and c stand for unknown quantities. Also, a and b are used together (ab, which means $a \times b$) to form another unknown. These unknowns are *terms*. The terms of the expression are the combinations of symbols that a sign does not separate—for example, $2a$, $3ab$, and $4c$ are the terms of the expression $2a + 3ab − 4c$. The plus (+) and minus (−) are operational signs—they tell you to add or subtract. Also, $2a$, $3ab$, and $4c$ indicate multiplication in that $2a = 2 \times a$, $3ab = 3 \times a \times b$, and $4c = 4 \times c$. Thus, algebraic expressions may be added, subtracted, multiplied, and divided.

Example Problem: If $a = 6$, $b = 4$, and $c = 3$, find the value of $2a + 3b + 5c$.

Solution: Substitute numerical values for letters:

$$(2 \times 6) + (3 \times 4) + (5 \times 3) = 12 + 12 + 15 = 39.$$

Grouping Terms

Another important aspect of algebraic expressions is that the terms within it may be grouped. Grouping terms indicates the order in which these operations should be carried out. Parentheses (), brackets [], and braces {} may be used for grouping terms. These symbols indicate that you should keep certain terms together; they also indicate the order in which you should perform the various operations in the expression. These grouping symbols must be removed before addition or subtraction can be performed. Removing parentheses, brackets, and braces means to perform the operation indicated inside them before working the rest of the expression. A few rules govern grouping symbols.

Rule 1. When an expression within the parentheses is preceded by the plus sign, the parentheses are removed without changing the signs within the parentheses.

Rule 2. When an expression within the parentheses is preceded by the minus sign, the parentheses are removed by changing the signs within the parentheses.

Rule 3. A number or symbol before the parentheses indicates that each term within the parentheses is to be multiplied by that number or symbol.

Example Problem: Simplify the expression, $3(a − b) + 4(b + c)$.

Solution: First remove the parentheses. Perform this calculation by understanding exactly what the expression states. In this case, the expression says to multiply a − b by 3, then add the result to 4 times b + c. So, to remove the parentheses:

$3 \times a$ and $3 \times b = 3a$ and $3b$; in the same way, $4 \times b = 4b$ and $4 \times c = 4c$. The result is

$$3a − 3b + 4b + 4c.$$

Then, combine like terms by performing the subtraction and addition:

$$3a + b + 4c.$$

Notice that only one part of the entire expression has like terms, which are $+4b$ and $−3b$. Subtracting $3b$ from $4b$ leaves $1b$, or simply b. Like terms are terms whose letters or symbols are the same. Just as you cannot add or subtract apples from oranges—they are unlike terms—you cannot add or subtract a from b, b from c, and so on. You can, of course, add or subtract like terms—apples and apples and oranges and oranges, as it were. In the case of the previous problem, subtract $3b$ from $4b$ to get b.

Example Problem: Simplify the following expression by removing the grouping symbols and by combining like terms:

$$3x + 4[2x - 3(x - 7y) + (x + 3y)]$$

Solution: Again, clearly understand the operations involved. In this case, the 4 before the first bracket ([) tells you to multiply the terms within the brackets by 4. However, the operations within the parentheses must be performed first. Performing the operations within the parentheses removes the brackets so that the expression becomes:

$$3x + 4(2x - 3x + 21y + x + 3y).$$

Now, remove the parentheses so that the expression becomes:

$$3x + 8x - 12x + 84y + 4x + 12y.$$

The result is:

$$3x + 96y.$$

Addition

Addition in algebra consists of collecting the terms of each expression and arranging them in columns. Place like terms in vertical columns. For example, to add $2a + 3b + c$ and $a + 6b + 8c$, arrange the terms in this manner:

$$\begin{array}{ccc} 2a & + 3b & + c \\ a & + 6b & + 8c \\ \hline 3a & + 9b & + 9c \end{array}.$$

Positive and negative quantities may be added. In other words, you can add a minus quantity to a plus quantity. Adding a negative quantity (a quantity with a minus sign before it) is equivalent to subtracting a positive quantity of the same value. For example, $-3z$ added to $5z$ is equal to $2z$, which is the same as subtracting $3z$ from $5z$. Adding negative quantities may also be expressed as $5z + (-3z) = 2z$.

Example Problem: Add $3a + 2ab - c$ and $3ab + 4c$.

Solution:

$$\begin{array}{ccc} 3a + 2ab & - & c \\ 3ab & + & 4c \\ \hline 3a + 5ab & + & 3c \end{array}.$$

Example Problem: Add $3x + 2xy - 3y$, $5x - 3xy + y$, and $3x + 4y$.

Solution: Arrange like terms in vertical columns and add:

$$\begin{array}{ccc} 3x + 2xy & - 3y \\ 5x - 3xy & + y \\ 3x & + 4y \\ \hline 11x - xy & + 2y \end{array}.$$

Subtraction

To subtract algebraic expressions, first change the sign of each term in the subtrahend. So, + in the subtrahend becomes −, and − becomes +. After changing the sign, then add the terms. Thus, $3x^2 + 5x + 10$ minus $x^2 + 2x - 4$ is carried out as follows:

$$
\begin{array}{r}
3x^2 + 5x + 10 \\
\underline{-x^2 - 2x + 4} \\
2x^2 + 3x + 14.
\end{array}
$$

If a term appears in the subtrahend and no like term appears in the minuend, bring down the term in the subtrahend into the result and change its sign. For example, $3a - 4ab + 3c$ is subtracted from $6a + 4c$ as follows:

$$
\begin{array}{r}
6a \qquad\quad + 4c \\
\underline{-3a + 4ab - 3c} \\
3a + 4ab + c.
\end{array}
$$

Example Problem: Subtract $2x^2 + 3x - 1$ from $5x^2 + x + 3$.

Solution: When writing down the subtrahend, change the signs; so, $2x^2 + 3x - 1$ becomes $-2x^2 - 3x + 1$. Then, add the minuend to the subtrahend in which the signs are changed:

$$
\begin{array}{r}
5x^2 + x + 3 \\
\underline{-2x^2 - 3x + 1} \\
3x^2 - 2x + 4.
\end{array}
$$

Example Problem: Subtract $6a^2 - 3a + 4ab$ from $12a^2 - b^2 + 3ab$.

Solution:

$$
\begin{array}{r}
12a^2 \qquad + 3ab - b^2 \\
\underline{-(6a^2 - 3a + 4ab \qquad\;)} \\
6a^2 + 3a - ab - b^2.
\end{array}
$$

In this example, notice the parentheses on either end of the subtrahend and the minus sign before the first parentheses. A minus sign in front of the parentheses is another way to denote the sign change in the subtrahend. The minus sign in this position indicates that all the terms inside the parentheses are multiplied by a minus. And, multiplying terms by a minus changes their signs (as you will see in the next section). Put another way, minus times minus is plus ($- \times - = +$) and minus times plus is minus ($- \times + = -$).

Multiplication

Algebraic multiplication can be indicated in four ways. For example, a times $b = c$ can be expressed as:

$$
\begin{aligned}
a \times b &= c. \\
a \cdot b &= c. \\
ab &= c. \\
a(b) &= c.
\end{aligned}
$$

When working with formulas and equations, the notation that expresses the operation in a formula or equation in the least confusing manner is the one used. The notation is the same in all cases.

Four simple rules govern algebraic multiplication.

Rule 1. A positive term multiplied by a positive term results in a positive product. For example:

$$3a \times b = 3ab.$$

Rule 2. A negative term multiplied by a negative term results in a positive product. For example:

$$-3a \times (-b) = 3ab.$$

Rule 3. A positive term multiplied by a negative term results in a negative product. For example:

$$3a \times (-b) = -3ab.$$

Rule 4. When like quantities are multiplied, their exponents are added in the product (when no exponent is present, it is considered to be 1). For example:

$$3a \times a^2 = 3a^{2+1} = 3a^3.$$

$$4b^2 \times 2b^3 = 8b^{2+3} = 8b^5.$$

Example Problem: Multiply $3a - 2b$ by $3b$.

Solution:

$$
\begin{array}{r}
3a - 2b \\
\times \quad 3b \\
\hline
9ab - 6b
\end{array}
$$

To multiply expressions when two or more terms are in the multiplier, follow these steps:

1. Multiply each term in the multiplicand by the last term.

2. Multiply each term in the multiplicand by the next to the last term and all other terms, working from right to left and placing like terms under each other. Unlike terms go in a separate column.

3. Then add like terms.

Example Problem: Multiply $-8a + 6b + 4c$ by $b + c$.

Solution: Multiply each term in the multiplicand by c, then by b, and add like terms:

$$
\begin{array}{l}
-8a + 6b + 4c \\
\times \qquad\quad b + c \\
\hline
-8ac + 6bc + 4c^2 \qquad\qquad\qquad \text{(Step 1)} \\
\qquad\quad 4bc - \quad 8ab + 6b^2 \quad \text{(Step 2)} \\
\hline
-8ac + 10bc + 4c^2 - 8ab + 6b^2. \ \text{(Step 3)}
\end{array}
$$

Division

Division of algebraic expressions involves the same processes of multiplication and subtraction that are used in solving arithmetic problems. The following rules apply to division of algebraic expressions.

Rule 1. When the signs of the divisor and the dividend are alike, the quotient is positive. For example:

$$-10x \div (-2x) = 5x.$$

Rule 2. When the signs of the divisor and dividend are different, the quotient is negative. For example:

$$4a \div (-2a) = -2a.$$

Rule 3. When a term with an exponent is divided by a like term with an exponent, the exponent of the divisor is subtracted from the exponent of the dividend. For example:

$$x^5 \div x^3 = x^{5-3} = x^2.$$

Example Problem: Divide $6x^2 + 9xy - 6x$ by $3x$.

Solution: Write down the dividend and divisor in the following manner:

$$3x \,\big)\, 6x^2 + 9xy - 6x \,.$$

Divide each term of the dividend by the divisor, write down each quotient, and place the appropriate sign before each quotient.

$$
\begin{array}{r}
2x\ \ + 3y\ \ - 2\ \ \ \ \ \\
\hline
3x \,\big)\, 6x^2 + 9xy - 6x \\
6x^2\ \ \ \ \ \ \ \ \ \ \ \ \ \ \ \ \\
\hline
9xy\ \ \ \ \ \ \ \\
9xy\ \ \ \ \ \ \ \\
\hline
6x \\
6x\,.
\end{array}
$$

Example Problem: Divide $x^2 - 2xy + y^2$ by $x - y$.

Solution: Arrange the dividend and divisor as for long division of numbers in arithmetic.

$$
\begin{array}{r}
x\ -\ y^2 \\
\hline
x - y \,\big)\, x^2 - 2xy +\ \ x \\
x^2 -\ xy \\
\hline
xy + y^2 \\
xy + y^2.
\end{array}
$$

Example Problem: Divide $a^4 + a^3 - 9a^2 - 16a - 4$ by $a^2 + 4a + 4$.

Solution: Arrange the dividend and divisor as in the previous example.

$$
\begin{array}{r}
a^2 -\ \ 3a - 1\ \ \ \ \ \ \ \ \ \ \\
\hline
a^2 + 4a + 4 \,\big)\, a^4 +\ \ a^3 -\ 9a^2 - 16a - 4 \\
a^4 + 4a^3 +\ 4a^2 \\
\hline
-\,3a^3 - 13a^2 - 16a \\
-\,3a^3 - 12a^2 - 12a \\
\hline
-\ \ a^2 -\ \ 4a - 4 \\
-\ \ a^2 -\ \ 4a - 4\,.
\end{array}
$$

Practice Problems

1. Find the sum of $2a + 3b$ and $a - 6b$.

   _____

2. Add $3b - 4bc$, $5a + 3bc + 2c$, and $6a + 4c$.

   _____

3. Find the sum of $6a + 20b + 4ab + c$, $6b - 8ab - 9c$, and $-4a + 6c$.

   _____

4. Add $5x + 3xy + 4y$ and $-8x - 2xy - 3y$.

   _____

5. Find the sum of $-8ab + 2a + c$, $7a + 9ab - 4c$, $8c - 12ab + 6a$, and $8b + 4c + 3ab$.

   _____

6. Subtract $3a^2 - 2b + c$ from $4a^2 - 6b + 2c$.

   _____

7. Take $6x + 3y$ from $4x^2 - 6xy - 4y$.

   _____

8. Subtract $6a(b + c) + 4(a + b)$ from $9a(b - c) - 8(b + c)$.

   _____

9. Subtract $4a^2 + 3ab + 6c^2$ from $9a^2 - 8c^2$.

   _____

10. Subtract $4(x + xy) - 6y$ from $6x^2 - 6(x + 2y)$.

    _____

11. Multiply $x^2 + 2xy + y^2$ by $4a$.

    _____

12. Multiply $4a + 6ab + 3c$ by $5ab$.

    _____

13. Multiply $3a + 4b$ by $a - b$.

    _____

14. Multiply $(a + 3b) - (a + b)^2$ by $4a$.

    _____

15. Simplify the expression $(x + y)^2 - (2x - 3y)$.

    _____

16. Divide $8a^2 + 4ab$ by $-2a$.

_____

17. Divide $27a^3 + 9a^2 + 12ab + 6b$ by $3a$.

_____

18. Divide $m^4 + 57m - 70$ by $m^2 + 3m - 5$.

_____

19. Divide $x^3 - 3x^2y + 3xy^2 - y^3$ by $x - y$.

_____

20. Divide $m^4 + 64$ by $m^2 + 4m + 8$.

_____

EQUATIONS

An *equation* may be simple or complicated, but it is always an expression of two quantities that equal each other. For example, $2 = 2$ is a simple equation. Another simple equation is $3 \times 4 = 2 \times 6$ because $3 \times 4 = 12$, $2 \times 6 = 12$, and $12 = 12$. If variables are used, an equation may read

$$3a = 2b$$

where

$$a = 4$$
$$b = 6.$$

In this case, $a = 4$ and $b = 6$; so, $3 \times 4 = 12$, $2 \times 6 = 12$, and $12 = 12$.

The two sides of an equation (the parts to the left and right of the equals sign) must remain equal, no matter what operation is performed to solve for the variable. If an arithmetical operation is performed on one side of the equation, the same operation must be performed on the other side of the equation.

Solving Problems with Equations

Solving algebraic equations usually involves letters or symbols that represent the numerical value of an unknown quantity. For instance, in the previous example, $a = 4$ and $b = 6$. Additional data in the problem make it possible to set up an equation. If enough data are given, the numerical value represented by the letter can be obtained. Following are the general steps for solving problems involving simple equations:

1. Let a letter or symbol represent one unknown number.

2. If a second number is given, represent it in terms of the first unknown, using the stated conditions in the problem.

3. Set up an equation, using the relations stated by the problem.

4. Solve for the unknown number.

5. Use the new number to solve for other unknown numbers, using the conditions given in the problem.

6. Check the numbers to see whether they meet the conditions of the problem.

Example Problem: The sum of two numbers is 92, and their difference is 4. Find the numbers.

Solution: Let x = the bigger number, and $x - 4$ = the smaller number. Set up the equation according to the conditions of the problem:

$$x + x - 4 = 92.$$

Solve for x:

$$2x - 4 = 92$$
$$2x = 92 + 4 = 96$$
$$x = 96 \div 2 = 48, \text{ which is the bigger number.}$$

Then solve for the other unknown:

$$48 - 4 = 44, \text{ the smaller number.}$$

Finally, check to see whether these numbers meet the conditions in the problem:

$$48 - 44 = 4.$$

This example not only shows how to set up a problem, it also shows the operations in solving an equation for an unknown. That is, to solve for x, you must move the -4 on the left side of the equation to the right side of the equation. In so doing, the sign changes from $-$ to $+$, because moving a term from one side of the equation to the other changes its sign. Put another way, moving subtraction to the equation's other side makes it addition. In the same way, moving addition to the equation's other side makes it subtraction.

Then, to solve for $2x$, which, remember, means $2 \cdot x$, you move the 2 to the right side of the equation. In so doing, you divide 96 by 2 to obtain 48 (the bigger number), because moving multiplication from one side to the other of an equation means that it becomes division. In the same way, moving division from one side to the other becomes multiplication.

Example Problem: Ten steel plates of a certain size weigh 60 pounds each. Their total weight is equal to the weight of eight plates of a different size. What is the weight of one of the different sized eight plates?

Solution: Choose y to equal the weight of one of the eight plates. Mathematically, it is stated as, "Let y = weight of one of the eight plates."

Because ten plates each weigh 60 pounds, then 10×60 lb = 600 lb, the weight of the ten plates.

Thus, $8y$ = the weight of the other eight plates.

But, the total weights are the same: 600 lb. Therefore, $8y = 600$ lb and $y = 600 \div 8 = 75$ lb, the weight of one plate.

Practice Problems

1. Let $a = 12$, $b = 4$, and $c = 2$. Find the value of the following:

   a. $2a + 3b$ _____

   b. $a + 2b + c$ _____

   c. $3b - c$ _____

2. In the equation $a + 2b + c = 14$, what is the value of a if $b = 2$ and $c = 8$?

   _____

3. Write the following equations, using the symbols a, b, c, d, and e where $a = 4$, $b = 1$, $c = 5$, $d = 2$, and $e = 3$.

   a. $4 + 5 + 1 = 3 + 4 + 1 + 2$ _____

   b. $1 + 5 + 3 = 3 + 1 + 2 + 3$ _____

   c. $5 + 4 + 2 = 5 + 3 + 1 + 2$ _____

4. A person receives $572 for 8 days of work. Using an algebraic equation, determine how much this person receives per day.

   _____

5. Five pounds of copper can be purchased at $1.07 per pound. An inferior grade may be bought for 72¢ per pound. By means of an equation, determine how many pounds of the inferior grade could be purchased for the same money that would be expended for 5 pounds of the better grade.

   _____

6. Divide 35 into two parts, so that 4 times the lesser number equals 3 times the greater number.

   Greater number: _____

   Lesser number: _____

7. A hundred and five feet of chain link fence material is available to build a square pen. An existing wall forms one side of the pen. What is the maximum area that can be enclosed by the wall and the fence material?

   _____

8. A small ship is chartered for a weekend excursion for 100 passengers at $150 each. The ship's owner agrees to reduce the rate $10 per person for each additional passenger. What is the total cost per person if 130 people participate in the cruise?

   _____

9. A student received A grades on 50% of one term's assignments and B grades on five other assignments for the term. The remaining term assignments, or $4/9$ of the total, were not handed in. How many assignments were there?

   _____

10. A speculator bought 100 shares of stock. She sold 25 shares at twice their cost, then the company went into bankruptcy. The remaining shares were sold for 20¢ each. Her net loss was $235. What was the initial price per share?

FORMULAS

A *formula* is a group of numerical symbols arranged to express briefly a single concept. A formula may be a simple term or expression, or it may be in the form of an equation. Formulas are used to save space and time when writing and working problems. For example:

$$T_m = D \times T_d$$

where

T_m = money earned in one month
D = number of days worked
T_d = money earned in one day.

Instead of a formula, the same information can be written: "The total amount of money earned in one month is equal to the number of days worked that month multiplied by the amount of money earned in one day." The formula is certainly quicker to write. In all formulas, when letters and symbols are used in place of numbers, the letters or symbols must be defined.

Example Problem: An oilfield worker earns $2,485 per month and works 12 months per year. His or her average living expenses are $2,050 per month. The balance is placed in a savings account. How much does he or she save in a year?

Solution: To reduce the statement to an algebraic formula, choose a letter for each factor to be considered in solving the problem. Let D stand for the number of dollars saved in a year, S for monthly earnings in dollars, and E for monthly expenses in dollars. Now using these letters, set up the algebraic formula:

$$D = (12 \times S) - (12 \times E)$$

or written in a simpler form,

$$D = 12S - 12E$$

where

D = number of dollars saved in a year
S = monthly earnings in dollars
E = monthly expenses in dollars.

Then solve for D by substituting numbers for the known factors:

S = $2,485; E = $2,050.
D = $(12 \times 2,485) - (12 \times 2,050)$
   = $29,820 - 24,600$
D = $5,220 saved in a year.

Transposition

Sometimes, to find the unknown quantity in a formula, it is necessary to *transpose*, or move, the terms. For example, consider the formula

$$T_m = D \times T_d$$

where

T_m = money earned in one month
D = number of days worked
T_d = money earned in one day.

This formula states that you can determine T_m if you know D and T_d. However, suppose you know the amount of the monthly check (T_m) and the number of days worked (D), and you need the daily rate of pay (T_d). In this case, the formula must be written to solve for T_d, the money earned in one day. That is, you must transpose, or move, T_d to one side of the equation so that it stands alone. And, you must not nullify or destroy the equation. Put another way, the goal is to move and isolate T_d so that it stands by itself on one side of the equation without destroying the equation. To keep from destroying the equation, whatever you do to one side of the equation, you must also do it to the other side. If you do not, you invalidate the equation—that is, the term or expression on one side does not equal the term or expression on the other side. In this example, to transpose, or solve for T_d, first divide both sides of the equation by D. The equation thus becomes:

$$\frac{T_m}{D} = \frac{D \times T_d}{D}.$$

The Ds on the right side of the equation cancel each other so that the equation becomes:

$$\frac{T_m}{D} = T_d$$

which is the same as

$$T_d = T_m \div D.$$

The daily rate (T_d) is equal to the monthly check (T_m) divided by the number of days worked during the month (D).

A commonly used formula that may also require transposing terms is:

$$\frac{a}{b} = \frac{c}{d}.$$

This formula sets up proportions and allows you to solve for an unknown when three factors are known. To understand this formula, substitute numbers for the variables; in this case, let $a = 12$, $b = 6$, $c = 4$, and $d = 2$.

$$12 \div 6 = 4 \div 2,$$

or

$$\frac{12}{6} = \frac{4}{2}.$$

Since both terms are equal to 2, they are equal to each other and constitute an equation. Note, too, that if the values are cross-multiplied—that is, if the number in the denominator on one side of the equation is multiplied by the numerator on the other side—the results equal each other. In this case, 12 is multiplied by 2 and 4 is multiplied by 6 and the result is 24 = 24. Cross-multiplying is one way to use the formula.

For instance, cross-multiplying transposes the terms in the equation $a/b = c/d$. Thus, cross-multiplying a by d and b by c results in a formula for direct proportion:

$$ad = bc.$$

The formula for direct proportion reads as, "the product of the means equals the product of the extremes." In a direct proportion, the first and fourth terms (a and d in the original formula, $a/b = c/d$, are called the extremes, and the second and third terms (b and c) are called the means. The formula $ad = bc$ can be further transposed into four variations:

$a = bc/d$ (both sides divided by d—remember: the ds on the left side cancel each other)

$b = ad/c$ (both sides divided by c—the cs on the left side cancel out)

$c = ad/b$ (both sides divided by b—the bs on the left side cancel out)

$d = bc/a$ (both sides divided by a—the as on the left side cancel out).

These transpositions can be utilized to solve proportional problems in various ways.

Example Problem: A photograph 8 inches wide by 10 inches deep is to be reduced to 3 inches wide to fit a magazine column. What is the depth of the reduced photo?

Solution: Set up the proportion by saying, "8 is to 10 as 3 is to x." Since the fourth term d is the unknown—the x—use the formula to solve for d (the unknown x becomes d) and substitute the known numbers for the letters: $a = 8$, $b = 10$, and $c = 3$.

$$d = bc/a$$
$$d = 10 \times 3 \div 8 = 3.75 \text{ in.}$$

Incidentally, simply cross-multiplying the appropriate values can solve the same proportion. In the case of the 8- by 10-in. photograph's being reduced so that the 8-in. width becomes 3 in., determining its reduced depth can be solved by setting up the proportion as:

$$\frac{8}{10} = \frac{3}{x} .$$

In this case, cross-multiplying results in $8x = 30 = 3.75$ inches.

Transposing terms and expressions of an equation can involve several steps. The important rule to remember is that the equation remains valid as long as each side is multiplied by, divided by, added to, or subtracted from equally. That is, the same mathematical operation must be done to both sides of the equation; if not, it is no longer an equation. Multiplication and division by the same term on the same side of the equation cancels that term, and addition and subtraction of the same term on the same side of an equation cancels the term. To illustrate, solve for b in the equation

$$a = \frac{20}{(b + 10)} .$$

Because you are solving for b, and the goal is to single out the b without nullifying the equation, the first step is to multiply each side by the expression $(b + 10)$. This step puts an expression with b in it on the left side of the equation. This step comes out as:

$$a(b + 10) = \frac{20}{(b + 10) \times (b + 10)}.$$

Since multiplying the same term on the same side of the equation cancels the terms, cancellation results in:

$$a(b + 10) = 20.$$

At this point, an a is on the left side of the equation along with a b. But, you only want a b on the left side because you are solving for b. So, to eliminate the a without destroying the equation, the next step is to divide both sides by a, which results in:

$$\frac{a(b + 10)}{a} = \frac{20}{a}.$$

The as on the left cancel to result in:

$$(b + 10) = \frac{20}{a}.$$

To eliminate the 10 on the left side without destroying the equation, the last step is to subtract 10 from each side:

$$b + 10 - 10 = \frac{20}{a} - 10,$$

which yields the desired result of:

$$b = \frac{20}{a} - 10.$$

Remember that when solving for an expression in an equation, the goal is to single out the expression without destroying the equation. As long as you add, subtract, multiply, or divide both sides of the equation by the same terms, you do not destroy the equation. The other factor to keep in mind is the cancellation rule: multiplication and division by the same term on the same side of the equation cancels that term, and addition and subtraction of the same term on the same side of an equation cancels that term.

Formula Applications

Formulas are used extensively in the petroleum and other industries. Many of the commonly used formulas deal with area and volume and therefore are given in chapter six, "Practical Geometry." Other established formulas and their applications in industry are presented here and in chapter eight.

Since most calculations are carried out on a calculator, be aware that a formula may have to be restated so that it gives the desired answer if you enter the numbers and operation signs one after the other on the calculator. In any case, it is vital to perform the operations in the proper sequence either by properly entering the values of the formula in the calculator or by rearranging the formula.

Well Control Formulas

Well control is a subject of concern in the petroleum industry. Briefly, well control involves recognizing and then properly handling an unwanted or unexpected entry of formation fluid into the wellbore. Such an entry is a *kick*. If rig crewmembers fail to recognize and take appropriate steps to control a kick, they can lose control of the well. The well can blow out. A *blowout* is the uncontrolled flow of formation fluids into the atmosphere or into another subterranean formation the wellbore has penetrated. Blowouts are not desirable because they can injure or kill personnel, destroy equipment, pollute the environment, and waste valuable resources.

An important formula in well control provides rig crewmembers with a way to determine the hydrostatic pressure of the drilling fluid (usually called mud) in the wellbore. The amount of hydrostatic pressure a drilling mud develops depends on its weight, or density, and on its depth, or height. The more a drilling mud weighs, the more hydrostatic pressure it develops. And, the deeper, or longer, the length of the mud column in the wellbore, the more hydrostatic pressure it develops. Usually, but not always, drilling mud completely fills the wellbore. So, often, the well's depth and the height, or depth, of the mud in the wellbore are the same.

A well's true vertical depth is critical to accurate calculations. Nowadays, many wells are deliberately not drilled vertically—that is, they deviate from vertical even to the point of being horizontal. (The wellbore actually runs parallel, rather than perpendicular, to the surface.) However, to calculate hydrostatic pressure correctly, the wellbore's true vertical depth and not its measured depth must be used. *Measured depth* is the depth of the well taken along the actual path it takes through the earth. Thus, a well's true vertical depth can be less than its measured depth, especially in horizontal wells.

In any case, the mud's weight and its true vertical height in the wellbore govern the amount of hydrostatic pressure it develops. And hydrostatic pressure plays a large role in controlling pressures in a well. The mud column's hydrostatic pressure must be the same as or, preferably, slightly greater than the pressure in formations the wellbore penetrates. The formula for determining hydrostatic pressure is:

$$HP = C \times MW \times TVD$$

where

HP = hydrostatic pressure, pounds per square inch (psi) or kilopascals (kPa)

C = a constant. Its value depends on how mud weight is expressed. In most of the U.S., mud weight is in pounds per gallon (ppg). On the Pacific Coast of the U.S., however, it may be in pounds per cubic foot (pcf). Other places express mud weight in kilograms per cubic metre (kg/m³). Where the mud weight is in ppg, the constant is 0.052. Where mud weight is in pcf, the constant is 0.0069. Where the mud weight is in kg/m³, the constant is 0.0098. In this case, hydrostatic pressure is in kilopascals (kPa) and true vertical depth is in metres.

MW = mud weight, ppg, pcf, or kg/m³

TVD = true vertical depth, feet or metres.

Example Problem: What is the hydrostatic pressure in a well that is full of mud that weighs 11.2 ppg and whose true vertical depth is 4,000 feet? (The constant for ppg is 0.052.)

Solution:

$$
\begin{aligned}
HP &= C \times MW \times TVD \\
&= 0.052 \times 11.2 \times 4,000 \\
&= 2,329.5999 \\
HP &= 2,330 \text{ psi. (For practical field use, round pressure to the nearest}
\end{aligned}
$$

whole number.) Thus, the pressure that an 11.2-ppg mud produces at the bottom of a 4,000-foot well is 2,330 psi.

Example Problem: What is the hydrostatic pressure in a well that is full of mud that weighs 83.8 pcf and whose true vertical depth is 4,000 feet? (The constant for pcf is 0.0069.)

Solution:

$$
\begin{aligned}
HP &= C \times MW \times TVD \\
&= 0.0069 \times 83.8 \times 4,000 \\
&= 2,312.88 \\
HP &= 2,313 \text{ psi.}
\end{aligned}
$$

Example Problem: What is the hydrostatic pressure in a well that is full of mud that weighs 1,342 kg/m^3 and whose true vertical depth is 1,219 metres? (The constant for kg/m^3 is 0.0098.

Solution:

$$
\begin{aligned}
HP &= C \times MW \times TVD \\
&= 0.0098 \times 1,342 \times 1,219 \\
&= 16,031.8004 \\
HP &= 16,032 \text{ kPa.}
\end{aligned}
$$

When a kick occurs, crewmembers stop pumping mud and close a large valve (a blowout preventer) and other, smaller valves at the top of the well. Shutting the blowout preventer and the other valves completely seals the well and prevents more formation fluids from entering. After a well is completely shut in on a kick, pressure appears on a gauge at the surface. This surface gauge is a drill pipe pressure gauge; it measures the amount of pressure the intruded fluids place on the drill string. (If the mud's hydrostatic pressure equaled or exceeded formation pressure, no kick could occur and, consequently, no pressure would appear on the gauge.)

Shut-in drill pipe pressure is vital to controlling the well. The amount of shut-in drill pipe pressure indicates the amount formation pressure exceeds hydrostatic pressure. For example, if shut-in drill pipe pressure is 300 psi, it means that hydrostatic pressure is 300 psi lower than formation pressure. Thus, crewmembers know that they must increase the mud's density enough to develop at least 300 more psi than it did when the well kicked. In short, they must pump a heavier mud into the well, a mud that weighs more than the original mud in use when the well kicked.

To determine how much to increase the mud weight, the following formula is used:

$$MWI = SIDPP \div TVD \div C$$

where

MWI = mud weight increase, ppg, pcf, or kg/m³
$SIDPP$ = shut-in drill pipe pressure, psi or kPa
TVD = true vertical depth, feet or metres
C = 0.052 for ppg, 0.0069 for pcf, and 0.0098 for kg/m³.

Example Problem: Given the following factors, find the mud weight increase needed to kill a well:

$SIDPP$ = 350 psi
TVD = 7,672 ft
C = 0.052.

Solution:

MWI = $SIDPP \div TVD \div C$
= 350 ÷ 7,672 ÷ 0.052
= 0.8773
MWI = 0.9 ppg.

In this case, the original mud weight was 10.2 ppg when the kick occurred; so, crewmembers must increase the mud weight to at least 11.1 ppg (10.2 + 0.9 = 11.1). Normally, they raise the mud weight about a decimal point or two higher to increase hydrostatic pressure to a value slightly greater than that needed to balance the formation pressure. (For in-depth coverage of well control, see: *An Introduction to Well Control*, catalog number 5.10020, ISBN 0-88698-185-9 and *Practical Well Control*, catalog number 2.80040, ISBN 0-88698-183-2. Both publications are available from Petroleum Extension Service, The University of Texas at Austin, www.utexas.edu/cee/petex.)

Ideal Gas Law Formula

In industry, it may be necessary to work with gas that is compressed and stored or expanded for transfer from one vessel to another. In such cases, knowing the pressure, volume, and temperature of a gas under various conditions may be required. For such calculations, the ideal gas law is used. This law combines Boyle's law and Charles's law and states that the volume of a quantity of gas varies inversely as the absolute pressure and directly as the absolute temperature. The formula for the ideal gas law is:

$$P_1 V_1 \div P_2 V_2 = T_1 \div T_2$$

where

P_1 = original pressure
P_2 = final pressure
V_1 = original volume
V_2 = final volume
T_1 = original temperature
T_2 = final temperature.

Pressures and temperatures in the formula are absolute. To convert gauge pressure to absolute pressure, add 14.7 psi to the gauge pressure. To convert Fahrenheit temperature to Rankine (absolute) temperature, add 460° to the Fahrenheit temperature. To convert Celsius temperature to Kelvin (absolute) temperature, add 273° to the Celsius temperature.

Example Problem: A volume of gas fills 1,000 cubic feet (ft³) at 80 pounds per square inch gauge (psig) and 80° Fahrenheit (F). What is the volume of the gas at 100 psig and 60°F?

Solution: In this case, V_1 is 1,000 ft³, P_1 is 80 psig, and T_1 is 80°F. What we are seeking is V_2 at P_2 (100 psig) and T_2 (100°F). So, first solve for V_2 by transposing the formula $P_1 V_1 \div P_2 V_2 = T_1 \div T_2$ so that it becomes:

$$V_2 = P_1 V_1 T_2 \div P_2 T_1.$$

Substitute figures for the factors:

$$
\begin{aligned}
P_1 &= \text{80 psig} + \text{14.7 psi} = \text{94.7 psia} \\
P_2 &= \text{100 psig} + \text{14.7 psi} = \text{114.7 psia} \\
T_1 &= \text{80°F} + \text{460°} = \text{540°R} \\
T_2 &= \text{60°F} + \text{460°} = \text{520°R} \\
V_1 &= \text{1,000 ft}^3.
\end{aligned}
$$

Then

$$
\begin{aligned}
V_2 &= (94.7 \times 1{,}000 \times 520) \div (114.7 \times 540) \\
    &= 49{,}244{,}000 \div 61{,}938 \\
V_2 &= 795.05 \text{ ft}^3.
\end{aligned}
$$

Incidentally, if using a calculator, the problem can be entered as:

$$94.7 \times 1{,}000 \times 520 \div 114.7 \div 540 = 795.05 \text{ ft}^3.$$

In any case, at 100 psig and 60°F, the volume of gas is about 795 ft³.

In the example, temperatures are in °F and to convert to the absolute temperature in °R, 460 was added to °F. If, however, the temperature needs to be in degrees Celsius (°C) and degrees Kelvin (K), then use the following formula to change °F to °C:

$$°\text{F} = {}^9/_5 °\text{C} + 32°$$

where

$$
\begin{aligned}
°\text{F} &= \text{temperature in degrees Fahrenheit} \\
°\text{C} &= \text{temperature in degrees Celsius.}
\end{aligned}
$$

Then, to obtain the absolute temperature in K, simply add 273 to the temperature in °C.

On the other hand, if the temperature is given in °C and you need °F, use the following formula:

$$°\text{C} = {}^5/_9(°\text{F} - 32°).$$

Ohm's Law Formula

In electrical calculations, the basis of many formulas is Ohm's law, which states that the current flowing in a circuit, in amperes, is directly proportional to the electromotive force (emf), in volts, and inversely proportional to the resistance of the circuit, in ohms. Expressed as a formula, this statement becomes

$$I = \frac{E}{R}$$

where

I = current, amperes
E = emf, volts
R = resistance, ohms.

The equation may be transposed to solve for the resistance if the current and voltage are known:

$$R = \frac{E}{I}$$

or to solve for voltage if current and resistance are known:

$$E = I \times R.$$

Unit Conversion Formulas

Many formulas are used to convert conventional units of measure to metric units and vice versa. For example, to convert miles to kilometres, the number of miles is multiplied by a conversion factor of 1.609 (see table 4.3). The formula for the conversion is:

$$km = mi \times C$$

where

km = kilometres
mi = miles
C = conversion factor, 1.609.

Since determining the constant in such calculations can be involved, tables are available for quick conversion of units (see tables 4.2, 6.1, and 6.3). Even with tables, however, some calculation is necessary to get the answer in the appropriate unit of measure.

Example Problem: If 1 cubic foot is equivalent to 7.48 gallons, what capacity tank would be needed to hold 50 gallons?

Solution: Set up a proportional equation, letting x equal the tank capacity needed to hold 50 gallons:

$1 : 7.48 = x : 50$
$x = 1 \times 50 \div 7.48$
$x = 6.684 \text{ ft}^3.$

Practice Problems

1. A formula for finding the permeability of a formation sample is

$$C = \frac{\mu QL}{A (P_1 - P_2)}$$

where
C = permeability, darcys
μ = viscosity of fluid used, centipoises
Q = volume of water passing through sample, cubic centimetres per second
P_1 = input pressure, atmospheres
P_2 = delivery pressure of fluid leaving sample, atmospheres

L = length of sample, centimetres

A = area of cross section of sample, square centimetres.

Given the following values for the variables, what is the permeability of the sample?

μ = 15

Q = 0.5

P_1 = 4

P_2 = 1

L = 12

A = 75.

(Note that units cancel out so that the answer is in darcys—the usual unit of measure used to describe permeability.)

_____

2. Convert the following metric measurements to conventional units: (Refer to table 4.3 for conversion factors.)

   a. 6 millimetres to inches _____

   b. 152 kilometres to miles _____

   c. 6 metres to yards _____

   d. 30 centimetres to inches _____

   e. 2 metres to feet _____

3. If 45 floor tiles cost $32.50, how much will 180 tiles cost?

   _____

4. If a litre of gasoline sells for 35.9 cents per litre, how many gallons can you buy for $5.00? (1 gallon = 3.785 litres)

   _____

5. Set up a formula to solve for each of the following variables of an automobile trip if the other two variables are known:

   a. The distance traveled (D), in miles _____

   b. The average speed (S), in miles per hour_____

   c. The time required for the trip (T), in hours _____

6. A formula useful in estimating the approximate number of barrels of oil in a storage tank is

$$V = d^2 \times 0.14 \times h$$

where

   V = volume of oil, barrels

   d^2 = diameter of tank, feet

   0.14 = a constant

   h = height of fluid, feet.

Using this formula, find the approximate number of barrels of oil in a tank when the height of the oil is 30 feet and the tank diameter is 114.5 feet.

7. Transpose the following formula for finding °API gravity (*API*) to solve for specific gravity (*sp gr*):

$$API = (141.5 \div sp\ gr) - 131.5.$$

sp gr = _____

8. Using the formula in problem 7, find the °API gravity of a crude oil whose specific gravity is 0.8311.

_____

9. Boyle's law expressed as a formula is

$$P_1 V_1 = P_2 V_2$$

where

P_1 = original pressure of gas (absolute)
V_1 = original volume of gas
P_2 = new pressure of gas (absolute)
V_2 = new volume of gas.

Transpose this equation to solve for each of the terms involved. Show the four formulas.

P_1 = _____

V_1 = _____

P_2 = _____

V_2 = _____

10. A volume of gas fills 2,000 cubic feet at 80 psig. If the temperature remains constant, what is the volume of 100 psig? (*Hint*: Convert psig to psia; then use the formula given in problem 9.)

_____

SELF-TEST

5. Principles of Algebra

Multiply each problem answered correctly by five to arrive at your percentage of competency.

1. Solve for x in the following equations:

   a. $3x = 12$

   _____

   b. $24 - x = 15$

   _____

   c. $x + 15 = 45$

   _____

   d. $2(2x - 4) + 4x = 152$

   _____

   e. $32x + 50x = x^2 + 3.5x + 10^3$

   _____

2. If $a = 6$, $b = 4$, and $c = 3$, find the value of each of the following:

   a. $3a + 6b + c$

   _____

   b. $a^2 + ab + 2c$

   _____

   c. $-3a^2 + 2ab - 2c^2$

   _____

   d. $2c^3$

   _____

   e. $3a + 2ab + 5b + bc + c^2$

   _____

3. Ten stands of pipe of a certain diameter weigh 900 pounds each. Their total weight is equal to the weight of six stands of pipe of a different diameter. What is the weight of one of the heavier stands of pipe?

   _____

4. Add the following algebraic expressions:

   a. $5x + 3y$ and $4x - 6y$

   _____

   b. $3x + 4y$ and $-4x - 8y$

   _____

    c. $5x - 3y + 4z$ and $8x - 2z + 3y$ and $7x - 5y - 9z$

---

    d. $3a + 4b - 6c + 8d$ and $7c + 4d$ and $-2a + b$ and $7c + 9d$

---

    e. $2x^2 + 4x^2y + 8xy^2$ and $3x^2 + 9xy^2$ and $-4x^2 + 3x^2y$ and $4x^2y - 9xy^2$

---

5. Subtract the following algebraic expressions:

    a. $2a - 2b + c$ from $a - 2b - 3c$

---

    b. $4a - 5b + 3c$ from $4a - 8b + 2c$

---

    c. $2x - y + 5z + 6$ from $3x - 2y + z$

---

    d. $4a^2b - 5ab^2 - b^3$ from $3a^2 - 3a^2b - 3ab^2$

---

    e. $x^3 - z^3 - 3x^2z + 3xz^2$ from $5x^2z + 8z^3 + 3x^3 + 2xz^2$

---

6. Simplify the following expressions by removing the grouping symbols and combining like terms.

    a. $(4a - 3b + 6) + (3a + 5b - 10)$

---

    b. $(2x^2 + 3x + 10) + (x^2 - 5x + 5) - (x^2 + x + 3)$

---

    c. $5x - [2x + (-3x + 4y) - (x - 8y) + 4x]$

---

    d. $(5a - 6b) - [4a - (4c - b) - 2c]$

---

    e. $-\{[(x^2 - a^2) + (y^2 - 2a^2)] - [(x^2 - a^2) - (y^2 - 2a^2)]\}$

---

7. Multiply the following expressions:

    a. $-5ab \times b$

---

    b. $2x^2 \times 3x^3$

---

c. $-4x^5 \times -6x^3$

_____

d. $(x + y)^2 \times 4xy$

_____

e. $(-x^3ay) \times (x^2 - 2ax + y^2)$

_____

8. Divide the following:

a. $8a^3 \div a^2$

_____

b. $(-18x^7) \div (6x^4)$

_____

c. $15x^2y^4 \div 75xy^4$

_____

d. $\dfrac{(a + x) - 2(a + x)^3}{a + x}$

_____

e. $(a^3 + 5a^2b + 5ab^2 - 3b^3) \div (a + 3b)$

_____

9. An owner sold $^2/_5$ of his farm to one person and 14 acres to a second person. Find the number of acres in the original farm if the owner had half of the farm left. (*Hint*: Let x equal the total number of acres in the original farm.)

_____

10. Using Ohm's law, $I = E \div R$, find the resistance (R) of a load if the voltage (E) across the load in a circuit is 120 volts and the current (I) through the load is 6 amperes.

_____

11. An electronic temperature process transmitter is delivering 16 mA at its output. If the transmitter was calibrated from 0 to 200°F to produce 4–20 mA, what is the current temperature at the input to the transmitter?

_____

12. Solve for the unknown in the following equations:

a. $A = \frac{1}{2}h(b + c)$, where $h = 13$, $b = 7$, $c = 11.2$.

_____

b. $h = r + \frac{1}{2}\overline{4\ r^2 - c^2}$, where $c = 6.5$, $r = 8.7$.

_____

c. $d = \dfrac{t}{3\sqrt{rs}}$, where $r = 4.2$, $t = 7$, $s = 5$.

_____

d. $a = bc \div d + c$, where $c = 6.8$, $d = 2.5$, $b = 9.2$.

_____

e. $S = {}^{1.157}/P - A$, where $S = 0.04764$, $A = 0.035$.

_____

13. Separate 55 into two parts so that two-thirds of the larger part is 5 more than the smaller part.

Larger part: _____

Smaller part: _____

14. A car that is 14 feet long overtakes a 28-foot truck that is going 50 miles per hour. How many feet per second must the car travel in order to pass the truck in 4 seconds?

_____

15. What is the speed of the car in problem 13, in miles per hour?

_____

16. A company has prepared a safety booklet that it will sell for $2.00 per copy. Under plan A, the company can get 15,000 copies printed for $5,000. Under plan B, it can get twice as many copies for one and a half times the cost.

a. What will be the cost per book under each plan?

Plan A: _____

Plan B: _____

b. How many books does the company have to sell before it starts making a profit?

Plan A: _____

Plan B: _____

c. After costs are recovered, how many books will be left to sell under each plan?

Plan A: _____

Plan B: _____

d. Assuming all books are sold, what is the total profit under each plan?

Plan A: _____

Plan B: _____

e. The company sold 25,000 copies and then decided to revise the booklet. Under plan A, the company would have to reorder 10,000 more copies at a printing cost of $4,500. Under plan B, the company would have to throw away 5,000 booklets that were out of date. With these statistics at hand, which plan would be the best choice for the company and why?

_____

17. Rewrite the following formula to solve for degrees Rankine, which is the absolute temperature used with Fahrenheit:

$$°C = {}^5\!/_9(°R - 492).$$

_____

18. Since the area (_A_) of a rectangle is found by multiplying its length (_l_) by its width (_w_), write a formula for finding the width of a rectangle when the area and length are known.

_____

19. If a 2" × 6" joist has 8 board feet in it, how long is the piece?

_____

20. What is the weight of a barrel of 40°API gravity oil?

_____

21. How many 8-inch square tiles are required to cover an entry that is 6 feet wide by 8 feet long?

_____

22. How much electrical power is delivered to a heater load requiring 240 volts at 200 amperes?

_____

23. What is the electrical resistance of a solenoid valve coil rated for 24 volts at 3 amperes?

_____

24. A standard electronic level transmitter, calibrated from 0 to 200", is determined to have 160" applied to its input. What is the signal current at its output?

_____

25. An electric heater is determined to have 12 ohms of resistance at a current level of 2 amperes. If two of these heaters are connected in electrical parallel, what is the total current from the supply?

_____

6
Practical Geometry

OBJECTIVES

Upon completion of chapter six, the student will be able to—

1. Find the perimeters of rectangles and other parallelograms, trapezoids, triangles, and polygons.

2. Find the circumferences of circles and ellipses.

3. Find the length of a triangle's side with the dimensions of two sides given.

4. Find the areas of parallelograms, trapezoids, triangles, regular polygons, circles, and ellipses.

5. Solve problems involving plane figures, by using geometrical formulas.

6. Find the volumes of rectangular solids, cylinders, elliptical solids, cones, pyramids, frustums of cones and pyramids, and spheres.

7. Find the lateral and total surface areas of some solid figures.

8. Solve problems involving solid figures, by using geometrical formulas.

9. Construct geometric figures, by using only a compass and a straightedge.

10. Use triangles and T-square properly for drawing geometric figures.

INTRODUCTION

Geometry deals with the measurements, properties, and relationships of points, lines, angles, surfaces, and solids. Plane geometry is concerned with plane, or two-dimensional, figures, such as squares, rectangles, triangles, and circles. Solid geometry deals with solid, or three-dimensional, objects, such as cubes, pyramids, cones, and spheres.

To solve advanced geometry problems, you must use deductive reasoning—that is, you must apply logical thinking along with special statements, which are called theorems, to solve problems. A *theorem* is a proposition or a formula that can be solved, or proved, by using basic assumptions and declarations called axioms.

An *axiom* is a statement or an idea that is accepted as true. For example, a mathematician may say that a statement or formula is axiomatic. By axiomatic, the mathematician means that the statement cannot be proven but that it is nevertheless accepted as true. An axiomatic equation, for instance, is $2 + 2 = 4$. It is axiomatic because the equation assumes that everyone agrees that figures we call numbers exist, that we agree what the numbers stand for, and that if we add two and two, we get four.

To solve complex geometry problems, you must clearly understand the terms, comprehend the theorems and their related formulas, and be able to draw conclusions based on given facts. Students of pure geometry spend a great deal of time proving theorems. That is, a theorem is proposed as true and, using

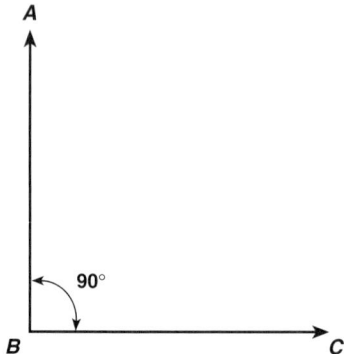

Figure 6.1   A right angle

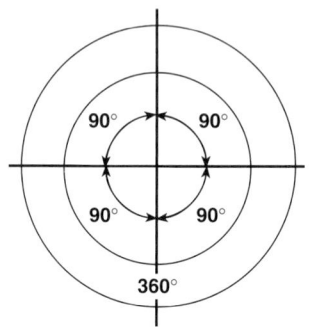

Figure 6.2   Circle

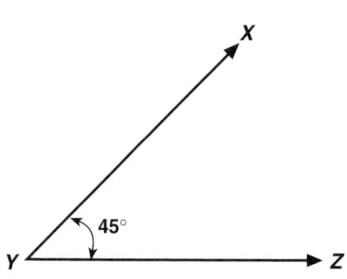

Figure 6.3   Acute angle

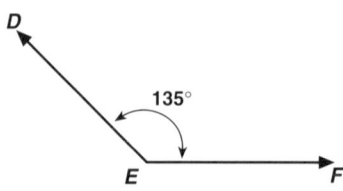

Figure 6.4   Obtuse angle

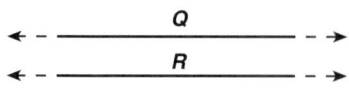

Figure 6.5   Parallel lines

axioms, students prove whether it is or is not true. This chapter, however, does not prove theorems. Rather, it covers practical geometry problems that you may encounter in the shop or field.

Geometric construction—that is, drawing accurate geometric figures—aids in visualizing a problem and its solution. So, you should also learn to use a geometric compass, straightedge, and basic drawing tools. (A geometric compass, unlike the direction finding instrument, is a V-shaped device for drawing circles or circular arcs.)

PLANE FIGURES

Lines and angles are the basic elements that make up plane geometric figures. A plane geometrical figure is a polygon or a circle. A polygon is any closed figure of three or more straight sides on the same plane. A polygon has length and width, but no depth. The most common mathematical problems involving plane figures are determining their perimeters and their areas. The *perimeter* of a polygon is the total distance of all its sides. The *area* of a polygon is the amount of space that it encloses.

Measurements of length and width (or base and altitude, or height) and angles determine the area and perimeter of a polygon. Polygons include squares, rectangles, multisided figures such as hexagons and octagons, trapezoids, parallelograms, and triangles. A circle is a closed plane curve, all points of which are equally distant from a point within called the center. Circle measurements include circumference, radius, diameter, and area.

Angles and Lines

An angle is formed when two straight lines, called sides, meet at a point, called the vertex. For example, in figure 6.1 sides *AB* and *BC* meet at point *B*, forming the angle *ABC*. The vertex of the angle is at *B*. The size of an angle is measured in degrees (°). One degree is a unit of measurement that is equal to $\frac{1}{360}$ of a circle. The angle in figure 6.1 contains 90° and is called a right angle. If a circle is divided into four equal parts, it contains four right angles of 90° each, or 360° (fig. 6.2). So, a circle contains 360°.

An angle that measures less than 90° is an acute angle (fig. 6.3), and an angle that measures more than 90° is an obtuse angle (fig. 6.4). The acute angle in figure 6.3 is a 45° angle. The obtuse angle in figure 6.4 is a 135° angle.

Lines may be straight or curved. Mathematically speaking, lines have length but not width and generally lie between two points. Parallel lines are straight lines in the same plane that do not meet, or intersect, however far extended. Lines *Q* and *R* in figure 6.5 are parallel lines; they are equidistant from all points on the lines.

Perpendicular lines meet at right angles to each other, as the lines *AB* and *BC* in figure 6.1. Put another way, line *AB* is perpendicular to line *BC*—that is, *AB* is perpendicular to *BC*.

Parallelograms

A parallelogram is a closed figure whose opposite sides are parallel (fig. 6.6). If the sides do not meet at right angles, as in figure 6.6, then the figure is a parallelogram. If, however, the sides meet at right angles—that is, if they are perpendicular to each other—then the parallelogram is a rectangle (fig. 6.7). If the sides are the same length and meet at right angles, the parallelogram is a square (fig. 6.8).

In the parallelogram in figure 6.6, line *CD* forms the base (length) and line *BD* is the height (width). In the rectangle (fig. 6.7), and in the square (fig. 6.8), line *CD* is the base and line *BD* is the height.

The perimeter of a plane figure is the distance around the figure. Put another way, the perimeter of a plane figure is the sum of the lengths of the sides. To find the perimeter of the rectangle shown in figure 6.7, add *AB*, which is 10 inches, *BD*, which is 4.5 inches, *DC*, which is 10 inches, and *CA*, which is 4.5 inches. So, the perimeter of the rectangle is:

$$10 \text{ in.} + 4.5 \text{ in.} + 10 \text{ in.} + 4.5 \text{ in.} = 29 \text{ in.}$$

To find the perimeter of a square, simply multiply the length of one side times 4. For example, the perimeter of the square in figure 6.8 is equal to line *AB* (which is 5 inches) × 4. Thus:

$$5 \text{ in.} \times 4 = 20 \text{ in.}$$

The distance around the square in figure 6.8—the square's perimeter—is 20 inches.

Example Problem: The length of two sides of a rectangle is 2 feet and the length of the other sides is 15 inches. What is the perimeter of this rectangle?

Solution: First, because the units of measurement must be the same, either convert 2 feet to inches or 15 inches to feet. To convert 2 feet to inches, multiply by 12 because 12 inches make up 1 foot. Thus:

$$2 \times 12 = 24 \text{ in.}$$

Then, add the length of the sides

$$24 + 24 + 15 + 15 = 78 \text{ in.}$$

Or, convert 15 inches to feet by dividing 15 by 12, the number of inches in 1 foot. Thus:

$$15 \div 12 = 1.25 \text{ ft.}$$

Then, 1.25 + 1.25 + 2 + 2 = 6.5 ft. You can confirm that this answer is correct because 78 in. ÷ 12 = 6.5 ft. Similarly, 6.5 ft × 12 = 78 in.

The area of a plane figure is the amount of surface space enclosed within the boundary lines of that figure (fig. 6.9). To find the area (*A*) of a parallelogram, multiply the number of units in its base (*b*) times the number of units in its height (*h*) to get the number of square units contained in the parallelogram (*A = bh*). To find the area of the rectangle in figure 6.9, multiply the base (13 inches) times the height (4.5 inches):

$$A = bh$$
$$A = 13 \text{ in.} \times 4.5 \text{ in.} = 58.5 \text{ in.}^2$$

As mentioned before, when computing areas, use the same units for base and height. For example, if the base is expressed in feet and the height in inches, change feet to inches or inches to feet before multiplying. Table 6.1 and the tables in chapter four will help with conversions in area measurement.

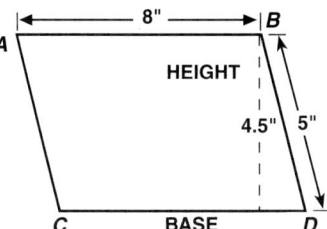

Figure 6.6   Parallelogram

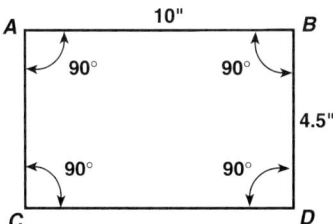

Figure 6.7   Rectangle

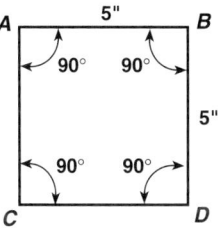

Figure 6.8   Square

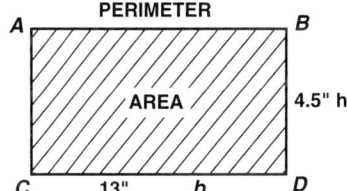

Figure 6.9   Area and perimeter of a rectangle

TABLE 6.1

Common Units of Area Measurement

1 ft²	=	144 in.²
1 yd²	=	9 ft²
1 rod²	=	30.25 yd²
1 acre	=	160 rods²
1 acre	=	4.840 yd²
1 mile²	=	1 section
1 section	=	640 acres
1 hectare	=	2.471 acres

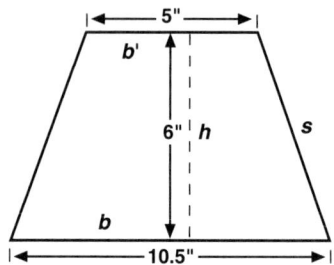

Figure 6.10   Trapezoid

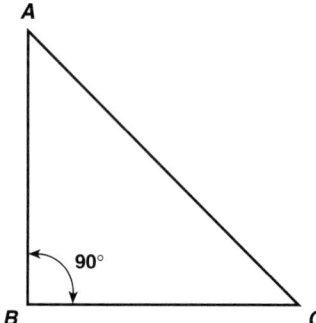

Figure 6.11   Right triangle

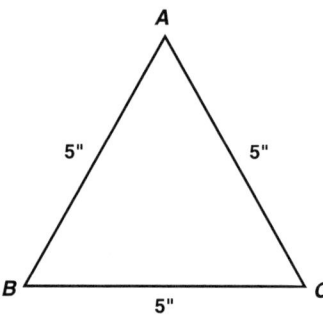

Figure 6.12   Equilateral triangle

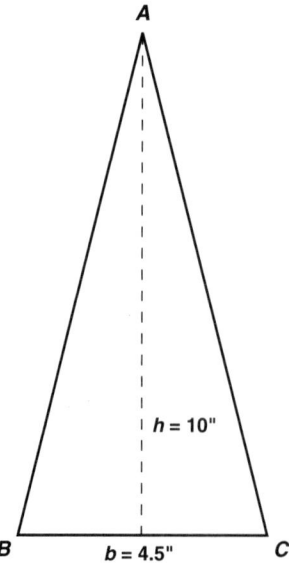

Figure 6.13   Isosceles triangle

Trapezoids

A trapezoid is a four-sided plane figure, two of whose sides are parallel (fig. 6.10). The area of a trapezoid is equal to one-half its height (h) times the sum of its bases (b and b'). To find the area of the trapezoid in figure 6.10, use the formula

$$A = \tfrac{1}{2}h \times (b + b').$$

Substituting the measurements in figure 6.10, the equation solves to:

$$A = \tfrac{1}{2}(6) \times (10.5 + 5) = 3 \times 15.5 = 46.5 \text{ inches}^2.$$

Triangles

A triangle is a closed figure formed by three straight lines connecting three points not in a straight line. The three points are the vertices of the triangle. The lines between the vertices are the sides. In the triangle in figure 6.11, sides AB and BC are legs. Side AC, which is the longest and is opposite the right angle, is the hypotenuse. Because angle ABC is 90°, this triangle is a right triangle. Only a right triangle has a hypotenuse.

An equilateral triangle is one with three equal sides (fig. 6.12), where AB equals BC equals CA. An isosceles triangle is one with two equal sides, or legs (fig. 6.13). Any one of the three sides of a triangle may be chosen as the base. In an isosceles triangle, however, base refers to the side that is not one of the legs. So, in figure 6.13, the base of the triangle is BC; the sides are AB and AC and are the same length.. A scalene triangle is one in which no two sides are equal (fig. 6.14).

The area of a triangle is equal to one-half the product of its base and altitude, or height. This statement, or theorem, can be written as a formula:

$$A = \tfrac{1}{2}bh$$

where

A = area
b = base
h = altitude, or height.

Using this formula, the area of the triangle in figure 6.13 is:

$$A = \tfrac{1}{2}(4.5 \text{ in.} \times 10 \text{ in.}) = \tfrac{1}{2}(45) = 22.5 \text{ in.}^2$$

The formula for the area of a triangle is derived from the fact that two identical triangles joined along equal sides form a parallelogram (fig. 6.15). And, the area

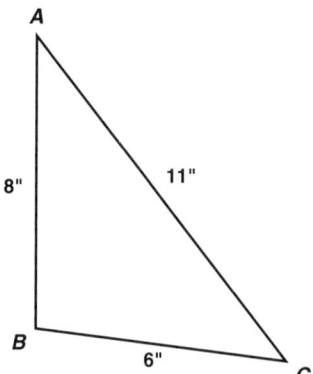

Figure 6.14   Scalene triangle

of a rectangle or parallelogram is equal to the product of its base and height (A = bh). A triangle is therefore one-half of a parallelogram. Thus, it is logical that the area of a triangle is found by taking one-half of the area of a parallelogram.

Example Problem: Find the area of the triangle in figure 6.16, the base of which is 18 inches and the altitude, or height, of which is 6 inches.

Solution:

A = ½bh.

A = 0.5(18 in. × 6 in.) = 108 ÷ 2 = 54 in.² (Remember: 0.5 is the same as the fraction ½.)

As stated before, a right triangle contains one 90° angle. The area of a right triangle is calculated in the same manner as that of any triangle. However, the sides of a right triangle bear a unique relationship to one another: in a right triangle, you can determine the length of one of its sides if you know the length of the other two sides. This relationship, which is a theorem, is stated as, the square of the hypotenuse of a right triangle is equal to the sum of the squares of the other two sides (legs). Thus, in a right triangle having the sides a and b and the hypotenuse c, $a^2 + b^2 = c^2$. Because the Greek mathematician Pythagoras of Samos discovered this relationship in the fifth century, B.C., it bears his name as the Pythagorean theorem.

Figure 6.17 illustrates the Pythagorean theorem when the sides a, b, and c of a right triangle are extended into squares from each side of the triangle. In the figure, one side of the triangle is 3 inches, another is 4 inches, and the other is 5 inches. Each square then contains 9 square inches, 16 square inches, and 25 square inches respectively. Because of this special relationship in right triangles, they are sometimes called three-four-five right triangles.

Example Problem: Find the length of the hypotenuse of the triangle in figure 6.18, where a = 3 inches, b = 4 inches, and c is unknown.

Solution:

c^2 = $a^2 + b^2$
c^2 = $3^2 + 4^2$ = 9 + 16 = 25
c = $\overline{25}$ = 5 in.

As long as the lengths of any two sides of a right triangle are known, you can determine the length of the third leg. To solve for an unknown leg length, transpose the equation to read

$$a^2 = c^2 - b^2$$

or

$$b^2 = c^2 - a^2.$$

Example Problem: Solve for leg b of the triangle in figure 6.19.

Solution:

b^2 = $c^2 - a^2$
b^2 = $5^2 - 3^2$ = 25 − 9 = 16 in.
b = $\overline{16}$ = 4 in.

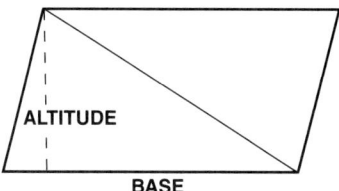

Figure 6.15   Parallelogram has two identical triangles

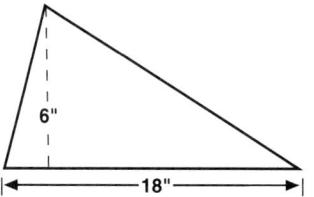

Figure 6.16   Dimensioned triangle

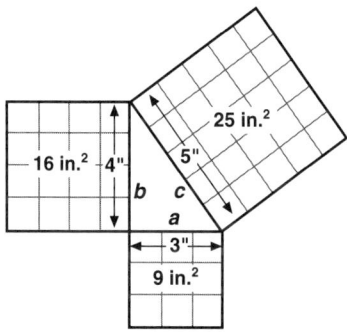

Figure 6.17   Diagram showing side relationships

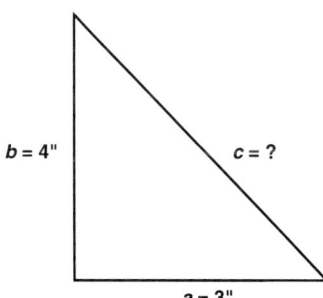

Figure 6.18   Triangle problem

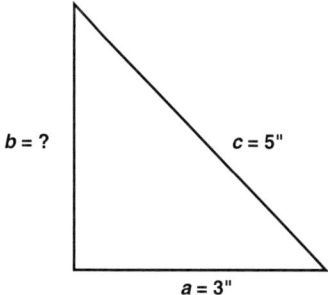

Figure 6.19   Triangle solution

Other Polygons

As stated earlier, a regular polygon is a closed figure with equal sides and equal angles. In addition to three-sided equilateral triangles and four-sided squares, regular polygons can be five-sided pentagons, six-sided hexagons, seven-sided heptagons, eight-sided octagons, and so forth. In any case, the perimeter (P) of a regular polygon, regardless of the number of sides, is determined by adding the measurements of each side or multiplying the number of sides times the length of one side. For example, figure 6.20 shows a hexagon, each side of which is 4 mm. So, the perimeter of this hexagon is

$$P = 6 \times 4 \text{ mm} = 24 \text{ mm.}$$

The area of a polygon is the number of square units contained in the surface bounded by its perimeter. To find the area of a regular polygon, use the formula:

$$A = \tfrac{1}{2}hP$$

where

A = area
h = altitude, or height, of the polygon
P = perimeter of the polygon.

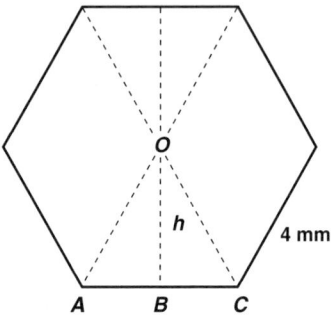

Figure 6.20   Regular six-sided polygon (hexagon)

In figure 6.20, the altitude (h) is the perpendicular line OB, drawn from the center O to bisect side AC.

Example Problem: Find the area of the octagon shown in figure 6.21, where the altitude (h) is 10 inches and the length of one of its sides is 6.6 inches.

Solution: First, find the perimeter of the octagon by multiplying the length of one side times the number of sides:

$$P = 6.6 \text{ in.} \times 8 = 52.8 \text{ in.}$$

Then solve for the area:

$$A = \tfrac{1}{2}hP$$
$$A = \tfrac{1}{2}(10 \times 52.8) = 528 \div 2 = 264 \text{ in.}^2$$

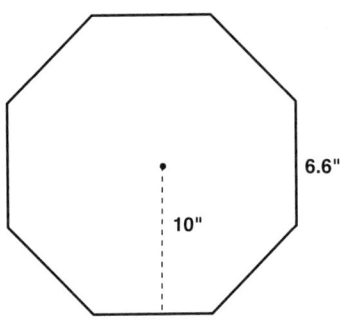

Figure 6.21   Octagon

An irregular polygon has sides of different lengths and different angles. Figure 6.22 shows an irregular polygon that has sides of 3, 5, 3, 4, and 5 inches. To find the perimeter of this, add the lengths of the sides together:

$$3 \text{ in.} + 5 \text{ in.} + 3 \text{ in.} + 4 \text{ in.} + 5 \text{ in.} = 20 \text{ in., or } 1 \text{ ft } 8 \text{ in.}$$

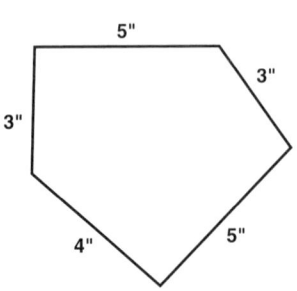

Figure 6.22   Irregular five-sided polygon (pentagon)

Finding the area of an irregular polygon is beyond the scope of this chapter. (However, Chapter 7 covers it.) For now, just be aware that calculating the area of irregular polygons involves drawing triangles within the polygon and then using special numbers that relate to right angles called trigonometric functions. A trigonometric function is obtained from certain ratios of the sides of a right triangle. For more information about trigonometric functions and determining the area of irregular polygons, refer to Chapter 7, "Trigonometry." Then, look in the section headed "Solving for Sides and Angles" for calculating areas of irregular polygons.

Circles

A circle is a closed curve all points of which are in the same plane and are equally distant from a point within it called the center. The distance around a circle is its circumference. A line from the center to any point on the circle is a radius. Figure 6.23 shows several parts of a circle. Lines *DB* and *DC* are radii, and all radii of the same circle are equal in length. A straight line drawn between any two points on the circle and passing through the center is the circle's diameter, which is line *AC* in figure 6.23. The length of two radii are the same as the diameter. Put another way, diameter (*d*) is equal in length to two radii (*r*), so $d = 2r$. All diameters of the same circle are equal. A straight line joining any two points on a circle is a chord (line *EF* in fig. 6.23). Any part of the curved line forming the circle is an arc. In figure 6.23, curves *AB*, *BC*, *CF*, and *AE* are arcs.

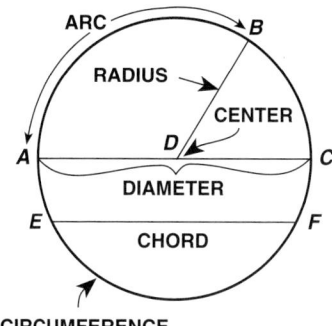

Figure 6.23   Parts of a circle

Two or more circles having the same center but different radii are concentric circles (fig. 6.24). A semicircle is half a circle (fig. 6.25).

The circumference of a circle is equal to its diameter (*d*) multiplied by the constant 3.141592654, which is generally shortened to 3.1416 or 3.14. This number is designated by the Greek letter π (*pi*). The value of π derives from the ratio of a circle's circumference to its diameter. For example, suppose a circle's diameter is 12 inches and its circumference is 37.7 inches. Expressed as a ratio of circumference to diameter, the figure is $^{37.7}/_{12}$, or $37.7 \div 12$. Dividing 37.7 by 12 equals 3.1416, the value of π. Another example: suppose a circle's diameter is 1.65 centimetres and its circumference is 5.18 centimetres. So, $5.18 \div 1.65 = 3.139$. In this case, the answer rounds to 3.14, the value of π. No matter the dimensions of the circle, dividing its circumference by its diameter always results in 3.1416 (π) or a number very close to it.

The equation for finding the circumference (*C*) of a circle is

$$C = \pi d, \text{ or } C = \pi 2r.$$

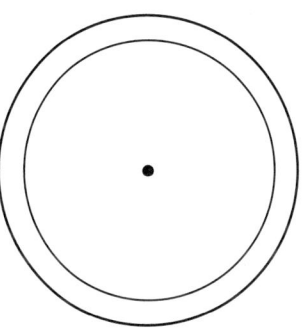

Figure 6.24   Concentric circles

Example Problem: The radius of a circle is 8 in. What is its circumference?

Solution:

$$C = \pi 2r$$
$$C = 3.1416 \times 2 \times 8 \text{ in.} = 50.2656 \text{ in., or } 50.27 \text{ in.}$$

By the way, some calculators have a π key, which means that instead of entering 3.14 or 3.1416, you merely press the key.

The formula $C = \pi d$ can also be used to find an unknown diameter or radius by transposing it to read

$$d = \frac{C}{\pi}.$$

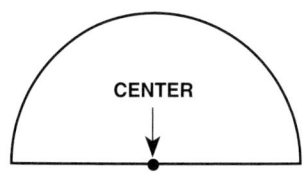

Figure 6.25   Semicircle

Example Problem: The circumference of a circle is 15.708 inches. What is its diameter?

Solution:

$$d = \frac{C}{\pi}$$
$$d = 15.708 \div 3.1416 = 5 \text{ in.}$$

The area of a circle (*A*) can be found by multiplying the square of its radius (*r*) by 3.1416 (π); that is,

$$A = \pi r^2.$$

Example Problem: What is the area of a circle whose radius is 6 inches?

Solution:

$$A = \pi r^2$$
$$A = 3.1416 \times 6 \times 6 = 113.1 \text{ in.}^2$$

If the area of a circle is known, then the formula $A = \pi r^2$ can be transposed to solve for the radius or diameter of a circle:

$$r^2 = \frac{A}{\pi}$$

$$r = \sqrt{\frac{A}{\pi}}$$

$$d = 2r.$$

Example Problem: Find the radius of a circle whose area is 254.4696 in².

Solution:

$$r = \sqrt{\frac{A}{\pi}}$$

$$r = \sqrt{176.715 \div 3.1416} = \sqrt{81} = 9 \text{ in.}$$

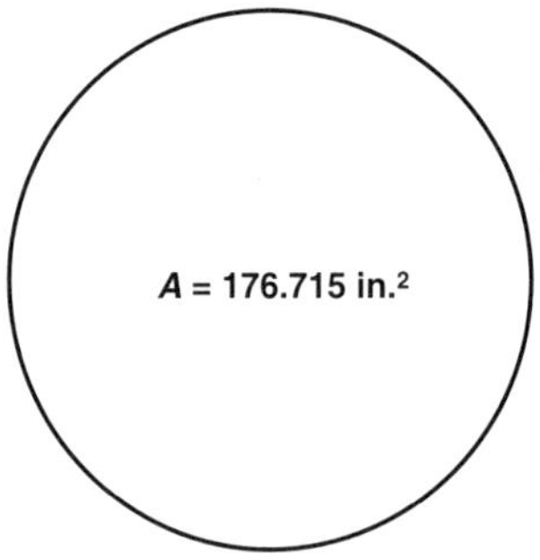

Example Problem: Find the diameter of the circle shown in figure 6.26, whose area is 176.715 in².

Solution:

$$r = \sqrt{\frac{A}{\pi}}$$

$$r = \sqrt{176.715 \div 3.1416} = \sqrt{56.25} = 7.5 \text{ in.}$$

$$d = 2 \times 7.5 = 15 \text{ in.}$$

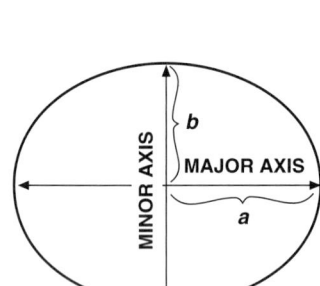

A = 176.715 in.²

Figure 6.26   Dimensioned circle

Ellipses

An ellipse is a kind of compressed circle. Technically, an ellipse is an oval-shaped closed curve with two axes (fig. 6.27). The longer axis is the major axis, and the shorter one is the minor axis. Letting *a* represent half of the major axis and *b* half of the minor axis, the formula for finding the perimeter (*P*) of an ellipse is

$$P = 2\pi \sqrt{(a^2 + b^2 \div 2)}.$$

Figure 6.27   Ellipse

Example Problem: Find the perimeter of an ellipse whose major axis is 10 inches and whose minor axis is 5 inches.

Solution: First, find half of the major axis (*a*) and minor axis (*b*). In this case:

$$a = \tfrac{1}{2} \text{ of } 10 = 5 \text{ in.}; \; b = \tfrac{1}{2} \text{ of } 5 = 2.5 \text{ in.}$$

Then, use the formula for finding the perimeter (*P*) of an ellipse,

$$P = 2\pi \sqrt{(a^2 + b^2 \div 2)},$$

and plug in the values.

$$P = 2 \times 3.1416 \times \sqrt{(5^2 + 2.5^2) \div 2}$$
$$= 6.2832 \times \sqrt{(25 + 6.25) \div 2}$$
$$= 6.2832 \times \sqrt{15.625}$$
$$P = 6.2832 \times 3.953 = 24.836 \text{ in.}$$

The area (A) of an ellipse is equal to one-half the major axis times one-half the minor axis times π. With a and b representing half the major axis and minor axis respectively, the equation for finding the area of an ellipse is

$$A = \pi ab.$$

This formula is similar to the formula for finding the area of a circle; however, the circle has one radius (r) while the ellipse has two dimensions (a and b) to deal with.

Example Problem: What is the area (A) of the ellipse shown in figure 6.28, where half of the major axis (a) is 2 inches and half of the minor axis (b) is 1.5 inches.?

Solution:

$$A = \pi ab$$
$$A = 3.1416 \times 2 \times 1.5 = 9.4248 \text{ in.}^2$$

In this example, and any problem like it, any two of the numbers may be multiplied together first. Then, the product of the two numbers is multiplied by the remaining number. For example,

$$3.1416 \times 2 = 6.2832, \text{ and } 6.2832 \times 1.5 = 9.4248.$$

Likewise,

$$2 \times 1.5 = 3, \text{ and } 3 \times 3.1416 = 9.4248.$$

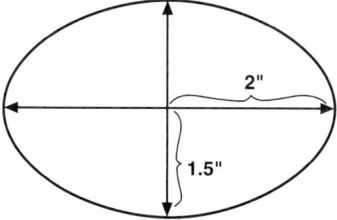

Figure 6.28   Dimensioned ellipse

CONVERSIONS AND FORMULAS TABLES

Tables 6.1 and 6.2 assist you in converting conventional units of area measurement and in finding areas of plane figures. In table 6.1, for example, to convert square

TABLE 6.2
Formulas for Finding Areas of Plane Figures

Figure	Formula	Meaning of Terms
Parallelogram, rectangle, square	$A = bh$	b = base h = height
Trapezoid	$A = \frac{1}{2}h(b + b')$	h = height b = bottom base b' = top base
Triangle	$A = \frac{1}{2}bh$	b = base h = altitude, or height
Other regular polygons (hexagon, octagon, etc.)	$A = \frac{1}{2}aP$	a = altitude P = perimeter
Circle	$A = \pi r^2$	π = 3.1416 r = radius
Ellipse	$A = \pi ab$	π = 3.1416 a = one-half of major axis b = one-half of minor axis

yards (yd²) to square feet (ft²), note that 1 square yard = 9 square feet. And, in table 6.2, note, for example, that the formula for finding the area of a triangle is $A = \frac{1}{2} bh$. Appendix C-1 in this manual is a bigger table of conversion factors. However, for the practice problems that follow these two tables, table 6.1 is adequate.

Practice Problems

Refer to tables 6.1 and 6.2 for area measurements and formulas. (If conversion factors given in Appendix C-1 are used, answers may vary.)

1. Find the number of square feet in the trapezoid below.

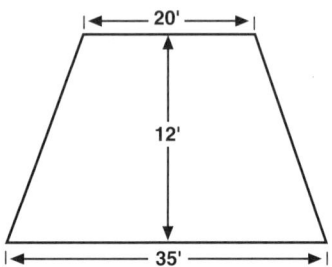

2. What is the area of the parallelogram below, in square feet?

3. What is the perimeter of a rectangle that measures 7 inches by 14 inches?

4. Find the number of acres in a field that is 2,700 feet long and 600 feet wide. (1 acre = 43,560 square feet.)

5. How many square yards are in a rectangle that is 125 feet long and 80 feet wide?

    What is the perimeter in yards?

6. A sheet of packing weighs 1.25 pounds per square foot (lb/ft²). How much does a piece of this packing that is 8 feet long and 30 inches wide weigh?

7. A saltwater pit is 250 feet long and 130 feet wide. How many acres does it cover? Round off the answer to the nearest common fraction.

_____

8. To reach a well location in a rice field, it was necessary to build a board road 1,400 feet long and 8 feet wide.

   a. How many square feet of lumber were required?

   _____

   b. If the boards were 1 foot wide, how many 8-foot boards were needed?

   _____

9. You are going to build a new bed for mounting on a pickup truck. The bed is to be 8 feet long × 54 inches wide × 22¼ inches deep. Without allowing for waste, how many square feet of sheet steel are needed to make the floor, two sides, and tailgate?

   _____

10. Find the areas of the triangles below.

   a.

   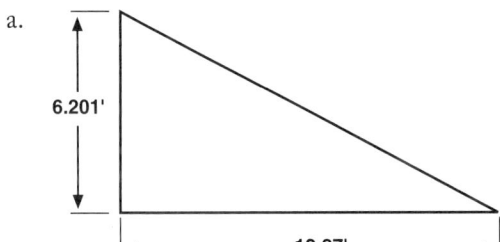

   6.201'

   12.27'

   _____

   b.

   3'

   6'

   _____

   c.

   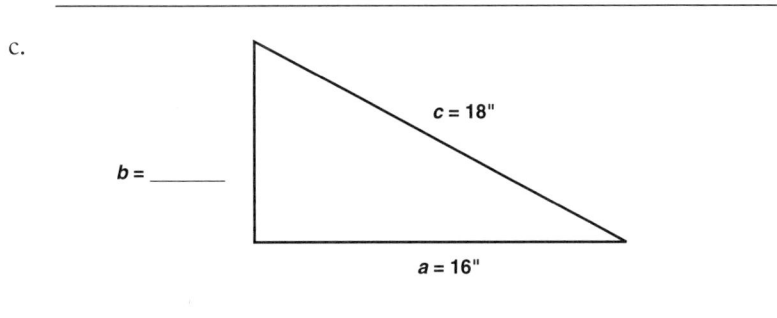

   $c = 18"$

   $b =$ _____

   $a = 16"$

   _____

11. Find the area of the shaded portion of the circle below.

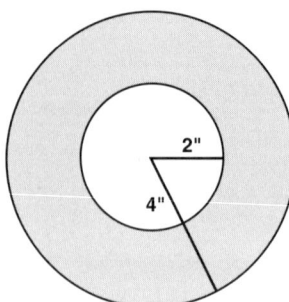

---

12. The tank below has a 7-foot outside diameter and is banded with eight bands that overlap 6 inches. How many feet of band iron were required to make these bands?

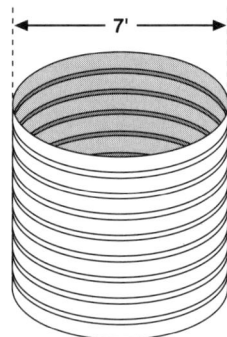

---

13. In landscaping a camp, an elliptical plot for a swimming pool is to be laid out. The major axis of the ellipse is 80 feet, and the minor axis is 60 feet. What part of an acre will the plot contain? (1 acre = 43,560 square feet.)

---

14. Calculate the total area of the irregular shape below.

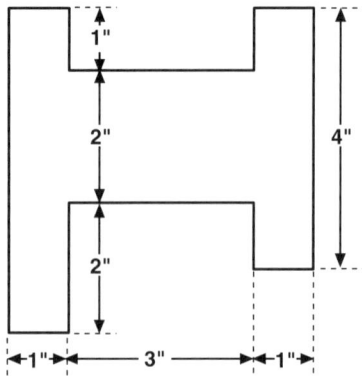

---

15. The tube sheet shown in the sketch below is 36 inches in diameter and has thirty 3-inch holes drilled in it.

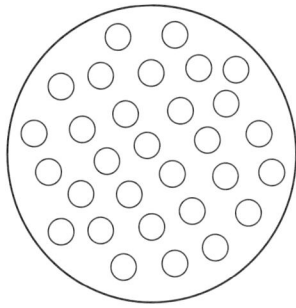

   a. How many square inches of metal were removed?

   _____

   b. How many square inches of metal are left in the sheet?

   _____

SOLID FIGURES

Plane surfaces have two dimensions: length and width. In solid figures, another dimension—thickness—occurs. Think of plane figures as two dimensional and solid objects as three dimensional. The amount of space a solid figure occupies is its volume. Volumes, which are obtained by dealing with all three dimensions, are expressed in cubic measures such as cubic inches (in.^{3}), cubic feet (ft^3), cubic yards (yd^3), and cubic metres (m^3). Just as in calculating dimensions of plane figures, all three dimensions in the calculation of volumes must be expressed in the same units. Feet cannot multiply inches nor can feet multiply yards or gallons multiply barrels.

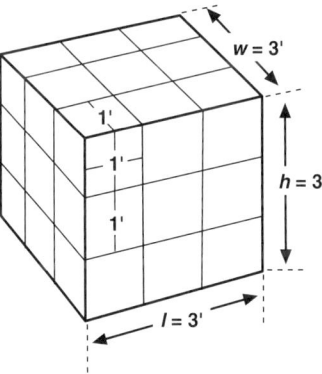

Figure 6.29   Cube

Rectangular Solids

The rectangular solid shown in figure 6.29, called a cube when all sides are equal, illustrates cubic measure. Multiplying the three dimensions of the solid—length (l) times width (w) times height (h)—gives its volume (V). The formula is:

$$V = lwh.$$

In figure 6.29, each dimension of the cube is 3 feet. Thus:

$$V = 3 \text{ ft} \times 3 \text{ ft} \times 3 \text{ ft} = 27 \text{ ft}^3.$$

Also notice in figure 6.29 that the cube is marked with lines 1 foot apart. If you sawed the cube apart along these lines, the cube would be divided into twenty-seven smaller blocks, each 1 foot high, 1 foot wide, and 1 foot deep. In short, 27 cubic feet are in a cube that is 3 feet on each side.

A cube is a rectangular solid whose sides are all equal. When the sides are not equal (fig. 6.30), the solid is simply a rectangular solid. You can determine the volume with the same formula for finding the volume of a cube, which is

$$V = lwh.$$

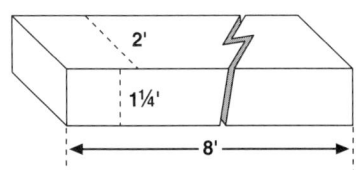

Figure 6.30   Rectangular solid

Example Problem: What is the volume of the rectangular solid shown in figure 6.30 where its length (*l*) is 8 feet, its width (*w*) is 2 inches, and its height (*h*) is 1¼ inches? Express the answer in cubic inches.

Solution: Using the formula $V = lwh$, plug in the values.

$$l = 8 \text{ ft} = 96 \text{ in.}, w = 2, \text{ and } h = 1^\circ \text{ in.}$$
$$V = 96 \text{ in.} \times 2 \text{ in.} \times 1.25 \text{ in.} = 240 \text{ in.}^3$$

Sometimes, volumes are expressed in gallons or barrels, and to solve for them, you sometimes have to convert measurements. Table 6.3 lists several conversions. These conversions change cubic feet into cubic inches and barrels; cubic metres into cubic feet, gallons, and barrels; cubic yards into cubic feet; and barrels into gallons and cubic feet.

Example Problem: How many barrels of water does it take to fill a rectangular tank that is 7 feet long, 4 feet wide, and 3½ feet high?

Solution:

$$V = lwh$$
$$V = 7 \text{ ft} \times 4 \text{ ft} \times 3\tfrac{1}{2} \text{ ft} = 98 \text{ ft.}^3$$

If 5.61 ft³ equals 1 barrel (bbl), then 98 ft³ equals 17.47 bbl because 98 ÷ 5.61 = 17.47 bbl.

To find the total surface area of all six sides of a rectangular solid, determine the area of each side and add the areas together. The answer is in square measurements. For example, to find the total surface area of the rectangular solid in figure 6.30, which is 8 feet long, 2 inches wide, and 1.25 inches high, first find the area of the top and multiply it by 2 to find the area of the top and the bottom:

$$A = lw$$
$$A = 96 \text{ in.} \times 2 \text{ in.} = 192 \times 2 \text{ sides} = 384 \text{ in.}^2$$

Then, find the area of the two sides:

$$A = 96 \text{ in.} \times 1.25 \text{ in.} = 120 \times 2 \text{ sides} = 240 \text{ in.}^2$$

Next, find the area of the two ends:

$$A = 2 \text{ in.} \times 1.25 \text{ in.} = 2.5 \times 2 \text{ sides} = 5 \text{ in.}^2$$

Finally, add the three dimensions to obtain the total surface area (*T*):

$$T = 384 + 240 + 5 = 629 \text{ in.}^2$$

Cylinders

In industry, it is necessary to find the volume of cylinders such as storage tanks, pipes, and pump cylinders. These volumes are usually expressed in gallons, barrels, or cubic inches and require the appropriate measurement conversions.

The volume of a cylinder (fig. 6.31) is equal to the product of the area of its base (*B*) times the height (*h*):

$$V = B \times h.$$

Remembering that the area of a circle is equal to πr^2, the area of the cylinder base is πr^2, and the formula for finding the volume of a cylinder is

$$V = \pi r^2 h.$$

TABLE 6.3

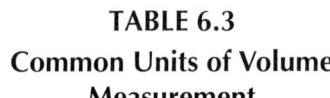

Common Units of Volume Measurement

1 ft³	=	1,728 in.³
1m³	=	35.31 ft³
1 yd³	=	27 ft³
1m³	=	264.12 gal
1 bbl	=	42 gal
1 m³	=	6.29 bbl
1 ft³	=	0.178 bbl
1 bbl	=	5.61 ft³

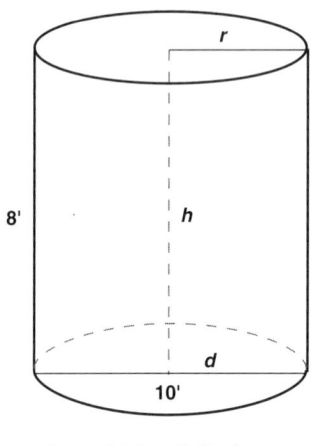

Figure 6.31   Cylinder

Example Problem: Find the volume of the cylinder shown in figure 6.31. Its diameter is 10 feet, and its height is 8 feet.

Solution:

r = ½ of 10 ft = 5 ft; h = 8 ft.
V = $\pi r^2 h$
V = 3.1416 × 5 ft × 5 ft × 8 ft = 628.32 ft³.

Example Problem: How many barrels of oil can be stored in a tank that is 20 feet in diameter and 10 feet high?

Solution:

r = ½ of 20 ft = 10 ft; h = 10 ft.
V = $\pi r^2 h$
V = 3.1416 × 10 ft × 10 ft × 10 ft = 3,141.6 ft³.

Because 1 cubic foot equals 7.48 gallons, then 3,141.6 cubic feet equals 23,499 gallons. And because 42 gallons equals 1 barrel, then

23,499 ÷ 42 = 559.5 bbl.

Elliptical Solids

An elliptical solid is similar to a cylinder except that its base is an ellipse instead of a circle. The equation for finding the area of an ellipse (A = πab) is used to determine the volume of an elliptical solid. Referring to figure 6.32, the formula for finding the volume of an elliptical solid can be written as

$$V = \pi abh$$

where

a = one-half the major axis
b = one-half the minor axis
h = height.

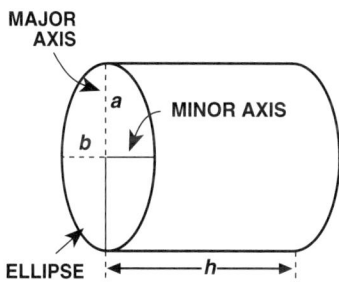

Figure 6.32   Elliptical solid

Example Problem: Find the volume of an elliptical tank whose major axis is 8 feet, whose minor axis is 6 feet, and whose height is 10 feet.

Solution:

a = 4 ft; b = 3 ft; h = 10 ft.
V = πabh
V = 3.1416 × 4 × 3 × 10 = 376.992, or 377 ft³.

Example Problem: A tank truck has an elliptical tank that is 14 feet long. The major and minor axes of the elliptical ends of the tank are 8 feet and 6 feet. How many gallons of gasoline does the tank hold?

Solution:

a = 4 ft; b = 3 ft; h = 14 ft.
V = πabh
V = 3.1416 × 4 × 3 × 14 = 527.789 ft³.

If 1 cubic foot equals 7.48 gallons, then

527.79 × 7.48 = 3,947.86 gal.

Cones

The most common type of cone—the right circular cone—is a solid generated by rotating a right triangle about one of its legs (fig. 6.33). The circle at the bottom of the cone is its base, and its altitude, or height, is the distance between the center of the base and the apex. It also has a slant height, which is measured from the apex to the edge of the cone's circular base.

The volume of a cone is equal to one-third of the area of the base times the altitude:

$$V = \frac{1}{3}(Bh)$$

or

$$V = \frac{1}{3}(\pi r^2 h).$$

Example Problem: Find the volume of a cone whose base has a 6-inch diameter and whose altitude is 7 in.

Solution:

$$\begin{aligned} V &= \frac{1}{3}(\pi r^2 h) \\ &= \frac{1}{3}(3.1416 \times 3^2 \times 7) \\ V &= 197.9208 \div 3 = 65.9736 \text{ in.}^3 \end{aligned}$$

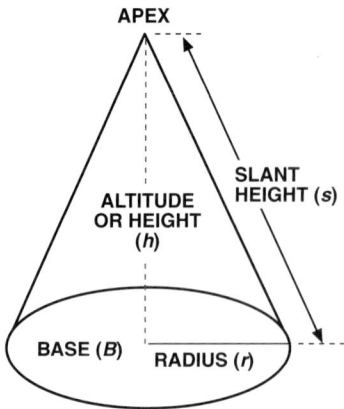

Figure 6.33   Cone

Another important cone measurement is its lateral area, or conical surface (fig. 6.34). Unfolding a cone and placing it flat on a surface shows its lateral area. The lateral area of a right circular cone equals half the product of the slant height and the circumference of the base:

$$L = \frac{1}{2}sC$$

where

L = lateral area
s = slant height
C = circumference of the base.

Since $C = 2\pi r$, then

$$L = \frac{s \times 2\pi r}{2} = \pi rs.$$

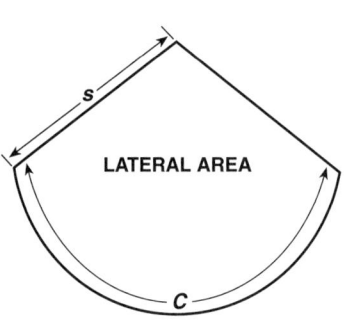

Figure 6.34   Lateral area of a cone or conical surface

Example Problem: Find the lateral area of a cone whose slant height is 5 millimetres and whose base radius is 2 millimetres.

Solution:

$$\begin{aligned} L &= \pi rs \\ L &= 3.1416 \times 2 \times 5 = 31.416 \text{ mm}^2. \end{aligned}$$

If you know the altitude and radius of the cone's base, you can calculate the slant height by using the formula for finding the hypotenuse of a triangle:

$$a^2 + b^2 = c^2.$$

Example Problem: What is the lateral area of the cone depicted in figure 6.35, where the height (h) is 12 inches and the radius (r) of the base is 6 inches?

Solution: First, find the slant height (s) of the cone, substituting h for a and r for b in the equation above:

$$\begin{aligned} s^2 &= h^2 + r^2 \\ s^2 &= 12^2 + 6^2 = 144 + 36 = 180 \\ s &= \overline{180} = 13.4164 \text{ in.} \end{aligned}$$

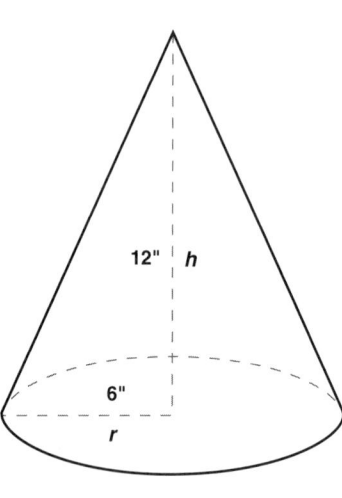

Figure 6.35   Dimensioned cone

Then solve for the lateral area:

$$L = \pi rs$$
$$L = 3.1416 \times 6 \times 13.4164 = 252.894 \text{ in.}^2$$

Pyramids

A pyramid is a solid figure that has a polygon for its base and triangles with a common vertex for its faces. Figure 6.36 shows three pyramids: one has a square as its base, another a pentagon, and the other a hexagon. These pyramids are regular pyramids; indeed, any pyramid that has a regular polygon as its base is a regular pyramid. Remember: a regular polygon has equal sides and equal angles.

The volume of a pyramid equals one-third the product of its base area (B) and its altitude (h):

$$V = \tfrac{1}{3}(Bh).$$

Example Problem: Find the volume of the pyramid shown in figure 6.37, which has a height (h) of 3 feet and whose base (b) is a square each side of which is 2 feet.

Solution:

$$B = 2 \text{ ft} \times 2 \text{ ft} = 4 \text{ ft}^2; h = 3 \text{ ft.}$$
$$V = \tfrac{1}{3}(Bh).$$
$$V = (4 \times 3) \div 3 = 4 \text{ ft}^3.$$

To find the lateral area of a pyramid, find the areas of the triangles that make up the lateral surface and add them together. Or, you can use a formula that involves the pyramid's slant height and perimeter. In a regular pyramid (one whose base is a regular polygon), the altitudes of these triangles from their common vertex to their bases are the same as the pyramid's slant height (fig. 6.38). The sum of the sides of the polygon that form the pyramid base equals its perimeter. Thus, using the slant height and the perimeter, the lateral area of a pyramid can be found using the formula

$$L = \tfrac{1}{2}(sP)$$

where

$$L = \text{lateral area}$$
$$s = \text{slant height}$$
$$P = \text{perimeter of the base.}$$

Example Problem: Find the lateral area of a pyramid whose slant height is 10 inches and whose base is a regular hexagon with 6-inch sides.

Solution:

$$L = \tfrac{1}{2}(sP)$$
$$L = 10 \text{ in.} \times 6 \text{ in.} \times 6 \text{ sides } \sqrt{2} = 180 \text{ in.}^2$$

The total area of a pyramid, like the total area of a cone, is found by adding the lateral area and the base area. For regular pyramids and right circular cones, the total area can be calculated with the formula

$$T = \pi r^2 + \pi rs$$

or

$$T = \pi r(r + s)$$

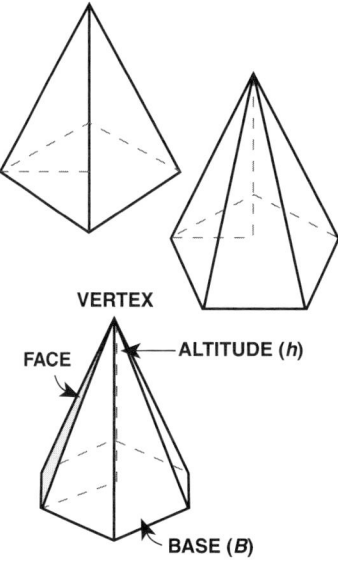

Figure 6.36   Pyramids

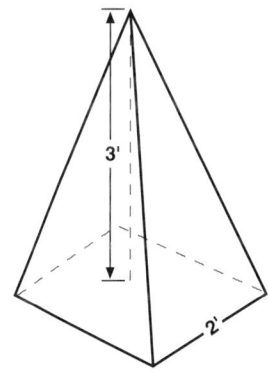

Figure 6.37   Dimensioned pyramid

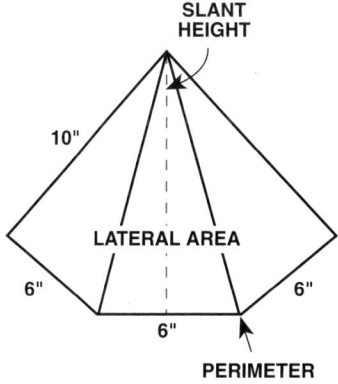

Figure 6.38   Lateral area of pyramid

where

T = total area
r = radius of the base
s = slant height.

This formula combines the formulas for calculating the lateral area and the base area.

Example Problem: Find the total area of a cone with a base radius of 1.5 feet and a slant height of 30 inches.

Solution: First, convert the measurements to inches or feet. Using feet as the measurement,

$$30 \text{ in. } \div 12 \text{ in. } = 2.5 \text{ ft.}$$

Then

$$T = \pi r(r + s)$$
$$T = 3.1416 \times 1.5 \times (1.5 + 2.5) = 18.8496 = 18.85 \text{ ft}^2.$$

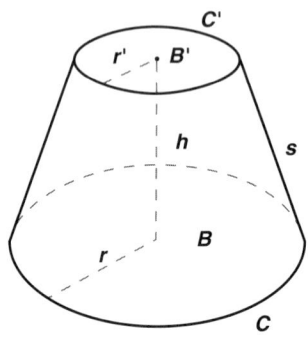

Figure 6.39   Frustrum of a cone

Frustums

When a pyramid or a cone is cut off at any point below the apex by a plane that is parallel to the base, the portion below the cutting plane is a frustum of a cone (fig. 6.39) or a frustum of a pyramid (fig. 6.40). (Frustum comes from the Latin word for piece or bit.) Frustums are commonly seen in concrete bases or in the welded portion of a pipe swedge.

The base of a pyramid or cone (B) and the section of the intersecting plane (B') are the bases of the frustum; the rest of the surface is the lateral surface (L). The altitude (h) of a frustum is the distance between the centers of the two bases. This altitude line is perpendicular to both bases.

The formula for finding the lateral area of a frustum of a cone is half the slant height (s) times the sum of the circumferences (C and C') of the bases:

$$L = \tfrac{1}{2}s(C + C')$$

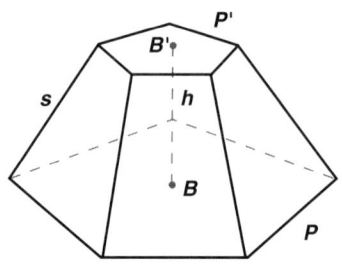

Figure 6.40   Frustrum of a pyramid

where

L = lateral area
s = slant height of the frustum
C = circumference of the cone base
C' = circumference of the section base.

Example Problem: Find the lateral area of the frustum shown in figure 6.41 where the slant height (s) is 52 millimetres, the radius of the larger base (C) is 21 millimetres, and the radius of the smaller base (C') is 12 millimetres.

Solution: Using the radii given, find circumferences C and C':

$$C = \pi 2r$$
$$C = 3.1416 \times 2 \times 21 \text{ mm} = 131.95 \text{ mm.}$$
$$C' = 3.1416 \times 2 \times 12 \text{ mm} = 75.40 \text{ mm.}$$

Then solve for lateral area (L):

$$L = \tfrac{1}{2}s(C + C')$$
$$L = 52 \times (131.95 + 75.40) \div 2 = 5{,}391.1 \text{ mm}^2.$$

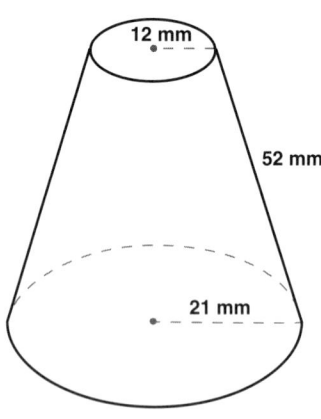

Figure 6.41   Frustrum cone calculations

The formula for finding the lateral area of the frustum of a pyramid is the same as that for a cone, except perimeters are used instead of circumferences:

$$L = \tfrac{1}{2}s(P + P')$$

where

L = lateral area
s = slant height of the frustum
P = perimeter of the pyramid base
P' = perimeter of the section base.

The volume of a frustum of a cone or a pyramid is obtained by the formula

$$V = \tfrac{1}{2}h(B + B' + \overline{B\,B'})$$

where

V = volume
h = height of the frustum
B = area of the cone or pyramid base
B' = area of the section base.

This formula can be used whether the frustum is regular or not—that is, the bases do not have to be perfect circles or even-sided polygons.

Example Problem: Find the volume of the frustum shown in figure 6.42, where the base (B) is a square that is 15 inches on each side, the section (B') is a square that is 6 inches on each side, and the height (h) is 24 inches. Give the answer in cubic feet.

Solution: First, solve for base areas:

B = 15 in. × 15 in. = 225 in.^{2}
B' = 6 in. × 6 in. = 36 in.^{2}

Then use the formula to find volume:

V = $\tfrac{1}{2}h(B + B' + \overline{B\,B'})$
V = 24 ÷ 3 × (225 + 36 + $\overline{225 \times 36}$)
V = 8 × (261 + 90) = 8 × 351 = 2,808 in.^{3}

Since 1 ft^3 equals 1,728 in.^{3}, then 2,808 ÷ 1,728 = 1.625 ft^3.

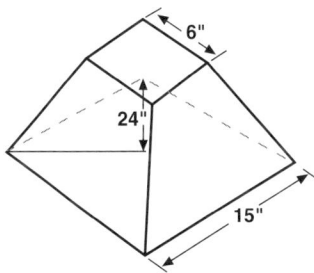

Figure 6.42   Frustrum pyramid calculations

Spheres

A sphere is a closed solid whose points are all equally distant from a point within called its center (fig. 6.43). A sphere is different from a circle in that a sphere exists in space while a circle is flat. It is helpful to think of a sphere as the shape of a basketball or a globe. Both the sphere and the circle have centers, diameters, and radii.

The volume of a sphere is equal to four-thirds π times the radius cubed:

$$V = \tfrac{4}{3}\pi r^3.$$

Example Problem: Find the volume of a sphere that has an 8-inch diameter.

Solution:

V = $\tfrac{4}{3}\pi r^3$, or 4 × π × r^3 ÷ 3
V = 4 × 3.1416 × 4 × 4 × 4 ÷ 3 = 268.0832 in.^{3}

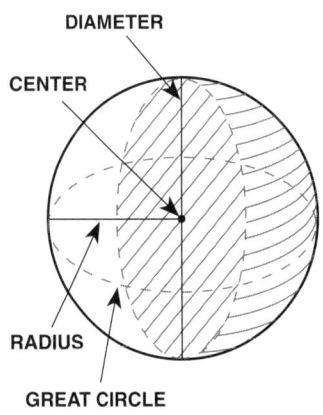

Figure 6.43   Sphere

A sphere is not a ruled surface. Thus, it cannot be rolled out into a flat surface without distorting its shape and area. Nevertheless, the area of a sphere can be determined by using the formula

$$S = \pi d^2$$

where

S = area of the sphere
d = diameter of the sphere.

Example Problem: Find the area of a sphere that has a 10-inch diameter.

Solution:

$S = \pi d^2$
$S = 3.1416 \times 10 \times 10 = 314.16$ in.^{2}

TABLES FOR VOLUMES AND SURFACE AREA

Table 6.4 summarizes the formulas for finding the volumes of the various solid figures covered in this chapter. Table 6.5 lists the formulas for finding the lateral or total surface areas of the solid figures discussed in this chapter.

TABLE 6.4
Formulas for Finding Volumes

Figure	Formula	Meaning of Terms
Rectangular solid, cube	$V = lwh$	l = length w = width h = height
Cylinder	$V = Bh$, or $V = \pi r^2 h$	B = area of base h = height π = 3.1416 r = radius of base
Elliptical solid	$V = \pi abh$	a = one-half the major axis b = one-half the minor axis h = height π = 3.1416
Cone	$V = \frac{1}{3}(Bh)$, or $V = \frac{1}{3}(\pi r^2 h)$	B = area of base r = radius of base h = height, or altitude π = 3.1416
Pyramid	$V = \frac{1}{3}(Bh)$	B = area of base h = height, or altitude
Frustum of cone or pyramid	$V = \frac{1}{3}h(B + B' + \overline{B B'})$	h = height of frustum B = area of cone or pyramid base B' = area of section base
Sphere	$V = \frac{4}{3}\pi r^3$	r = radius of sphere

TABLE 6.5
Formulas for Finding Lateral or Total Surface Areas of Solid Figures

Figure	Formula	Meaning of Terms
Rectangular solid, cube	$T = 2A + 2A'$ $+ 2A''$	T = total area A = area of side 1 A' = area of side 2 A'' = area of side 3
Cylinder (right)	$L = 2\pi rh$ $L = hC$ $T = 2\pi r(r + h)$	L = lateral area r = radius of base h = altitude C = circumference of base π = 3.1416
Cone (right circular)	$L = \frac{1}{2}sC$	s = slant height C = circumference
Pyramid (regular)	$L = \frac{1}{2}sP$	s = slant height P = perimeter
Frustum of a cone (right circular)	$L = \frac{1}{2}s(C + C')$ $T = L + B + B'$	s = slant height C = circumference of cone base C' = circumference of section base B = area of cone base B' = area of section base
Frustum of a pyramid (regular)	$L = Hs(P + P')$ $T = L + B + B'$	s = slant height P = perimeter of pyramid base P' = perimeter of section base B = area of pyramid base B' = area of section base

Practice Problems

Refer to tables 6.3, 6.4, and 6.5 for help in solving these problems.

1. Cast iron weighs 450 pounds per cubic foot. What is the weight of the casting shown in the sketch below?

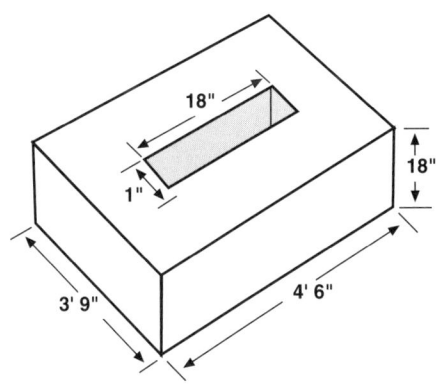

2. A ditch for a pipeline is to be 1,500 feet long, 12 inches wide, and 30 inches deep. How many cubic yards of dirt must be removed?

_____

3. How many gallons of water does the rectangular trough shown below hold?

_____

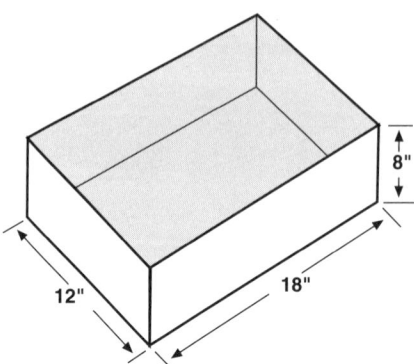

4. A cylindrical gasoline tank is 5 feet in diameter and 10 feet long. How many gallons of gasoline can it hold?

_____

5. A gauger gauged a cylindrical oil tank 15 feet in diameter and found that the oil's height was 6 feet 6 inches. How many barrels of oil are in the tank?

_____

6. A single-cylinder pump has a stroke of 16 inches, and the cylinder is 12 inches in diameter. How many gallons of oil does the pump deliver at each stroke?

_____

7. A company makes metal signs out of ⅛-inch metal plate. Workers cut the signs in the form of an ellipse whose major axis is 3 feet and minor axis is 2 feet. If the plate weighs 489 pounds per cubic foot, what does each sign weigh?

_____

8. An engine foundation is to be poured 3 feet wide by 8 feet long by 5 inches deep. How many cubic feet of concrete will be needed?

_____

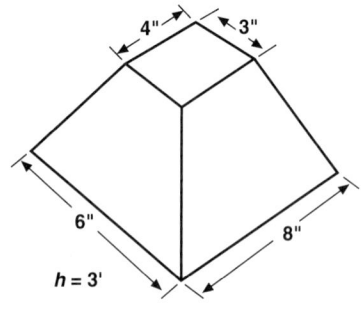

9. Calculate the number of cubic yards of concrete required to pour the footing represented in the sketch to the left.

_____

10. Find the volume of the oil heater shown in the sketch below.

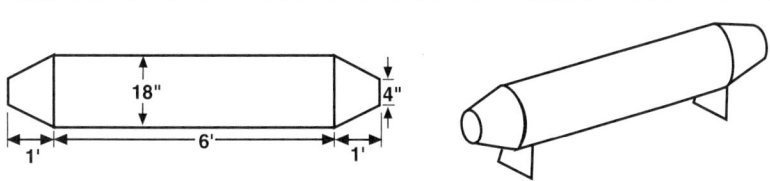

11. The casting shown in the sketch is made of brass that weighs 530 pounds per cubic foot. The casting has a hole that is 1½ inches in diameter through its center. Calculate its weight.

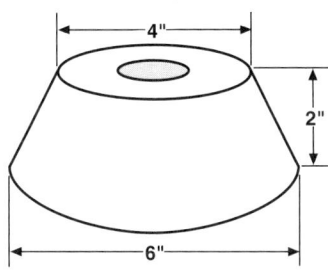

12. The exhaust hood shown to the right is connected to an exhaust fan that has a capacity of 500 cubic feet per minute.

    a. What is the volume capacity of the hood?

    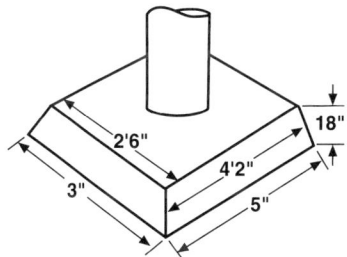

    b. After finding the hood's volume capacity, determine how many times per minute the fan moves that volume of air.

13. The compressor building whose floor is shown in the sketch below has a basement 8 feet deep. It is necessary to exhaust the air from the basement once each 3 minutes to prevent the accumulation of gases. How many fans, each with a capacity of 12,000 cubic feet, are required?

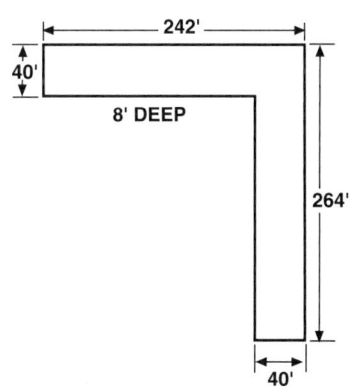

14. What is the weight of the brass ball shown in the sketch if the brass of which it is cast weighs 534 pounds per cubic feet?

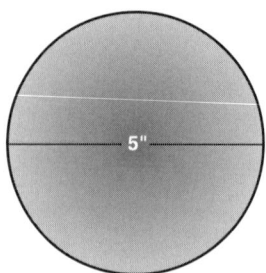

15. Calculate the lateral area of the doghouse sketched below. The door is a semicircle. *Note*: Divide it into regular shapes, find the areas of each shape, and add them to determine the lateral area. The lateral area does not include the floor and roof.

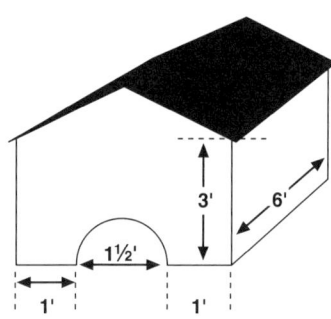

GEOMETRIC CONSTRUCTIONS

Geometric construction involves drawing lines, angles, and curves in regular patterns that are easily recognizable. Careful geometric construction during the design of a structure ensures that the structure, whether it is a building, a pipeline, or a piece of machinery, fits together when it is built. Normally, the original design and layout is done on paper. However, when the structure is built, the job usually requires tape measures, rulers, stakes, surveying instruments, or other tools and equipment. The same measurement principles apply with chalk and stakes as with pencil and paper.

Construction Tools

Many mathematicians distinguish between constructing and drawing geometric figures. A compass and straightedge are the only tools used to construct figures, while many instruments are used in drawing them, the more common being a ruler, drawing board, T-square, and triangles. True constructions are accomplished without using a calibrated protractor for angles or a ruler for measuring lines. (A protractor is a semicircular-shaped instrument that is marked with the degrees of a circle and measures and marks angles.)

A geometric compass is a hinged instrument with a point on one arm and a drawing implement on the other. The angle between the two arms can be made wider or narrower as needed. The point of the compass is placed on a given point on the figure. The other end of the compass—the drawing end—is swung around without moving the point or the angle between the two arms of the compass. By swinging the drawing end of the compass, an arc is created. If the drawing end of the compass is swung all the way around, a circle is created.

The distance from the point end to the drawing end of the compass is the compass radius. Figure 6.44 shows the compass radius as line *AB*. The curved line *CD* is the arc created by a compass with a radius of *AB*.

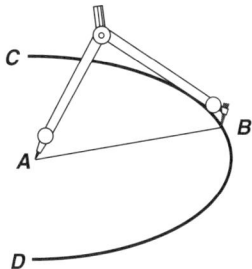

Figure 6.44   Compass drawing

Bisecting

Bisecting is dividing something in half with a line. Using only a compass and a straightedge, you can bisect a line, an arc, or an angle.

Bisecting a Line. Place the point of the compass at one end of the line, and swing the compass so that the drawing end lightly makes one arc above the line and another arc below the line (fig. 6.45). Then, place the point of the compass at the other end of the line and repeat the previous procedure without changing the angle between the compass arms (fig. 6.46). The arcs above and below the line should cross at a single point. Use a compass radius that is greater than one-half the length of the line *AB* in order for the arcs to cross.

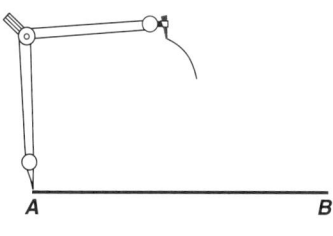

Figure 6.45   Bisecting a line

Using the straightedge, connect the points where the arcs cross with a line (fig. 6.47). Line *AB* now has been bisected by line *CD*. Line *CD* is a bisector and is perpendicular to line *AB* (fig. 6.48).

Bisecting an Arc. Use the same procedure for bisecting a line with points *A* and *B* occurring at the arc ends (fig. 6.49). The bisector is perpendicular to the chord joining the ends of the arc (chord *AB*).

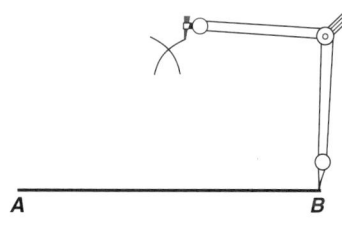

Figure 6.46   Bisecting a line

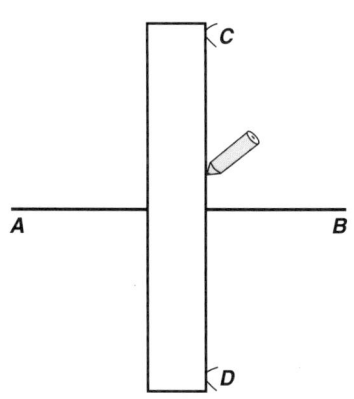

Figure 6.47   Bisecting a line

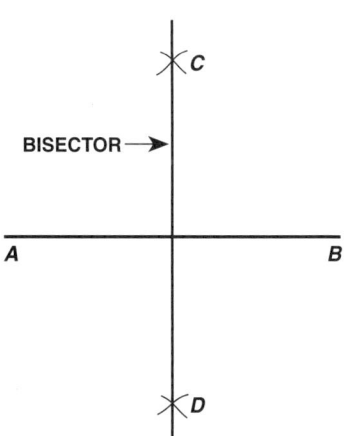

Figure 6.48   Bisecting a line

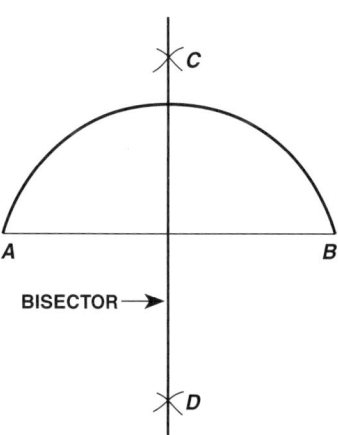

Figure 6.49   Bisecting an arc

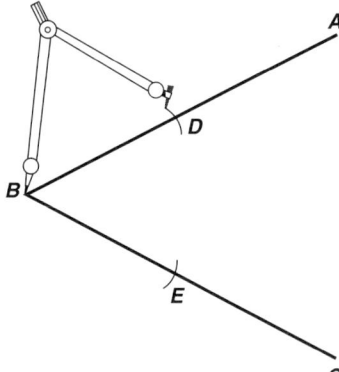

Figure 6.50   Bisecting an angle

Bisecting an Angle. First, look at figure 6.50. Set the compass at a radius smaller than the angle sides, and then place the point of the compass at the angle vertex (*B* in fig. 6.50). Swing the compass so that two arcs lightly cut across the sides of the angle (at *D* and *E*).

Now, without changing the compass radius, place the compass point at *E* and draw an arc through the middle of the angle. Move the compass point to *D* and draw an intersecting arc at *F* (fig. 6.51).

Using the straightedge, draw a line joining *B* and *F*. Line *BF* bisects angle *ABC* (fig. 6.52).

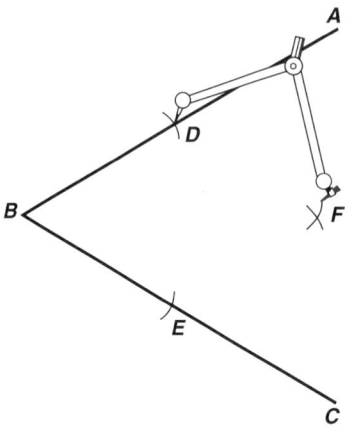

Figure 6.51   Bisecting an angle

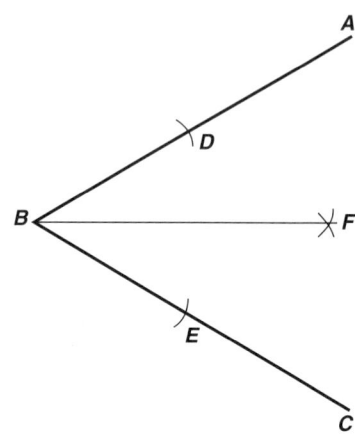

Figure 6.52   Bisecting an angle

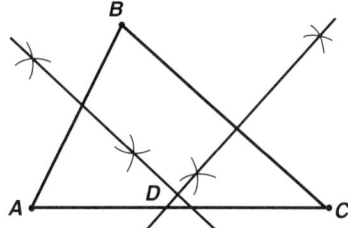

Figure 6.53   Constructing a circle

Constructing Figures

Many figures can be constructed using only a compass and straightedge.

Constructing a Circle Passing through Any Three Points Not in a Straight Line. As figure 6.53 shows, first connect the three points *A*, *B*, and *C* with straight lines to form a triangle. Then, bisect any two sides of the triangle, such as *AB* and *BC*. The intersection of these two bisectors at point *D* is the center of the circle. Place the compass point at *D*, adjust the compass radius to *DA*, *DB*, or *DC*, and draw the circle. It passes through all three points (fig. 6.54).

Constructing an Angle Equal to a Given Angle. Note the angle *ABC* in figure 6.55. You can reproduce this angle with a compass and straightedge. First, draw a line *B'C'* to simulate line *BC* in angle *ABC* in figure 6.55. Next, place the point of the compass at the vertex *B* of angle *ABC* and draw an arc that intersects sides *AB* at *D* and *BC* at *E* (fig. 6.56). Without changing the compass radius, place the compass point at *B'* and draw an arc that intersects *B'C'* at *E'*.

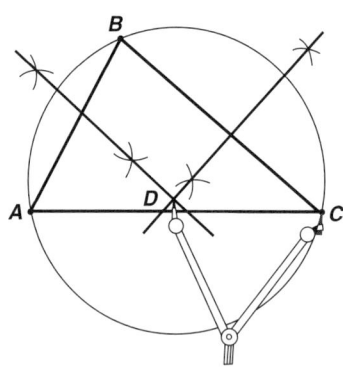

Figure 6.54   Constructing a circle

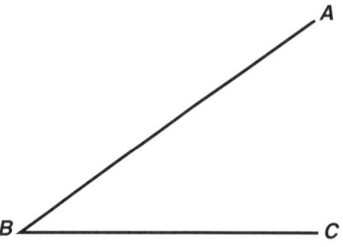

Figure 6.55   Constructing an angle

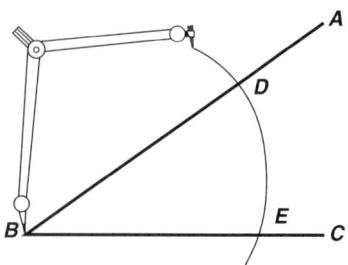

Figure 6.56   Constructing an angle

Return to angle *ABC*, place the compass point at *E*, and adjust the compass angle so that the writing end will be on *D*. With this compass radius, and with *E'* as the center, make an arc that intersects the existing arc at *D'* (fig. 6.57). Using a straightedge, draw a line connecting points *B'* and *D'*. The resulting angle *A'B'C'* is equal to angle *ABC* (fig. 6.58).

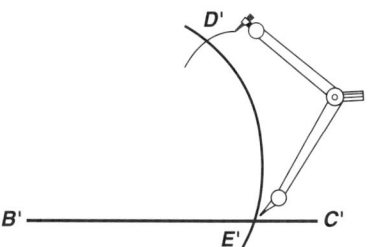

Figure 6.57   Constructing an angle

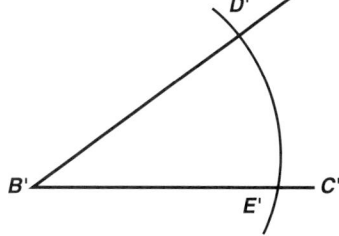

Figure 6.58   Constructing an angle

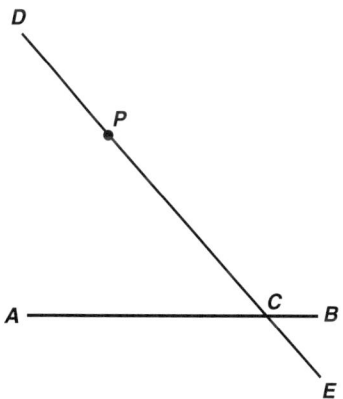

Figure 6.59   Constructing a parallel line

Constructing a Line through a Given Point That Is Parallel to a Given Line. To construct a line parallel to line *AB* that passes through point *P* in figure 6.59, draw line *DE* that connects point *P* with line *AB* at *C*. The next step is to draw another angle with the same measure as angle *ACD* located above point *P*. To draw this angle, draw an arc (*FG*) through angle *ACD* and, without changing the compass radius, draw the same size arc using point *P* as the center (fig. 6.60).

With the compass, measure the distance between *F* and *G*. Using this compass radius and *H* as the center, draw a small arc that intersects the existing arc at point *I* (fig. 6.61). With the straightedge, draw line *JK* through points *I* and *P*. Line *JK* is parallel to line *AB* and passes through point *P* (fig. 6.62).

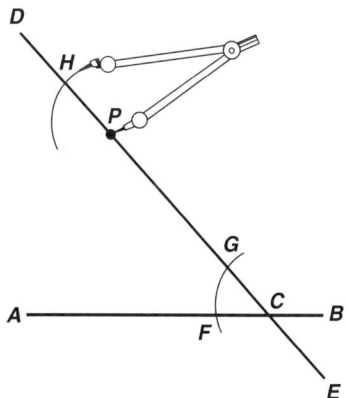

Figure 6.60   Constructing a parallel line

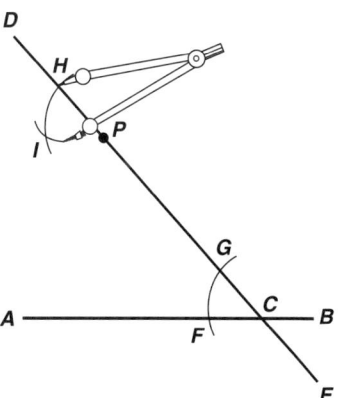

Figure 6.61   Constructing a parallel line

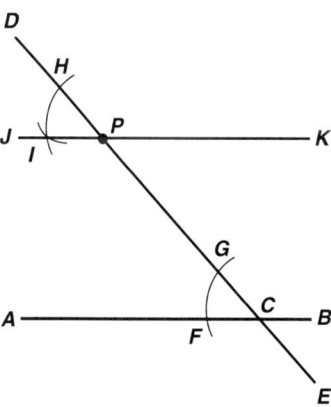

Figure 6.62   Constructing a parallel line

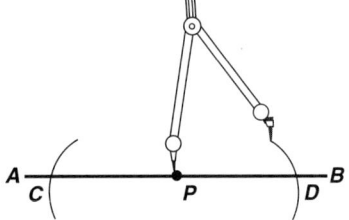

*Figure 6.63   Constructing a
perpendicular line*

Constructing a Perpendicular to a Line from a Point on the Line. To draw a line perpendicular to a given line *AB* at point *P*, place the compass point on *P*, and draw arcs intersecting the line at *C* and *D* on either side of *P* (fig. 6.63). Now enlarge the compass radius slightly, and with the compass point on *C*, draw a large arc above the line. Without changing the compass radius, move the compass point to *D* and draw another arc that intersects the first arc at *E* (fig. 6.64). Using a straightedge, draw a line connecting *E* and *P*. Line *EP* is perpendicular to line *AB* (fig. 6.65).

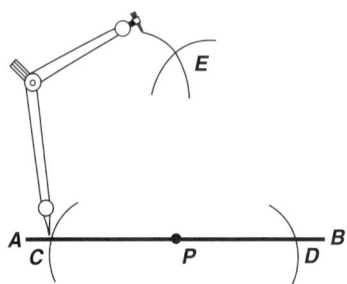

*Figure 6.64   Constructing a
perpendicular line*

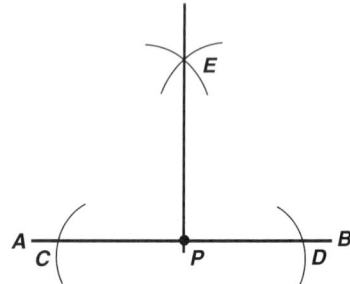

*Figure 6.65   Constructing a
perpendicular line*

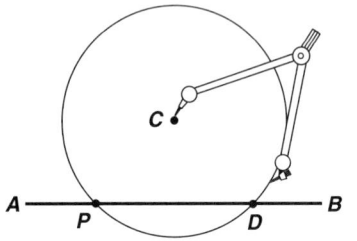

*Figure 6.66   Constructing a
perpendicular line*

Constructing a Perpendicular to a Line from a Point Near the End of the Line. Another method can be used to draw a perpendicular line when the desired point is very close to the end of the line. To draw a line perpendicular to line *AB* at point *P* (fig. 6.66), select a second point *C* somewhere above line *AB*. Place the compass point at *C* and the compass drawing end at *P*. With the compass radius at this setting, draw a circle.

This circle intersects line *AB* and points *P* and *D*. With a straightedge, draw a line connecting *C* at the circle's center with the intersection of circle and line at *D*. Extend this line all the way to the other side of the circle through *C* to intersect the circle at *E* (fig. 6.67).

The new line *DCE* is the circle's diameter. Point *E* is directly above the original point *P* on line *AB*. Using a straightedge, draw a line connecting points *E* and *P*. Line *EP* is perpendicular to the given line *AB* (fig. 6.68).

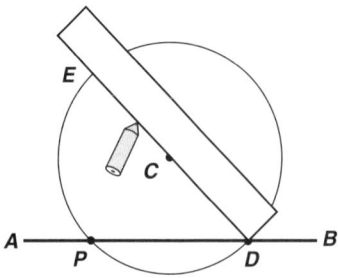

*Figure 6.67   Constructing a
perpendicular line*

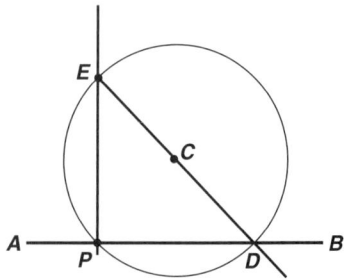

*Figure 6.68   Constructing a
perpendicular line*

Constructing a Perpendicular to a Line from a Point Not on the Line. To draw a perpendicular to a line *AB* from a point *P* that is not on the line, put the compass point on *P* and set the compass at an angle wide enough to draw an arc that intersects line *AB* in two places (points *C* and *D* in fig. 6.69). Then set the compass radius so that it is larger than half the distance between points *C* and *D* (fig. 6.70). Placing the compass point on *C*, draw an arc below the line. Without changing the compass radius, move the compass point to *D* and draw another arc below the line. The arcs below the line should intersect at *E* (fig. 6.71). With the straightedge, draw a line connecting point *P* above the line with point *E* below the line. The new line *PE* is perpendicular to line *AB* (fig. 6.72).

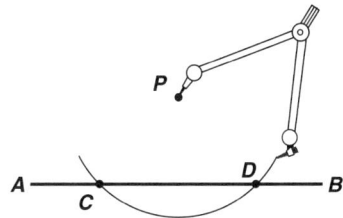

Figure 6.69   Constructing a perpendicular line

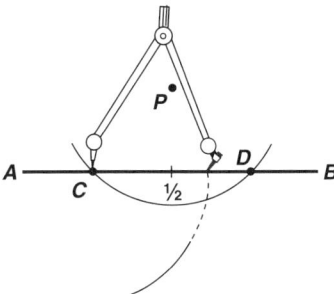

Figure 6.70   Constructing a perpendicular line

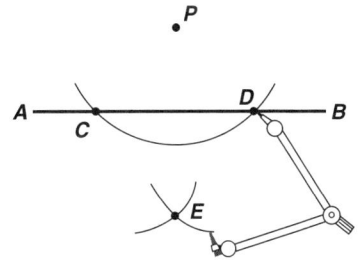

Figure 6.71   Constructing a perpendicular line

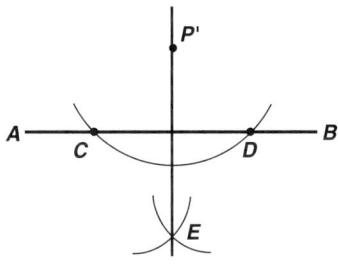

Figure 6.72   Constructing a perpendicular line

Dividing a Given Line into Any Number of Equal Parts. To divide line *AB* into five equal parts, draw line *AC* at any convenient angle, starting at point *A* (fig. 6.73). From *A*, mark off on line *AC* five equal divisions of any length using dividers or a compass. From point 5 on *AC*, draw a line to *B* (fig. 6.74). Then construct lines from 1, 2, 3, and 4 that are parallel to line *5B*. (Use the procedure for drawing parallel lines given earlier.) These lines divide line *AB* into five equal parts.

Laying Out a Half Circle into Six Equal Parts. Draw line *CD* perpendicular to the semicircle diameter *AB* that intersects at *D* (fig. 6.75). (Use the method of constructing perpendiculars described earlier.) Using the distance between *C* and *D* as a compass radius, construct arcs intersecting the semicircle to the left and right of the perpendicular at points *E* and *F*. Without changing the compass radius, move the compass point to *A* and construct an arc intersecting the semicircle at point *G* (fig. 6.76). Move the compass point to *B* and draw another arc intersecting the semicircle at point *H*. The circumference is now divided into six equal parts at points *E*, *G*, *C*, *H*, and *F*. Connecting these points to point *D* with a straightedge divides the semicircle into six equal parts.

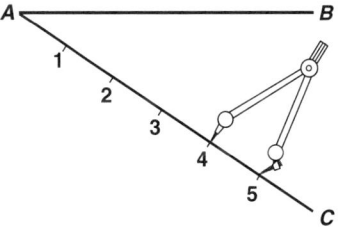

Figure 6.73   Dividing a line into parts

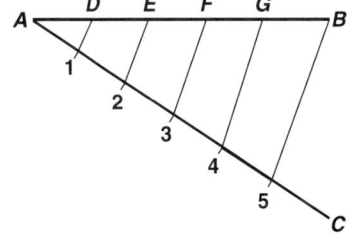

Figure 6.74   Dividing a line into parts

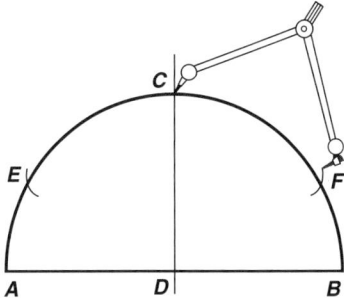

Figure 6.75   Dividing a circle into equal parts

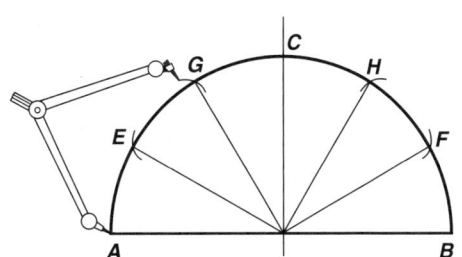

Figure 6.76   Dividing a circle into equal parts

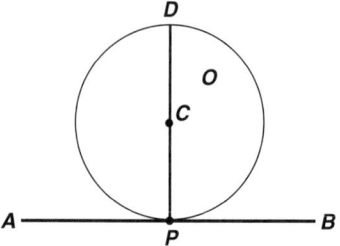

Figure 6.77 Constructing tangents to circle

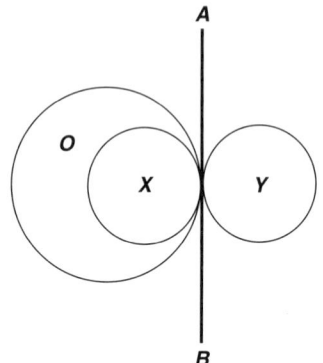

Figure 6.78 Constructing tangent to circle

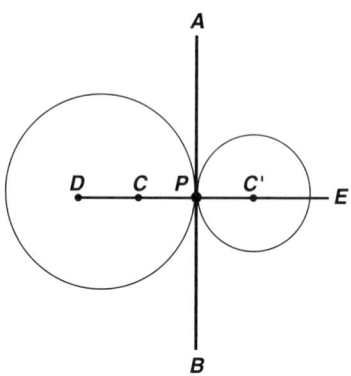

Figure 6.79 Constructing tangent to circle

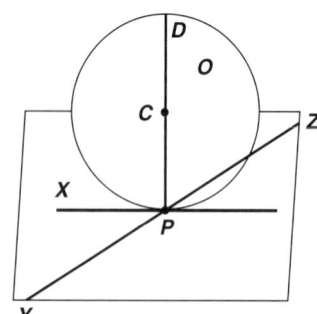

Figure 6.80 Constructing sphere to plane

Constructing Tangents

Tangent means touching, and a line that touches a circle at only one point is tangent to the circle. Likewise, the circle is tangent to the line (fig. 6.77). The point where the two touch is the point of contact, or point of tangency. Several theorems apply to lines and circles that are tangent to each other.

1. A straight line perpendicular to a radius at its outer extremity is tangent to the circle. (*AB* is perpendicular to *CP* and therefore is tangent to circle *O*.)

2. A line perpendicular to a tangent at its point of contact passes through the center of the circle. (Line *PD* is perpendicular to *AB* and therefore passes through center *C* of circle *O*.)

3. A line from the center of a circle and perpendicular to a tangent passes through the point of contact. (Line *CP* is perpendicular to *AB* and therefore passes through point *P*.)

Two circles are tangent to each other if they are both tangent to the same line at the same point, either internally or externally. In figure 6.78, circles *O* and *X* are tangent internally, and circles *O* and *Y* are tangent externally. Both are tangent to line *AB*. If two circles are tangent, a straight line drawn through their centers (known as the line of centers) passes through their point of contact. In other words, line *DE* drawn through *C* and *C'* passes through *P* (fig. 6.79). Also, line *DE* is perpendicular to line *AB*.

Solid figures may also be tangent to each other (fig. 6.80). A line can be tangent to a sphere if it touches the sphere at only one point. Like circles and lines, spheres and planes have similar theorems.

1. A plane perpendicular to a diameter of a sphere at one end is tangent to the sphere. (Plane *X* is perpendicular to diameter *DP* and therefore is tangent to sphere *O*.)

2. If a plane is tangent to a sphere at the end of a diameter, it is perpendicular to the diameter. (If plane *X* is tangent to sphere *O* at *P*, then it is perpendicular to diameter *DP*.)

3. If a sphere is tangent to a plane, it is tangent to any line in the plane passing through the point of contact. (Sphere *O* is tangent to plane *X* and therefore tangent to line *AB*, line *YZ*, and all lines passing through point *P*.

Like tangent circles, the line of centers of tangent spheres pass through the point of contact. These facts should help in constructing the following tangents.

Constructing a Line Tangent to a Circle from a Point on the Circle. Using a straightedge, draw a line connecting the circle's center *C* with the given point *P* on the circle (fig. 6.81). Then construct a line perpendicular to line *CP* at *P*, using the procedure for constructing perpendiculars given earlier. Line *AB* is tangent to circle *O*.

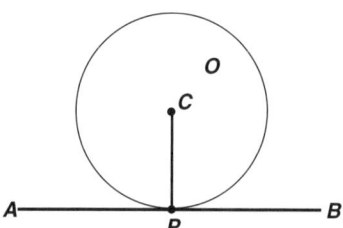

Figure 6.81 Constructing sphere to plane

Constructing Lines Tangent to a Circle from a Point Outside the Circle. With the straightedge, connect the circle's center C with the given point P outside the circle (fig. 6.82). Using the bisecting method described earlier, bisect line CP, producing point A that lies halfway between the end points of line CP. Measure line CA with the compass, and with this compass radius and the compass point at A, draw a second circle, which should go through the center C of circle O and point P outside the circle. The new circle intersects circle O at B and D (fig. 6.83). Now, with a straightedge, draw lines BP and DP, which are tangent to circle O (fig. 6.84).

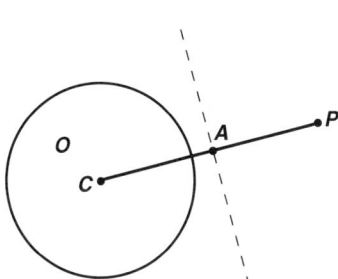

Figure 6.82   Constructing sphere to plane

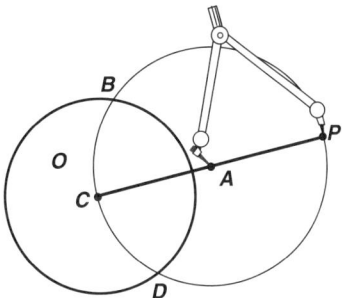

Figure 6.83   Constructing sphere to plane

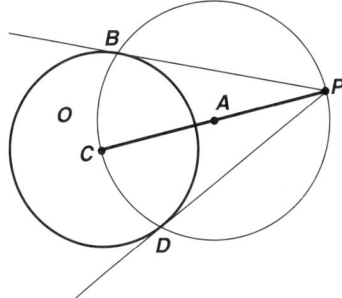

Figure 6.84   Constructing sphere to plane

Constructing a Circular Arc of Given Radius, Tangent to Two Nonparallel Lines. Draw perpendicular lines EF and EG to the two given lines AB and CD using the method described earlier (fig. 6.85). Set the compass radius equal to XY. With the compass point at the intersection of lines AB and FE, draw an arc that intersects the perpendicular line FE at point K. With the same compass radius, repeat the procedure using lines CD and EG, creating an arc at point L.

Now using the method described for constructing parallel lines, draw a line IH parallel to AB at point K and line IJ parallel to DC at point L. The new lines intersect at point I. With the compass radius set the same and with point I as the center, construct an arc that touches the original lines AB and CD at M and N (fig. 6.86). This arc is tangent to the nonparallel lines AB and CD at points M and N.

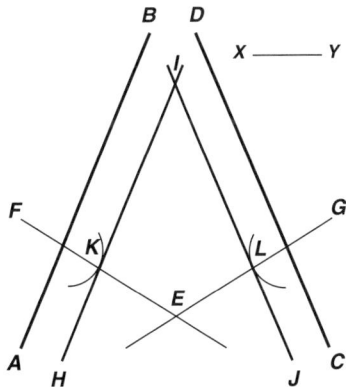

Figure 6.85   Constructing sphere to nonparallel lines

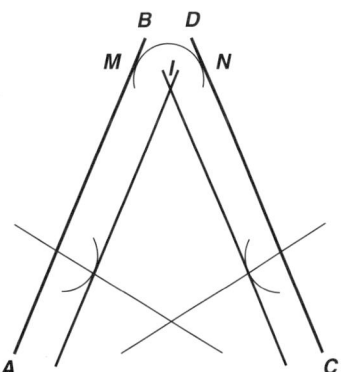

Figure 6.86   Constructing sphere to nonparallel lines

Drawing Procedures

Geometric figures can be drawn instead of constructed by using a drawing board, a T-square, and drafting triangles. A drawing, or drafting, triangle is a flat plastic right triangle that is available in many sizes and angles. It allows you to draw certain angles easily. The more commonly used triangles are the 45°–45° and the 30°–60°, meaning that the triangles have 45° and 45° angles or 30° and 60° angles opposite the right angle (fig. 6.87).

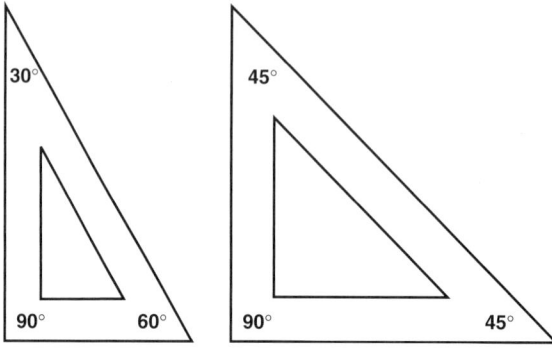

Figure 6.87   Drawing triangles

The proper use of triangles and T-square for drawing lines is illustrated in figure 6.88. Arrows indicate the direction in which to draw the lines.

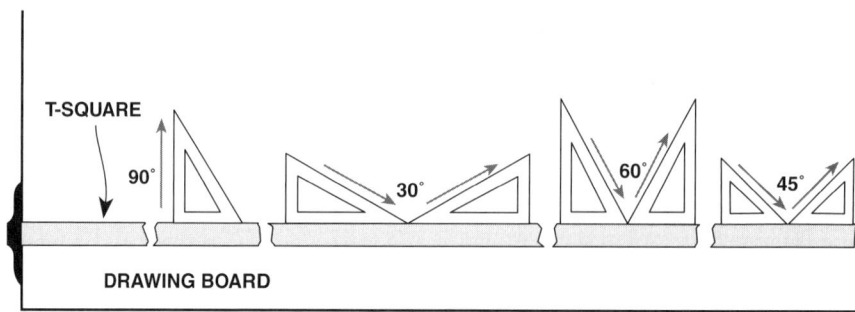

Figure 6.88   Proper drawing strokes

Drawing a Line Parallel to a Given Line. To draw a line parallel to the given line *AB* (fig. 6.89), place a triangle in position so that one edge coincides with line *AB*. Place the T-square blade against the other edge of the triangle. Hold the T-square firmly in place, slide triangle to desired new position, and draw parallel line *CD* (fig. 6.90).

Figure 6.89   Using drawing triangle

Figure 6.90   Using drawing triangle

Drawing a Line Perpendicular to a Given Line. To draw a line perpendicular to the given line *AB*, place a 30°–60° triangle so that its hypotenuse coincides with the line (fig. 6.91). Hold the T-square firmly in place and turn the triangle about its 90° angle to the position shown in figure 6.92, then draw line *CD*. Line *CD* is perpendicular to line *AB*.

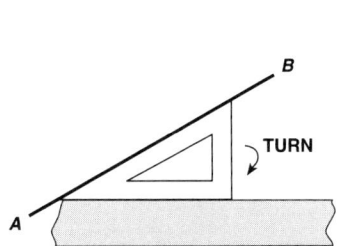

Figure 6.91   Drawing a line perpendicular to a line

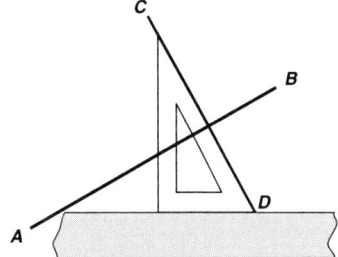

Figure 6.92   Drawing a line perpendicular to a line

Dividing a Circle into Equal Parts. By using both triangles, you can divide a circle into as many as twenty-four equal parts, each part having a 15° angle. Figure 6.93 shows the steps involved in achieving the final sunburst pattern. Arrows indicate the direction in which to draw the lines.

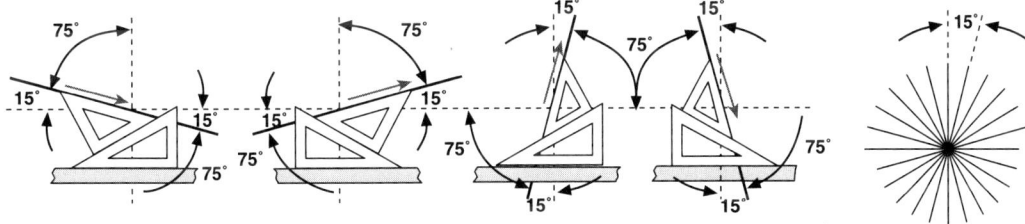

Figure 6.93   Dividing a circle into equal parts

Practice Problems

(Note: Since these are construction exercises, no answers are given.)

1. Erect a perpendicular to a line that is 2¾ inches long at a point that is 1⅛ inches from the end of the line.

2. Erect a perpendicular to a line 3⅜ inches long at one end of the line.

3. Draw a line across the boiler plate in the sketch from point *B* and perpendicular to the line *AB*. Erect perpendiculars to line *BC* at each end of the line.

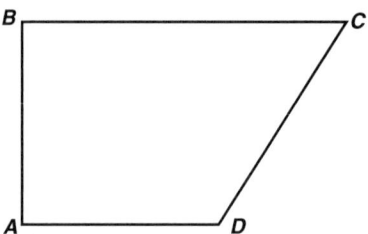

4. Bisect a line that is 3½ inches long.

5. Lay out an angle of 75° with a protractor. Then bisect this angle using only a compass and a straightedge. Show all construction lines.

6. Divide a line 2⅝ inches long into seven equal parts.

7. Draw a line *AB*. Draw a perpendicular line *CD* at point *D* on line *AB*. Angle *CDB* is a 90° angle. Construct an angle of 45° by bisecting angle *CDB*.

8. The drawing below is of a pipe that is to be installed in the corner of a building. With a compass and straightedge, determine graphically the length of the pipe joining the two 45° ells.

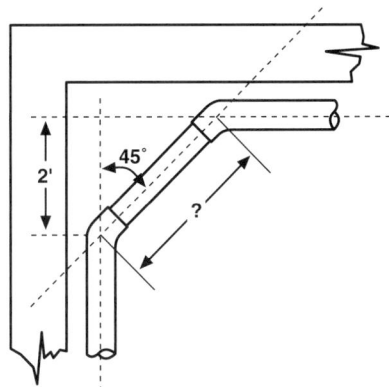

9. Draw a triangle. Then, construct another triangle so that two angles and one side will be equal respectively to two angles and the side of the first.

10. Draw a circle. Construct a tangent to the circle at a point on the circle.

SELF-TEST

6. Practical Geometry

*Multiply each question/problem answered correctly by five
to arrive at your percentage of competency.*

Use tables 6.1–6.5 as a reference for solving these problems. Using conversion factors given in Appendix C-1 will alter answers somewhat.

1. Find the area of a circle with a radius of 2 millimetres.

   _____

2. Three holes have been drilled into a small metal plate. The diameter of each hole is ⅜ of an inch. What is the total area of the holes?

   _____

3. How many gallons will the tank shown below hold?

   _____

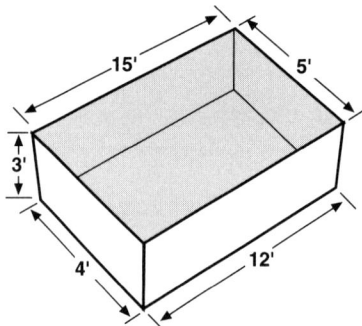

4. How many square inches of braking surface are there on the clutch disk shown in the sketch below?

   _____

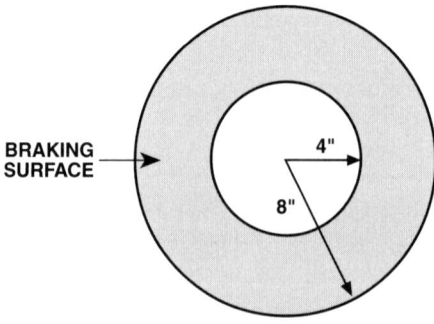

5. Find the volume of a sphere 14 inches in diameter.

   _____

6. How many barrels will the cylindrical tank in the sketch below hold?

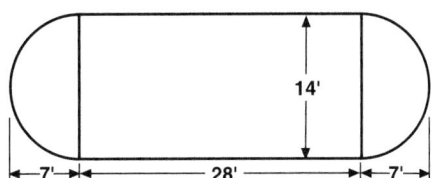

7. How many square inches are contained in a triangular gusset plate if its base is 8 inches and its height is 4½ inches?

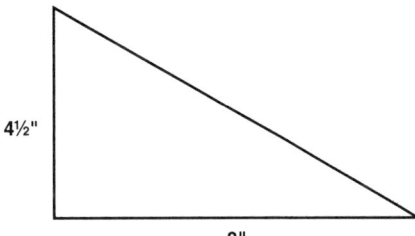

8. A farm shaped in the form of a parallelogram is cut in two by a highway as shown below.

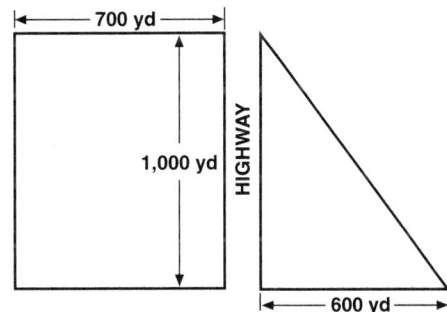

How many acres are there on the left of the road?

9. How much paint will it take for the two gable ends of the roof shown in the sketch if 1 gallon covers 350 square feet?

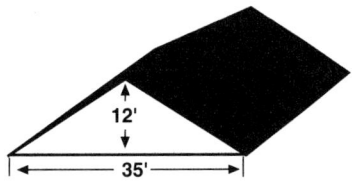

10. A water tower is constructed as shown in the sketch. How many gallons of water are needed to fill the tank to the top of the conical section?

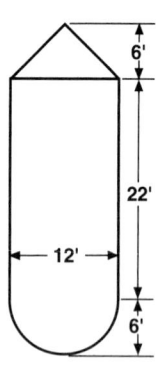

_____

11. The tank in the sketch below is in the shape of a half ellipse. How many gallons does this tank hold?

_____

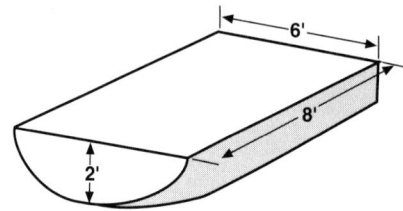

12. In the five right triangles shown to the right, solve for x.

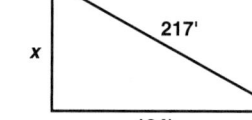

a. _____

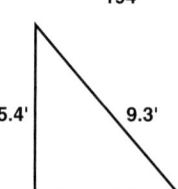

b. _____

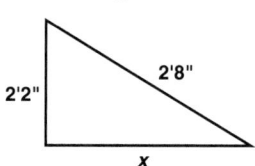

c. _____

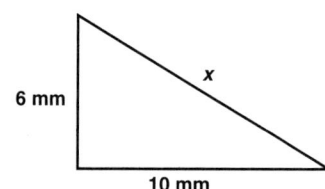

d. _____

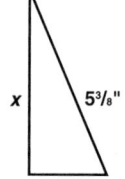

e. _____

13. How many bolts are required for the joint shown on the sketch of the tank if the bolt holes are spaced 3 inches apart? The outside diameter of the tank is 10 feet 6 inches.

_____

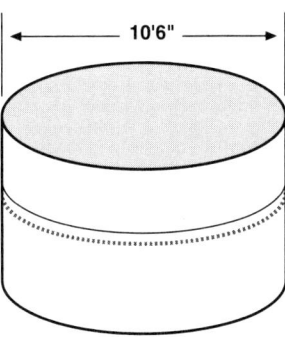

14. Find the number of square feet in a floor that is 28 feet 6 inches by 18 feet 9 inches.

_____

15. If 1 gallon of paint covers 400 square feet, how many gallons are needed to paint the walls of a room that is 18 feet long, 12 feet wide, and 8 feet high?

_____

16. Find the area of a trapezoid whose parallel sides are 10 inches and 20 inches and whose height is 7 inches.

_____

17. Find the number of cubic feet in nine pieces of plastic that are 2 inches thick, 6 inches wide, and 3 feet long. Round off to two decimal places.

_____

18. The diameter of a fully inflated truck tire is 30 inches. How many revolutions per minute will the wheel make if the car travels at the rate of 40 mph?

_____

19. Find the lateral area of a cone whose slant height is 6 inches and whose base diameter is 4 inches.

_____

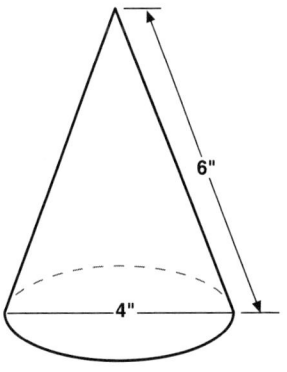

20. Find the volume of the cone frustum shown below.

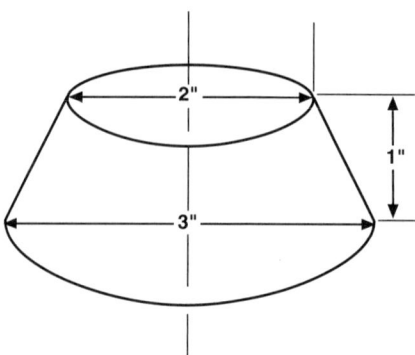

7

Trigonometry

OBJECTIVES

Upon completion of chapter seven, the student will be able to—
1. Explain the principles of right triangle trigonometry.
2. Define the three most common trigonometric functions, or ratios: sine, cosine, and tangent.
3. Use trigonometric functions to find the unknown measurement of one side of a right triangle when the other two sides or one acute angle and one side are known.
4. Use a table of trigonometric functions or a scientific calculator to quickly solve for missing values in triangles.
5. Name the reciprocals of the sine, cosine, and tangent.
6. Determine which trigonometric function to use when solving problems involving triangles.
7. Solve for unknown measurements of oblique triangles by using trigonometric formulas.
8. Find the area of triangles by using trigonometric functions.
9. Apply trigonometric formulas to everyday situations by constructing triangles to represent the situations.
10. Solve problems requiring the use of inverse functions.

INTRODUCTION

Trigonometry is the study of triangles and their use in solving problems. Indeed, the word trigonometry derives from the Greek words for triangle measurement. All triangles have sides and angles, and trigonometry deals with the relationship these sides and angles have to each other. Using trigonometry, unknown data can be found from given, or known, data. For example, surveyors can compute a distance that they cannot physically measure by determining angles and lengths with their surveying instruments. Then, using these known values and trigonometric ratios, they can find the unknown measurement.

To understand the basis for trigonometry, consider the two triangles in figure 7.1. Although triangle $A'B'C'$ is larger than triangle ABC, the angles are the same—90, 60, and 30 degrees in this case. Because the angles are the same, the ratios of the sides of the triangles are the same. That is, the length of the sides of both triangles is proportional to each other because the angles are the same. The principles of trigonometry come from these ratios.

The *right triangle* (a triangle with a right, or 90-degree, angle) is the basis for all trigonometry calculations. The relationships between a right triangle's angles and sides are simple and well known. Even so, trigonometric formulas can also solve problems with *oblique triangles*, which are triangles without right angles, as you will learn later in this chapter.

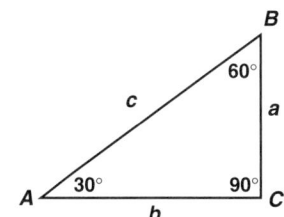

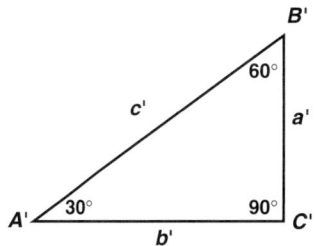

Figure 7.1   Triangle ratios

187

RIGHT TRIANGLE TRIGONOMETRY

Right triangle trigonometry allows you to solve problems involving triangles that have a right angle. When you know the values of two of the sides, or the value of one angle and one side of a right triangle, you can indirectly calculate the values of the remaining angles and sides.

Angles are measured from an arbitrary radial line of a circle. A *radial line* is a straight line that begins at the center of the circle and runs, or radiates, outward from the center. Draw several radial lines inside a circle and the spaces between the lines are angles. Angles are measured in degrees, minutes, and seconds. A circle has 360 degrees (360°). Each degree is made up of 60 minutes (60') and each minute is made up of sixty seconds (60"). Thus, a 360° circle contains 21,600' and 1,296,000".

A *quadrant* is one-fourth of a circle, or 90°, and a 90° angle is a right angle. A right triangle has one right angle (a 90° angle) and two acute angles, which are angles of less than 90°. Regardless of the size of each of the two acute angles, their sum is 90°. Therefore, the sum of all three angles in a right triangle is 180°. This fact can be helpful when determining the value of the acute angles in a right triangle. For example, if you know that a right triangle has one acute angle of 60°, you also know that the other acute angle is 30° because $180 - 90 - 60 = 30°$ and $90 + 60 + 30 = 180°$.

Trigonometric Functions

A trigonometric function expresses the relationship between the angles and sides of a right triangle. Trigonometric functions, or ratios, involve two sides and an acute angle of a right triangle. One side of a right triangle is the hypotenuse. The other two sides of a right triangle are generally referred to as being opposite or adjacent to an angle.

Figure 7.2 shows the sides of a right triangle. Side c is the hypotenuse and is opposite right angle C. Side a is opposite angle A, and side b is opposite angle B. Side a is also adjacent to angle B, and side b is adjacent to angle A. Although side c is also adjacent to angles A and B, the side opposite the right angle is always the hypotenuse. Understanding this terminology is important because trigonometric functions are defined in terms of these sides. The three most commonly used trigonometric functions are sine (sin), cosine (cos), and tangent (tan).

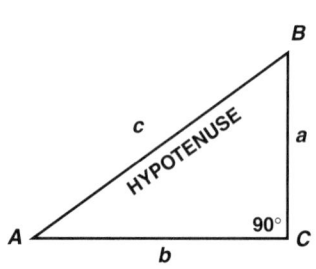

Figure 7.2   Sides of a triangle

Sine

The sine of an acute angle in a right triangle is a trigonometric function equal to the length of the side opposite the angle divided by the hypotenuse. Put another way, sine is found by dividing the side opposite the acute angle by the hypotenuse:

$$\text{sine} = \frac{\text{opposite side}}{\text{hypotenuse}}.$$

As mentioned earlier, figure 7.2 shows a right triangle with angles A, B, and C and sides a, b, and c. Angles A and B are acute angles. Side a is opposite angle A and side b is opposite angle B. Angle C is 90°. Side c is opposite angle C and is the hypotenuse. The sine of angle A (referred to as sin A) is

$$\text{sine } A = \frac{a}{c}$$

and the sine of angle B is

$$\text{sine } B = \frac{b}{c}.$$

In any right triangle, the ratio is constant and changes only when the acute angle changes. The values of sine range from 0.0175 for a 1° angle to 1.000 for a 90° angle. Thus, as the size of the acute angle increases, the value of the angle's sine increases. Put another way, the sine of a small angle is small, and the sine of a large angle is large. As an angle approaches 90°, its sine becomes larger and is equal to 1 at 90° (as shown in table 7.1). The values of the trigonometric functions are based on acute angles, those lying between 0° and 90°. However, the functions exist for angles greater than 90°. For angles between 90° and 180° (obtuse angles), the sine values repeat, being 1 at 90° and decreasing to 0 at 180°. For angles between 180° and 360°, the values of the sines are negative; for example, sin 270° = –1.00.

Table 7.1 gives the sine and other functions for any angle from 0° to 90°. As you will learn soon, these values are needed to solve problems.

Scientific calculators with trigonometric function keys can easily find the functions of an angle. On most calculators with trigonometric keys, you simply enter the angle on the number keys and then press the desired function key. Each key is labeled sin, cos, or tan.

Example Problem: Find the numerical value of sin 26°.

Solution: Using table 7.1, locate 26° in the left-hand column. Then, read the answer under the sine column, which is 0.4384. Or, using a calculator with trigonometric functions, press 2, 6, and sin to get 0.4383711.

Example Problem: Find the sine of angle B in the right triangle shown in figure 7.3.

Solution: First, determine the value of angle B. To do so, recall that the angles of a right triangle add up to 180°. Thus, because angle C is 90° and angle A is 39°, then angle B is

$$90° - 39° = 51°.$$

Then, from table 7.1, or by using a calculator, sin 51° = 0.7771.

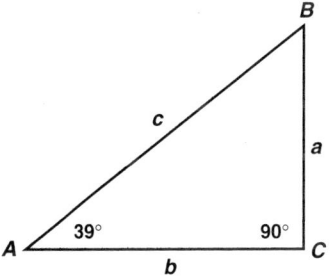

Figure 7.3 Angles of a triangle

Cosine

Cosine (cos) is the trigonometric function for a given angle in a right triangle that is equal to the length of the side adjacent to the angle divided by the hypotenuse. Put another way, the cosine of an acute angle in a right triangle is obtained by dividing the side adjacent to the angle by the hypotenuse:

$$\text{cosine} = \frac{\text{adjacent side}}{\text{hypotenuse}}.$$

So, referring to figure 7.2, the cosine of angle A is the adjacent side b divided by the hypotenuse, c. Likewise, the cosine of angle B is the adjacent side a divided by the hypotenuse, c. The formulas are:

$$\cos A = \frac{b}{c}$$

TABLE 7.1
Natural Trigonometric Functions

Angle	Sine	Cosine	Tangent	Angle	Sine	Cosine	Tangent
1°	0.0175	0.9998	0.0175	46°	0.7193	0.6947	1.0355
2°	0.0349	0.9994	0.0349	47°	0.7314	0.6820	1.0724
3°	0.0523	0.9986	0.0524	48°	0.7431	0.6691	1.1106
4°	0.0698	0.9976	0.0699	49°	0.7547	0.6561	1.1504
5°	0.0872	0.9962	0.0875	50°	0.7660	0.6428	1.1918
6°	0.1045	0.9945	0.1051	51°	0.7771	0.6293	1.2349
7°	0.1219	0.9925	0.1228	52°	0.7880	0.6157	1.2799
8°	0.1392	0.9903	0.1405	53°	0.7986	0.6018	1.3270
9°	0.1564	0.9877	0.1584	54°	0.8090	0.5878	1.3764
10°	0.1736	0.9848	0.1763	55°	0.8192	0.5736	1.4281
11°	0.1908	0.9816	0.1944	56°	0.8290	0.5592	1.4826
12°	0.2079	0.9781	0.2126	57°	0.8387	0.5446	1.5399
13°	0.2250	0.9744	0.2309	58°	0.8480	0.5299	1.6003
14°	0.2419	0.9703	0.2493	59°	0.8572	0.5150	1.6643
15°	0.2588	0.9659	0.2479	60°	0.8660	0.5000	1.7321
16°	0.2756	0.9613	0.2867	61°	0.8746	0.4848	1.8040
17°	0.2924	0.9563	0.3057	62°	0.8829	0.4695	1.8807
18°	0.3090	0.9511	0.3249	63°	0.8910	0.4540	1.9626
19°	0.3256	0.9455	0.3443	64°	0.8988	0.4384	2.0503
20°	0.3420	0.9397	0.3640	65°	0.9063	0.4226	2.1445
21°	0.3584	0.9336	0.3839	66°	0.9135	0.4067	2.2460
22°	0.3746	0.9272	0.4040	67°	0.9205	0.3907	2.3559
23°	0.3907	0.9205	0.4245	68°	0.9272	0.3746	2.4751
24°	0.4067	0.9135	0.4452	69°	0.9336	0.3584	2.6051
25°	0.4226	0.9063	0.4663	70°	0.9397	0.3420	2.7475
26°	0.4384	0.8988	0.4877	71°	0.9455	0.3256	2.9042
27°	0.4540	0.8910	0.5095	72°	0.9511	0.3090	3.0777
28°	0.4695	0.8829	0.5317	73°	0.9563	0.2924	3.2709
29°	0.4848	0.8746	0.5543	74°	0.9613	0.2756	3.4874
30°	0.5000	0.8660	0.5774	75°	0.9659	0.2588	3.7321
31°	0.5150	0.8572	0.6009	76°	0.9703	0.2419	4.0108
32°	0.5299	0.8480	0.6249	77°	0.9744	0.2250	4.3315
33°	0.5446	0.8387	0.6494	78°	0.9781	0.2079	4.7046
34°	0.5592	0.8290	0.6745	79°	0.9816	0.1908	5.1446
35°	0.5736	0.8192	0.7002	80°	0.9848	0.1736	5.6713
36°	0.5878	0.8090	0.7265	81°	0.9877	0.1564	6.3138
37°	0.6018	0.7986	0.7536	82°	0.9903	0.1392	7.1154
38°	0.6157	0.7880	0.7813	83°	0.9925	0.1219	8.1443
39°	0.6293	0.7771	0.8098	84°	0.9945	0.1045	9.5144
40°	0.6428	0.7660	0.8391	85°	0.9962	0.0872	11.4301
41°	0.6561	0.7547	0.8693	86°	0.9976	0.0698	14.3007
42°	0.6691	0.7431	0.9004	87°	0.9986	0.0523	19.0811
43°	0.6820	0.7314	0.9325	88°	0.9994	0.0349	28.6363
44°	0.6947	0.7193	0.9657	89°	0.9998	0.0175	57.2900
45°	0.7071	0.7071	1.0000	90°	1.0000	0.0000	∞

and

$$\cos B \ = \ \frac{a}{c} .$$

Note in table 7.1 that as the angle increases, the cosine decreases. The cosine of an acute angle is never equal to or greater than 1 because 1 would be 0°. The cosine of a small angle is large. For example, referring to table 7.1, the cosine of 2° is 0.9994; but, the cosine of 89° is 0.0175.

Example Problem: Find the numerical value of cos 16°.

Solution: Table 7.1 shows that the cosine of 16° is 0.9613. Or, enter 1, 6, and cos on a scientific calculator to get 0.9612617.

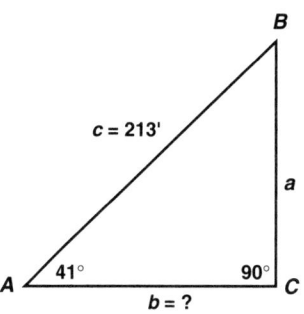

Figure 7.4   Dimensioned triangle

Example Problem: In the right triangle in figure 7.4, find b if c = 213 feet and angle A = 41°.

Solution: Because the length of the side adjacent to the hypotenuse (side b) is desired and the value of angle A is known, use the cosine function:

$$\cos A \ = \ \frac{b}{c} .$$

Next, solve for side b. To solve for b, transpose and multiply both sides of the formula by c. (Remember: $^c/_1 = c$.) Thus:

$$\frac{b}{c} \ = \ \cos A$$

$$\frac{b}{c} \times \frac{c}{1} \ = \ \cos A$$

$$b \ = \ \cos A \times c.$$

Now substitute the known values: (To find the value of the cosine of 41°, use table 7.1 or a scientific calculator.)

$$b \ = \ \cos 41° \times 213 = 0.7547 \times 213$$
$$b \ = \ 160.75 \text{ ft.}$$

Tangent

The tangent of an acute angle in a right triangle is equal to the length of the side opposite the angle divided by the length of the adjacent side. That is, to find the tangent of an angle, divide the side opposite the angle by the side adjacent to the angle:

$$\text{tangent} \ = \ \frac{\text{opposite side}}{\text{adjacent side}} .$$

Referring to figure 7.2, the tangent of angle A (tan A) is the ratio of the side opposite angle, which is a, to the side adjacent to angle A, which is b; or, in equation form:

$$\tan A \ = \ \frac{a}{b} .$$

Likewise, the tangent of angle B is the ratio of the side opposite angle B, which is b, and the side adjacent to angle B, which is a; or, in equation form:

$$\tan B \ = \ \frac{b}{a} .$$

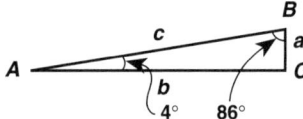

Figure 7.5   Triangle side ratios

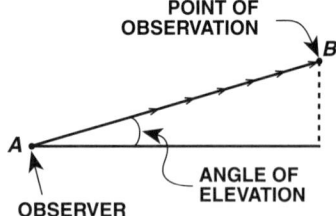

Figure 7.6   Triangle angle
definitions

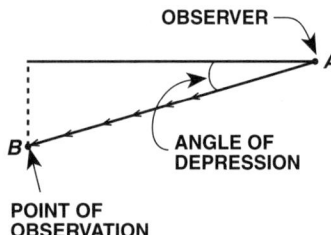

Figure 7.7   Triangle angle
definitions

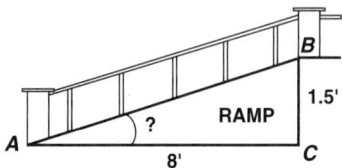

Figure 7.8   Dimensioned triangle

The tangent of an angle changes when the angle changes. As the angle increases, the tangent increases. For example, the tangent of 15 degrees is 0.2479, while the tangent of 75 degrees is 3.7321. Figure 7.5 shows a right triangle with a very small acute angle—angle A, which is only 4 degrees. Consequently, the side opposite angle A (side a) is also very short. The tangent of a small angle is also small—0.0699 in this case. Also notice in figure 7.5 that angle B is close to 90 degrees—86 degrees in this case. When an angle comes close to 90 degrees, the hypotenuse (c) is nearly parallel to the side opposite the angle (b).

The tangent function is often used to find the angle of elevation and the angle of depression. In figure 7.6, angle A is the angle of elevation. The observer's line of sight is from A to B. When the observation is made from a high point (A in figure 7.7) down to a low point (B), the angle at A is the angle of depression.

Example Problem: Find the angle of elevation (to the nearest degree) of a ramp built from the ground up to a building doorway. As figure 7.8 shows, the ramp rises 1.5 feet in a horizontal distance of 8 feet.

Solution: Use the tangent function and substitute known values:

$$\tan A \ = \ 1.5 \text{ ft} \div 8 \text{ ft} = 0.1875.$$

Using table 7.1, the nearest number to 0.1875 under the tangent column is 0.1944, which is the tangent for 11 degrees. Thus, the ramp rises at about an 11-degree angle from the ground; or, its angle of elevation is about 11 degrees.

Reciprocals

Three other trigonometric functions are the cosecant (csc), secant (sec), and cotangent (cot). These functions are the reciprocals of the sine, cosine, and tangent, respectively. The reciprocal of any number is 1 divided by that number—for example, the reciprocal of 20 is $\frac{1}{20}$, or 0.05 ($20 \div 1 = 0.05$). The formulas for the reciprocals are:

$$\text{cosecant} \ = \ \frac{\text{hypotenuse}}{\text{opposite side}}$$

$$\text{secant} \ = \ \frac{\text{hypotenuse}}{\text{adjacent side}}$$

$$\text{cosecant} \ = \ \frac{\text{adjacent side}}{\text{opposite side}}.$$

Table 7.2 compares the reciprocal functions with the three main functions.

It is interesting to note that in a right triangle, the sine of one acute angle is the cosine of the other, and that the tangent of one angle is the reciprocal of the tangent (the cotangent) of the other. For example, in a right triangle with acute angles of 30° and 60°, the tangent of 30° is 0.5774 (refer to table 7.1 for this value). The other acute angle in a right triangle is 60°, and the tangent of 60° is 1.7321, which is very close to $\frac{1}{0.5774}$, or the reciprocal of 0.5774 ($1 \div 0.5774 = 1.732$). Incidentally, some scientific calculators have a reciprocal key that can quickly find the reciprocal functions of an angle. The reciprocal key is usually designated as $1/x$. To use it, enter the number you wish to find the reciprocal of and press $1/x$.

TABLE 7.2
Reciprocal Functions

Function	Reciprocal
$\sin A = \dfrac{a}{c}$	$\csc A = \dfrac{1}{\sin A}$ or $\csc A = \dfrac{c}{a}$
$\cos A = \dfrac{b}{c}$	$\sec A = \dfrac{1}{\cos A}$ or $\sec A = \dfrac{c}{b}$
$\tan A = \dfrac{a}{b}$	$\cot A = \dfrac{1}{\tan A}$ or $\cot A = \dfrac{b}{a}$
$\sin B = \dfrac{b}{c}$	$\csc B = \dfrac{1}{\sin B}$ or $\csc B = \dfrac{c}{b}$
$\cos B = \dfrac{a}{c}$	$\sec B = \dfrac{1}{\cos B}$ or $\sec B = \dfrac{c}{a}$
$\tan B = \dfrac{b}{a}$	$\cot B = \dfrac{1}{\tan B}$ or $\cot B = \dfrac{a}{b}$

Inverse Functions

When a function of an angle is known but the angle is not, the inverse function is used to find the value of the angle. The inverse function represents an angle. Thus, an angle whose function is x is the inverse function of the angle. For example, the angle whose sine is 0.7071 is 45 degrees (see table 7.1). Put another way, $\sin x = 0.7071 = 45°$. Arc sin x or $\sin^{-1} x$ denotes an angle whose sine is x.

On scientific calculators with an inverse function key, the inverse sine key is likely to be labeled as $\sin^{-1}$. (However, refer to the instructions that accompany your calculator because differences exist.) The inverse cosine and tangent keys may be labeled $\cos^{-1}$ and $\tan^{-1}$. To make the inverse function keys work on most scientific calculators, you use the calculator's second-level function key, which is usually labeled 2nd or 2ndF. (Again, refer to your calculator's instructions.) The inverse function may be labeled on the body of the calculator above the keys or in small letters of a different color on the keys themselves. In any case, to determine an inverse function, first enter the known value of the function. Then, press 2nd or 2ndF. Finally, press the desired inverse function key—for instance, $\sin^{-1}$. A calculator shows the angle whose sine is 0.7071 as 44.999451, which is very close to 45°— indeed, so close as to be considered the same.

Example Problem: Find the angle whose tangent is 2.1546.

Solution: Using table 7.1, find the nearest numbers in the tangent column to 2.1546. These numbers are 2.1445 (65 degrees) and 2.2460 (66 degrees). Interpolate to find the exact angle.

2.1546 − 2.1445 = 0.0101
2.2460 − 2.1546 = 0.0914
0.0101 ÷ 0.0914 = 0.11.

Since 2.1445 is closer to 2.1546 than 2.2460, the angle is 66 degrees + 0.11 = 65.11 degrees. A calculator finds the answer to be 65.1028 degrees.

Applying Trigonometric Functions

Trigonometric functions are very helpful in problem solving. For example, if the hypotenuse and one acute angle are known, the sine is used to find the opposite side, and the cosine is used to find the adjacent side.

Example Problem: In the right triangle in figure 7.9, angle A = 34° and side c = 150 feet. Find the lengths of sides a and b in feet.

Solution: To find side a use

$$\sin A = \frac{a}{c}.$$

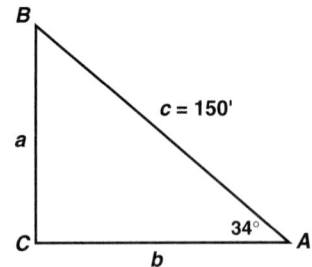

Figure 7.9   Dimensioned triangle

Transpose to

$$a = c \times \sin A.$$

From table 7.1, the sine of 34° is 0.5592. Substitute numerical values:

$$a = 150 \times 0.5592 = 83.88 \text{ ft.}$$

To find side b, use

$$\cos A = \frac{b}{c}$$

$$b = c \times \cos A.$$

From table 7.2, the cosine of 34° is 0.8290.

$$b = 150 \times 0.8290 = 124.35 \text{ ft.}$$

If an acute angle and the opposite side are known, the hypotenuse can be found by using the sine, and the adjacent side can be found by using the tangent.

Example Problem: In the right triangle in figure 7.10, angle A = 40° and side a = 37 in. Find the lengths of sides c and b in inches.

Solution: To find side c:

$$\sin A = \frac{a}{c}$$

$$\sin 40° = 0.6428$$

$$c = a \div \sin A = 37 \div 0.6428 = 57.56 \text{ in.}$$

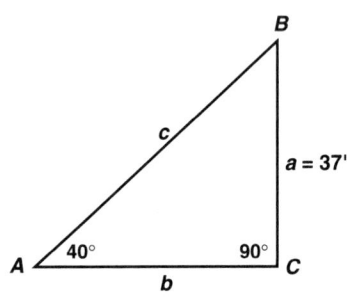

Figure 7.10   Dimensioned triangle

To find side b:

$$\tan A = \frac{a}{b}$$

$$\tan 40° = 0.8391$$

$$b = a \div \tan A = 37 \div 0.8391 = 44.1 \text{ in.}$$

These functions may be used in many combinations to solve for missing measurements. To choose the correct function for any problem, find the function that contains two known quantities and the unknown quantity desired.

The size of an angle may be found if two of the sides of the triangle are known. Using the values of the sides, the value of the trigonometric function of the angle can be found. Then, using the inverse function, the angle that matches the value of the function can be found in table 7.1 or by using a scientific calculator.

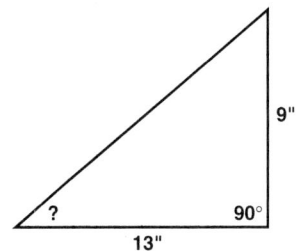

Example Problem: In a right triangle, two legs are 13 inches and 9 inches long. Find the smaller acute angle.

Solution: First, draw the triangle as in figure 7.11 and note that the smaller angle is opposite the shorter side. Then, the tangent of the unknown angle is 9/13, or 0.6923 (9 ÷ 13 = 0.6923). Using the inverse function, find an angle whose tangent is 0.6923. A calculator shows that the angle is 34.6949. The table of functions shows that tan 34° = 0.6745 and tan 35° = 0.7002. Since 0.6923 is closer to 0.7002 than it is to 0.6745, the unknown angle is found to be about 34.6°, or 34°36'.

Figure 7.11   Dimensioned triangle

Incidentally, to convert degrees (°) to minutes ('), set up a proportion. Since 60' are in 1° and since 0.6° is the same as 6/10, determine the number of minutes in 0.6° by the proportion, 6 is to 10 as x is to 60, or:

$$\frac{6}{10} = \frac{x}{60}.$$

Then, cross-multiply to determine that

$$10x = 360°$$
$$x = 36'.$$

Practice Problems

Round off answers to two places.

1. A right triangle has one 15-degree angle. How many degrees does the other acute angle contain?

    _____

2. Find the length of side c in the triangle below.

    _____

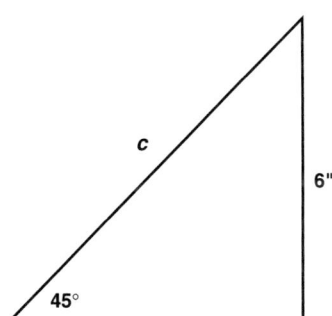

3. The guy wire to the top of a gin pole that is 30 feet high makes an angle of 45 degrees with the ground. What is the length of the wire?

    _____

4. Find the length of the hypotenuse in the accompanying sketch.

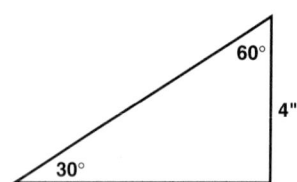

5. Find the length of the side opposite the 60-degree angle in the sketch in problem 4.

6. The side opposite a 36-degree angle in a right triangle is 12 inches long. How long is the hypotenuse?

7. Find the length of the pieces of galvanized roofing needed to cover the roof in the sketch below. (Note: Treat as two right triangles.)

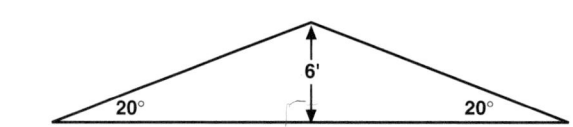

8. A 30-foot ladder is placed against a house at an angle of 48 degrees to the ground.

   a. How high is the top of the ladder above the ground?

   b. How far is the bottom of the ladder from the house?

9. A guy wire to a derrick is 200 feet long and is anchored to the ground 150 feet from the derrick. What angle does the wire make with the ground?

10. In the wedge shown below, find the value of angle *A* between the two sides.

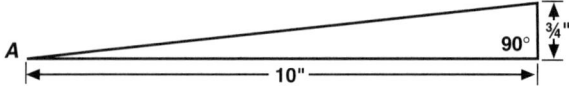

OBLIQUE TRIANGLE CALCULATIONS

Right triangle calculations in trigonometry solve most of the problems encountered in petroleum or other industrial work. Usually, oblique triangles, as well as other shapes, can be divided into right triangles, and the problems can be solved using trigonometric functions. A good approach to everyday situations is to find the triangle or triangles in the application and then solve for missing sides and angles.

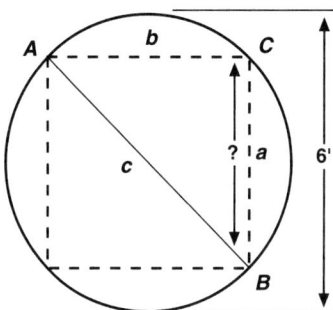

Figure 7.12   Oblique triangle

Example Problem: Determine the largest square that can be milled from a 6-inch circular disk.

Solution: To visualize the problem, draw a circle of any size and inscribe a square inside it (fig. 7.12). Then, passing through the center of the circle, connect two opposite corners of the square, making a right triangle. Label the right angle C and the other angles and sides as shown in figure 7.12. Determine the known dimensions: side c = 6 inches, angle $C = 90°$, and angles A and $B = 45°$ (half of a right angle). Now, use the sine function,

$$\sin A = \frac{a}{c}$$

and, because side a is the unknown side, transpose to solve for a:

$$a = \sin A \times c$$
$$a = \sin 45° \times 6 \text{ in.} = 0.7071 \times 6 = 4.24 \text{ in.}$$

Solving for Sides and Angles

Unknown measurements of oblique triangles that have two sides and one angle, two angles and one side, or all three sides known can be found by using the following trigonometric formulas. Reference points for the formulas are shown in figure 7.13.

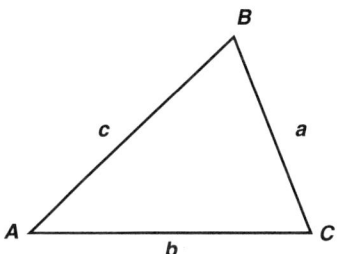

Figure 7.13   Dimensioned oblique triangle

When two angles and one side are known, the other angle and the other two sides can be found. Assuming the known side is b and the known angles are B and C, the formulas are

$$A = 180° - (B + C)$$
$$c = \frac{b \sin C}{\sin B}$$
$$a = \frac{b \sin A}{\sin B}.$$

Example Problem: Figure 7.14 shows an oblique triangle with angles A, B, and C and sides a, b, and c. Angle A is 35°, angle C is 40°, and side a is 10 feet long. Find angle B and the lengths of sides b and c.

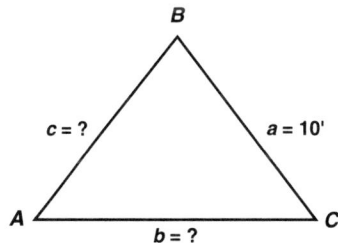

Figure 7.14   Dimensioned oblique triangle

Solution: Transpose the formula $A = 180° - (B + C)$ to solve for B:

$$B = 180° - (A + C)$$
$$B = 180° - (35° + 40°) = 105°.$$

Substitute the known side a in the formulas given:

$$c = \frac{a \sin C}{\sin A}$$
$$c = 10 \text{ ft} \times \sin 40°) \div \sin 35°$$
$$c = 10 \text{ ft} \times 0.6428) \div 0.5736 = 11.2064, \text{ or } 11.21 \text{ ft}.$$

To solve for side b, use the sine of angle B. Angle B is more than 90° (it is an obtuse angle). To find the sine of an obtuse angle, subtract the value of the obtuse angle (in this case, 105°) from 180° and use the sine of the difference (75°) in the equation:

$$b = \frac{a \sin B}{\sin A}$$

$$b = 10 \text{ ft} \times \sin 75°) \div \sin 35°$$
$$b = (10 \text{ ft} \times 0.9659) \div 0.5736 = 16.8392, \text{ or } 16.84 \text{ ft.}$$

Thus, angle $C = 40°$; side $c = 11.21$ ft; and side $b = 16.84$ ft.

When two sides and the included angle are known, the other two angles and side can be found. Assume the known sides are b and c and the known angle is A. Then

$$a^2 = b^2 + c^2 - 2bc \cos A$$
$$\sin B = \frac{b \sin A}{a}$$
$$C = 180° - (A + B).$$

When two sides and the angle opposite one of these sides is known, the other two angles and side can be found. Assume the known angle is B and the known sides are b and c. Side b is opposite angle B. The formulas are

$$\sin C = \frac{c \sin B}{b}.$$
$$A = 180° - (B + C).$$
$$a = \frac{b \sin A}{\sin B}.$$

When all three sides are known, the angles can be found. With sides a, b, and c known, angles A, B, and C can be found using these formulas:

$$\cos B = \frac{c^2 + a^2 - b^2}{2ca}.$$
$$\sin C = \frac{c \sin B}{b}.$$
$$A = 180° - (B + C).$$

Finding Areas

When two sides and the included angle of a triangle are given, the area of the triangle can be computed by using the sine of the angle. Assuming sides b and c and angle A (fig. 7.15) are known, the formula for finding the area is

$$\tfrac{1}{2}bc \sin A.$$

This formula works for acute angles. If the included angle is obtuse, the formula becomes

$$\text{area} = \tfrac{1}{2}ac \sin (180° - B)$$

or

$$\text{area} = \tfrac{1}{2}bc \sin (180° - A).$$

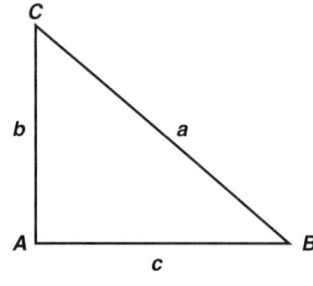

Figure 7.15   Area of a triangle

Practice Problems

Round off answers to two places.

1. The equal sides of a roof make an angle of 35° with the horizontal. If each side of the roof is 50 feet, how long is the span of the roof?

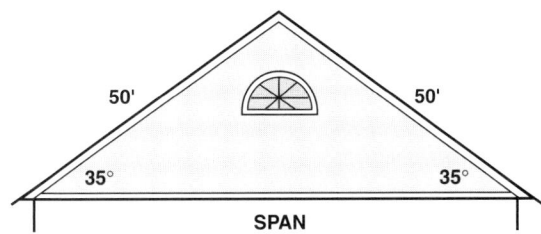

2. Find the length of side *a* in the triangle shown below.

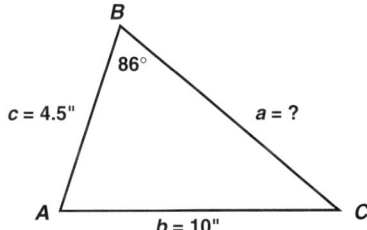

3. Find the angle whose cosine is 0.5240.

4. An oblique triangle has a 4-inch side and a 6-inch side that join to make a 65° angle. What is the measurement of the other side?

5. Using the sine and cosine, find the dimensions of *X* and *Y* in the sketch below.

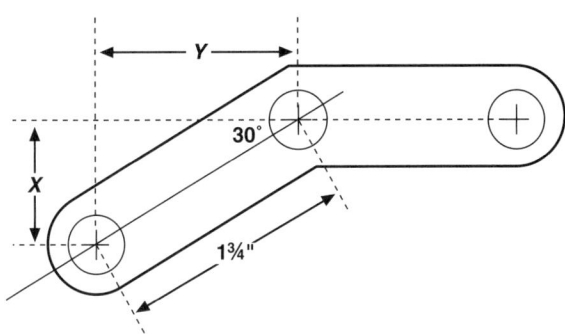

6. Using the sine, find dimension *c* in the plate sketched below.

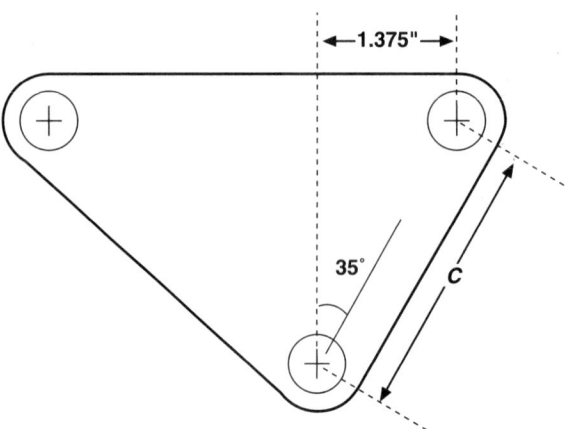

7. Find the area of the triangle below, using trigonometric functions.

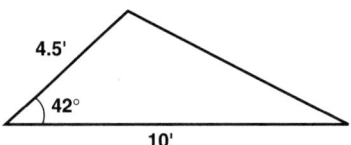

8. Approximately how many square feet are in a triangular plot of ground that is 20 feet × 30 feet × 35 feet?

9. The derrick legs in the sketch below make an angle of 10° with the vertical. Calculate their length.

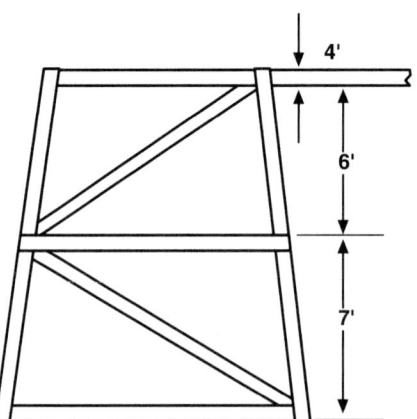

10. What is the length of each side of the largest pentagon that can be cut from a circular disk that is 6 inches in diameter?

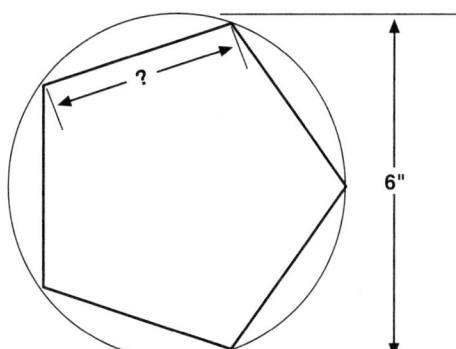

SELF-TEST

7. Trigonometry

Multiply the number of problems solved correctly by five to arrive
at your percentage of competency.

Round off answers to one place.

1. What is the length of side *b* in the sketch below?

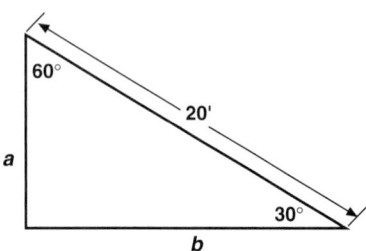

2. Find the lengths of the two guy wires shown in the accompanying sketch.

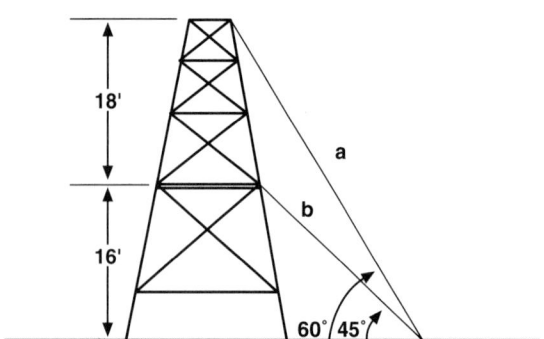

3. Find the length of the guy wire in the sketch below.

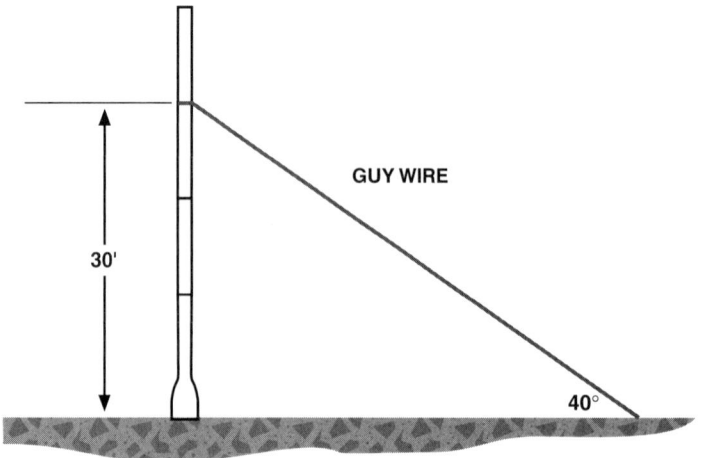

4. A right triangle has angles of 28° and 62°. The hypotenuse is 12 inches long. How long is the side opposite the 28° angle?

_____

5. Find the length of the hypotenuse of the triangle shown below.

_____

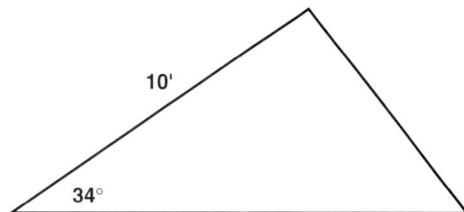

6. A balloon that is 600 feet high when straight up is driven by the wind until its tie rope is at an angle of 80° to the ground. What is the height of the balloon at this point?

_____

7. Calculate the following distances on the sketch below:

   a. Length of the guy wire _____

   b. Height from top of the pole to the ground _____

   c. Height of the pole _____

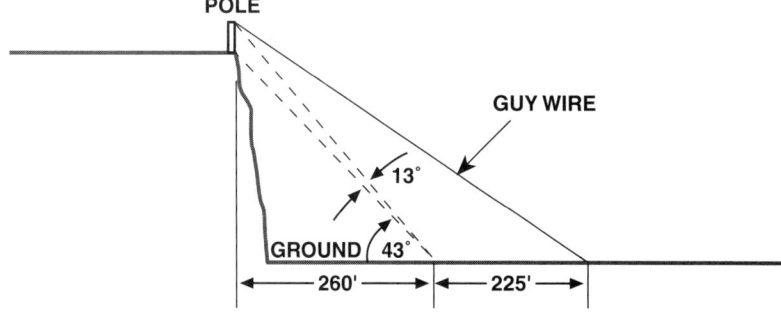

8. Find the angles of the triangle shown in the sketch below.

Angle *A*: _____

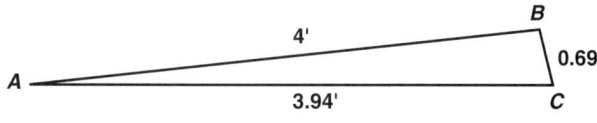

Angle *B*: _____

Angle *C*: _____

9. Find the length of side *a* on the sketch to the right.

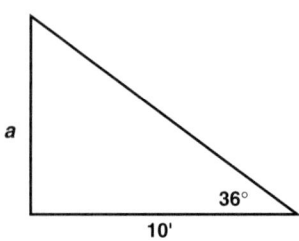

_____

10. If a right triangle has an angle of 46° and the side opposite the angle is 9 feet, what is the length of the adjacent side?

_____

11. Find the length of the center section of the **A**-frame shown in the sketch.

_____

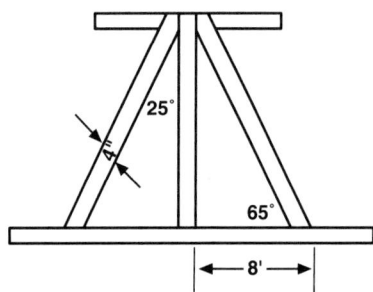

12. The plot of ground shown in the sketch below is cut in two by a river. Side *b* and the 32° and 90° angles are known, but it is impossible to reach point *X*, which is marked by a stake across the river. Calculate lengths of sides *a* and *c*.

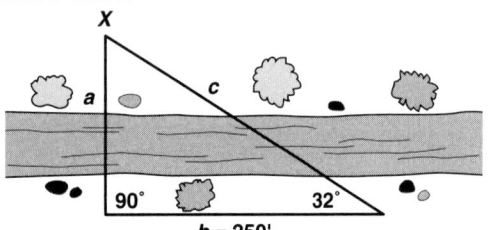

Side *a*: _____

Side *c*: _____

13. Find the distance between the centers for the pulleys shown in the sketch.

_____

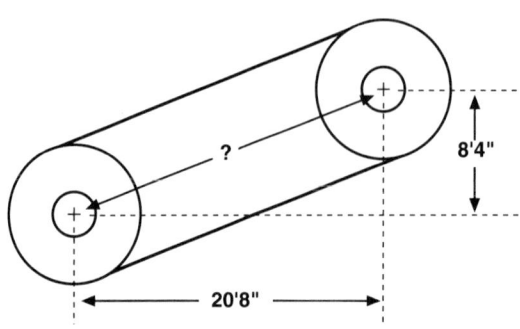

14. One of the acute angles in a right triangle is 39°.

    a. What is the cotangent of the 39° angle?

    _____

    b. What is the cotangent of the other acute angle in the triangle?

    _____

15. What is the length of each side of the largest hexagon that can be milled from a 6-inch circular disk?

_____

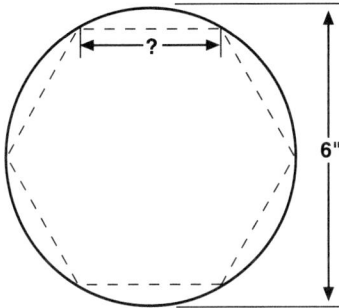

16. What is the length of each side of the largest octagon that can be milled from a 6-inch circular disk?

_____

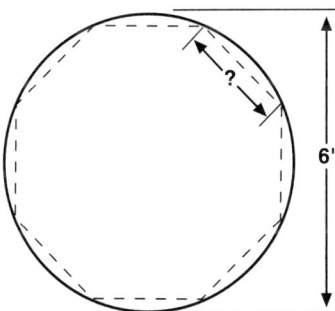

17. Find the radius of a circle circumscribed around an octagon with 10-inch sides.

_____

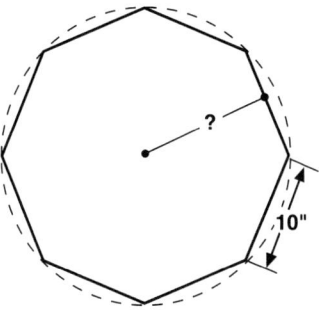

18. Nine holes are to be drilled on the circumference of a circle that is 7.25 inches in diameter. What should be the distance between the centers of the holes?

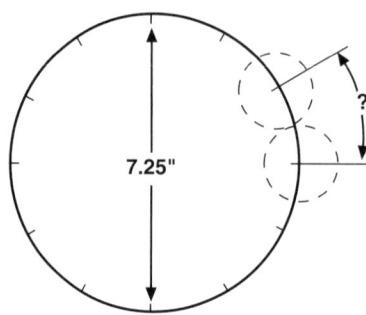

19. Find the distance at which to set dividers for laying off twelve equally spaced holes on a 9-inch diameter circle.

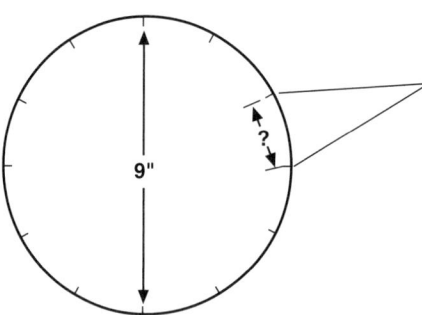

20. What is the diameter of a circle circumscribed around a square that is 7 inches on a side?

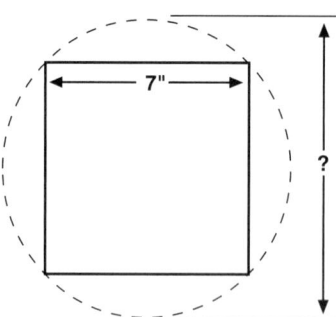

8

Advanced Math Concepts

OBJECTIVES

Upon completing chapter eight, the student will be able to—

1. Identify binary devices that have only two states.
2. Write numbers with value in binary format.
3. Compare various number systems with specific base numbers.
4. Add, subtract, multiply, and divide logic numbers.
5. Create octal numbers from binary numbers.
6. Create hexadecimal numbers from binary numbers.
7. Prepare numbers in ASCII format.
8. Write numbers in BCD and Gray Code format.

INTRODUCTION

The first chapter of this manual pointed out that our numbering system is based on the ten fingers on our hands. Because we have ten fingers and it is easy to count on them, it is logical that we base our numbering system on 10. Later, Arabic speaking scholars developed this primitive system into what we now term the base 10, or decimal, numbering system. In the base 10 system, characters begin at 0 and end at 9. To count above 9, we create the double-digit number of 10. In the number 10, 1 represents the value ten and the 0 represents the value one. Put another way, the number 10 represents zero ones and one ten. We commonly refer to this number as representing the quantity ten. When we say the number, we normally don't say, "one ten and zero ones"; instead, we abbreviate it simply to ten.

Now, consider the number 203. In this case, 2 is in the hundreds column, 0 is in the tens column, and 3 is in the ones column. So, the number 203 represents 2 hundreds, 0 tens and 3 ones. However, we simply say that this number is two hundred-three. Notice that the place of the numbers in the base 10, or decimal, numbering system shows their relative value.

Thus, the ones are in the right-most column. Then, moving one column at a time from right to left, come the tens, the hundreds, the thousands, the ten thousands, and so on. For example, the number 34,895 has 5 ones, 9 tens, 8 hundreds, 4 thousands, and 3 ten thousands. Notice that each succeeding column is a multiple of 10—that is, 10 is ten ones, 100 is 10 tens, 1,000 is 10 hundreds, and so on. Since the base number in the decimal system is 10, number values can be shown in columns (table 8.1).

The example decimal number is written as 3,634,209 and is read as three million, six hundred thirty four thousand, two hundred-nine. In reality, this number results from adding seven numbers: 3,000,000 + 600,000 + 30,000 + 4,000 + 200 + 00 + 9. The placement of the numbers signifies their relative value in this system. When saying or writing a complete number, we abbreviate it.

TABLE 8.1
Base Ten Numbers

Base number with exponent	10^6	10^5	10^4	10^3	10^2	10^1	10^0
Equivalent decimal value	1,000,000	100,000	10,000	1,000	100	10	1
Example decimal number	3,	6	3	4,	2	0	9

BINARY ARITHMETIC

Most people take the decimal number system for granted because the numbers are easy to see, speak, and write. However, computers are a big part of most people's lives. And, a digital computer does not have the ability to see, speak, or write decimal numbers. Decimal numbers are characters or symbols of varying shapes that the core of a computer system cannot recognize directly. Special programs must be written before a computer can deal with the numbers that are so familiar to us. On the other hand, computers easily recognize the presence or absence of an electrical signal. That is, a computer can recognize a signal that has two states: either present or absent. A two-state signal is referred to as a binary signal. You can think of a computer as having only one finger that it is either present or absent depending on which of the two states is occurring.

Binary logic is the study of statements, devices, or symbols that can be represented in two distinct states or conditions. Examples of binary conditions are on, off; true, false; yes, no; black, white; and high, low. An example of a device that exhibits two states is a light switch because it is either on or off. Many devices exhibit a two-state condition and are referred to as binary devices. No contradiction exists between these two states and they are considered to be absolute, but opposite, conditions. Just as the decimal system is termed the base-ten system, the binary system is termed a base-two system.

The binary numbering system uses the symbols 1 and 0 to represent the only two numbers in the system. Keep in mind that 1 and 0 do not have numerical value, although they resemble the decimal system's 1 and 0: 1 and 0 are only symbols that represent one of two binary states.

The binary number system has several characteristics, which include—

1. It is a base 2 numbering system.
2. It uses 1 and 0 as symbols to make discrete decisions or to show states of devices.
3. The largest valued symbol is 1. The lowest valued symbol is 0.
4. The decimal number 0 is a valid number when converting binary numbers to decimal.

A binary number can be converted to decimal and decimal to binary by using table 8.2. Table 8.2 counts through 4 binary digits. Incidentally, binary digit is often shortened to bit; so, 4 binary digits can also be called 4 bits. The table shows the binary digits, or bits, as 2 raised to the exponents 0, 1, 2, and 3. Larger binary numbers can be achieved by adding more columns with values such as 2^4, 2^5, 2^6, 2^7, and 2^8.

To use the table to convert binary numbers to decimal numbers, be aware that any 0 in the columns of binary numbers is not valid. On the other hand, a 1 in the columns of binary numbers indicates the value is valid, or, as a mathematician might

TABLE 8.2
Decimal Equivalents

Base number with exponent	2^3	2^2	2^1	2^0	Equivalent decimal value
Equivalent decimal	8	4	2	1	
value	0	0	0	0	0
	0	0	0	1	1
	0	0	1	0	2
	0	0	1	1	3
	0	1	0	0	4
	0	1	0	0	5
	0	1	1	0	6
	0	1	1	1	7
	1	0	0	0	8
	1	0	0	1	9
	1	0	1	0	10
	1	0	1	1	11
	1	1	0	0	12
	1	1	0	1	13
	1	1	1	0	14
	1	1	1	1	15

say, it exists. A zero's not being valid simply means that it is not used to convert a binary number to a decimal number. A one's being valid, or existing, means that the number is used to convert a binary number to a decimal number. Thus, by adding the values that are shown with a 1 in the table, the equivalent decimal number can be obtained.

Example Problem: What is the equivalent decimal value of the binary number 1010?

Solution: First, find the binary number 1010 in the table. Notice that the 0 in the rightmost column of the binary number is positioned under the 1 column of the equivalent decimal value column. However, because the binary number is 0, it is not valid. (It is not used to determine the decimal equivalent.) The next number in the binary number is 1 and it is in the 2 column of the equivalent decimal value column and it exists, or is valid. (It is used to determine the decimal equivalent.) The third binary number is 0 and it lies under the 4 column of the equivalent decimal number column. It is not valid because it is 0. The fourth binary number is 1 and it lies in the 8 column. Because it is a 1, it is valid. Therefore, the equivalent decimal number is 0 + 2 + 0 + 8 = 10.

Example Problem: Convert the binary number 101 into a decimal number.

Solution:

$$101 = 4 + 0 + 1 = 5.$$

Although table 8.2 only carries the numbers to 2^3 (8), the columns can be carried onward to the left as far as desired—for example, 2^4 (16), 2^5 (32), 2^6 (64), 2^7 (128), 2^8 (256), and so on. So, to convert the binary number 110011 to its decimal equivalent, start with the rightmost number and proceed to the left. In this case, the binary 1 is in the decimal 1 column, a binary 1 is in the decimal 2 column, a binary 0 is in the decimal 4 column, a binary 0 is in the decimal 8 column, a binary 1 is in the decimal 16 column, and a binary 1 is in the decimal 32 column. So, the decimal equivalent of the binary number 110011 is 32 + 16 + 0 + 0 + 2 + 1 = 51. Fifty-one is the decimal equivalent of 110011.

Example Problems: Find the decimal equivalents of the binary numbers 100001, 1010101, and 111000111.

Solution:

> 100001 = 32 + 0 + 0 + 0 + 0 + 0 + 1 = 33.
> 1010101 = 64 + 0 + 16 + 0 + 4 + 0 + 1 = 85.
> 111000111 = 256 + 128 + 64 + 0 + 0 + 0 + 4 + 2 + 1 = 455.

Binary Numbers

As explained earlier, binary numbers are different from decimal numbers—that is, base 2 is used for binary numbers and base 10 is used for decimal numbers. All decimal numbers can be represented as the number 10 raised to some power, and all binary numbers can be represented as the number 2 raised to some power.

The decimal system uses the digits ranging from 0 to 9 while the binary system uses only the digits 1 and 0. Only two digits fit well with electronic systems, because 0 can represent the absence of voltage while 1 can represent the presence of voltage. Switches or solid-state devices can provide voltage or not provide voltage as required.

Regardless of the numbering system, a number's value depends on its position relative to a reference point. In binary systems, the reference point is the binary point. In decimal systems, the reference point is the decimal point.

In the base 10 decimal system, a number becomes 10 times larger each time it moves one position to the left of the decimal point. For example, the number 3,747 can be represented in weighted columns as—

10^3	10^2	10^1	10^0
1,000	100	10	1
3	7	4	7

The number 3,747 also means 3,000 + 700 + 40 + 7 = 3,747. In this case, the decimal point is not written. If it were, the number would be written as 3,747.0. Normally, numbers that only exist to the left of the decimal point do not show the decimal point.

In base ten numbers with a decimal point, numbers to the right of the decimal point can be shown as—

10^{-1}	10^{-2}	10^{-3}	10^{-4}
0.1	0.01	0.001	0.0001
6	2	1	9

In this case, the number is written as 0.6219 and it means it is made up of six tenths (0.6), two hundredths (0.02), one thousandth (0.001), and nine ten thousandths (0.0009).

Example Problem: Write the decimal number 98.38 in columns showing powers of 10.

Solution:

10^1	10^0		10^{-1}	10^{-2}
9	8	‖	3	8

The double line between the positive and negative exponents indicates the decimal point.

In base 2 binary systems, a number doubles each time it is moved one position to the left of the binary point. The binary number 1111 (base 2) can be written as 1,000; 100; 10; and 1; or as $1(2^3) + 1(2^2) + 1(2^1) + 1(2^0)$. This binary number can be converted to a decimal number by writing $8 + 4 + 2 + 1 = 15$ because 2^3 is 8, 2^2 is 4, 2^1 is 2, and 2^0 is 1. The columns below demonstrate this process in a different format where the numbers increase in weight to the left of the binary point.

2^3	2^2	2^1	2^0	Decimal No.
8	4	2	1	
1	1	1	1	15.0
1	0	1	0	10.0
0	1	0	1	5.0

Binary numbers with negative powers of base two are those that are to the right of the binary point. These numbers decrease in half for each column to the left. For example—

2^{-1}	2^{-2}	2^{-3}	2^{-4}	Decimal No.
0.5	0.25	0.125	0.0625	
1	1	1	1	0.9375
1	0	1	0	0.6250
0	1	0	1	0.3125

The decimal equivalent for the binary number 0.1111 is 0.9375 because $0.5 + 0.25 + 0.125 + 0.0625 = 0.9375$. Similarly, the decimal equivalent of the binary number 0.0101 is 0.3125 because $0.25 + 0.0625 = 0.3125$.

Binary Addition

Binary numbers can be added, subtracted, multiplied, and divided using similar methods as those for decimal numbers. To identify binary numbers, this text uses the subscript 2 ($_2$) after 1 and 0. So, a binary 1 is 1_2 and a binary 0 is 0_2. Similarly, decimal numbers use the subscript 10 ($_{10}$). First, remember that $0_2 + 0_2 = 0_2$ and that $0_2 + 1_2 = 1_2$. However, $1_2 + 1_2$ becomes more complicated because, in the binary system, $1_2 + 1_2$ is not 2. This number does not exist in the binary system. So, $1_2 + 1_2 = 0_2 + c$. When adding binary numbers larger than 1, c symbolizes the phrase "to carry." That is, in binary notation, c means to carry 1_2 to the next column to the left. In column format, it shows as:

$$c$$
$$1_2$$
$$+ \quad 1_2$$
$$\overline{10_2}$$

Thus, $1_2 + 1_2 = 10_2$ because $1_2 + 1_2 = 0_2$ with a 1_2 carried to the next column. You can verify this answer by converting to the equivalent decimal number. First, $1_{10} + 1_{10} = 2_{10}$—that is, the decimal equivalent of the binary number 10_2 is 2_{10}, and the binary equivalent of $10_2 = 2_{10}$.

Example Problem: Add the following binary numbers to arrive at a binary sum.

Solution:

	c	*c c*	*cccc*
0101	0011	0110	1111
+ 0010	+ 1010	+ 0111	+ 1111
0111	1101	1101	11110

You can check the results by converting the binary numbers to their decimal equivalent numbers and adding the decimal numbers. Thus—

0101 = 5	0011 = 3	0110 = 6	1111 = 15
+ 0010 = 2	+ 1010 = 10	+ 0111 = 7	+ 1111 = 15
0111 = 7	1101 = 13	1101 = 13	11110 = 30

In the following examples, the binary numbers added on the left are equivalent to the decimal numbers added on the right. *Augend*, the quantity to which another quantity is added, and *addend*, one of a collection of numbers to be added, is the number to be added. The binary numbers to be added are 110.11 and 011.01, while the equivalent decimal numbers to be added are 6.75 and 3.25.

Function	Binary						Decimal
No. carried		*1*	*1*	*1*	*1*	*1*	*1 1*
Augend		1	1	0.	1	1	6.75
Addend		0	1	1.	0	1	3.25
Sum	1	0	1	0.	0	0	10.00

Referring to the binary columns, begin the operation with the right-hand column and work to the left, just as when adding decimal numbers. Beginning at the right-hand column, 1 is added to 1, and the result is 2. But, because 1 is the highest number in the binary system, a 0 is written for the sum of 1 plus 1, just as 0 is put down for the sum of 9 plus 1 in the decimal system. The 1 is carried over to the next column and placed in the no. carried row.

Adding the second column, the 1 in the augend is added to the 1 in the no. carried row; so, again, a 0 is written for the sum of 1 plus 1, and a 1 is carried to the third column in the no. carried row. When the third column is added, the result is again 1 plus 1; so, a 0 is placed in the sum, and the 1 is carried to the fourth column.

The fourth column now has 1 plus 1 plus 1. As before, 1 plus 1 gives 0 and 1 to carry; so, the remaining 1 and 0 are added to get 1 for the sum in the fourth column. Going to the fifth column, 1 plus 1 gives the sum of 0 and 1 is carried into the sixth column. The final sum is 1010.00, or 10.00 in decimal form. In binary addition, then, the rule is that 1 plus 1 equals 0 with 1 to carry to the next column.

Subtraction

Subtraction of binary numbers is similar to subtraction of decimal numbers. In the examples that follow, the binary numbers subtracted at left are equivalent to the decimal numbers at right. Minuend is the number from which another number, the subtrahend, is to be subtracted. The first example subtracts the binary number 010.11 from the binary number 110.10 and compares it with subtraction of the equivalent decimal numbers 6.50 from 2.75.

Function	Binary						Decimal
No. borrowed	*−1*	*1*	*1*	*1*			*−1 1 1*
Minuend	1	1	0.	1	0		6.50
Subtrahend	− 0	1	0.	1	1		−2.75
Difference		1	1.	1	1		3.75

Start subtraction with the right-most column of the binary number. In this column, 1 in the subtrahend cannot be subtracted from 0 in the minuend; so, borrow 1 from the next column to the left. A 1 is moved one column to the right and becomes worth 2. So, subtracting 1 from 2 leaves a 1 for the right-hand column. Enter this number in the bottom row, which is the difference row.

In the second column, a −1 is in the subtrahend, a 1 in the minuend, and a −1 in the number borrowed column. These add up to a −1. A 1 is borrowed from the third column, which is treated as a 2 when adding up the second column so a 1 is placed in the difference level for the second column.

The operation continues in the like manner for the remaining columns. Remember that, in subtraction, the borrows column has the same sign as the subtrahend, which is a minus (−).

Multiplication

Multiplication of binary numbers is the simplest of the four mathematical operations. Binary multiplication is the same as decimal multiplication. Even the placement of the binary point in the product is determined in the same way as placement of the decimal point. In the following example, 10.11 is multiplied by 1.01 with its decimal equivalent of 2.75 multiplied by 1.25.

Function			Binary				Decimal
Multiplicand			1	0.	1	1	2.75
Multiplier			×	1.	0	1	× 1.25
			1	0	1	1	1325
		0	0	0	0		550
	1	0	1	1			275
Product	1	1.	0	1	1	1	3.4375

Division

Division of binary numbers is also simple. Because a trial divisor must be either a 1 or a 0, there is little likelihood of failing on the first trial. In the following examples, borrowing was done mentally during all intermediate subtraction steps. The placement of the binary point and the placement of the decimal point are determined in the same manner.

As an example, the binary number 11.0111 is divided by 1.01 and the decimal number 3.4375 is divided by 1.25.

$$
\begin{array}{r}
10.11 \\
1.01 \overline{)\ 11.0111} \\
\underline{101} \\
111 \\
\underline{101} \\
101 \\
\underline{101} \\
\end{array}
$$

$$
\begin{array}{r}
2.75 \\
1.25 \overline{)\ 3.4375} \\
\underline{250} \\
937 \\
\underline{875} \\
625 \\
\underline{625.} \\
\end{array}
$$

TABLE 8.3
BCD Code and Binary and Decimal Equivalents

BCD	Binary	Decimal
0000	0	0
0001	1	1
0010	10	2
0011	11	3
0100	100	4
0101	101	5
0110	110	6
0111	111	7
1000	1000	8
1001	1001	9

SPECIAL BINARY CODES

Scientists have developed several special binary codes, which are based on binary concepts. These special codes are used in applications to speed up a process or create a standard to which users can adhere.

Binary Coded Decimal (BCD) Code

Binary coded decimal (BCD) code is used in input devices such as thumbwheel switches and seven-segment displays in electronic equipment. BCD code is the four-digit binary equivalent of the ten decimal numbers of 0 through 9. To show a decimal number in BCD, the equivalent binary number for each digit is written. For example, the decimal number 72 becomes 0111 0010 in BCD, because the binary number for 7 is 0111 and the binary number for 2 is 0010. The space between the four digits signifies a BCD coded number. Table 8.3 shows BCD code numbers and the equivalent binary and decimal numbers.

American Standard Code for Information Interchange (ASCII)

Programmable logic controllers (PLCs) use American standard code for information exchange (ASCII) to represent numbers, letters, and symbols. PLCs are digital devices used in process control systems to run and manage the variables that occur in such systems. For example, a PLC could maintain a constant level of liquid in a tank or a vessel by reading level sensors and sending corrective signals to devices to adjust the liquid flowing into and out of the vessel.

ASCII is also used to send and receive alphanumeric data such as the letters and symbols found on a computer keyboard. Each character or symbol uses up to 7, or sometimes 8, binary digits to represent it. Table 8.4 is a partial list of ASCII characters and symbols. They can also be found in digital handbooks and on the World Wide Web (the Internet).

As an example of using the table, note that the letter F has the binary ASCII code of 100 0110 and that the + sign has the ASCII code of 010 1011.

TABLE 8.4
ASCII Characters and the Equivalent Binary Numbers

Character	Binary No.	Character	Binary No.
@	100 0000	0	011 0000
A	100 0001	1	011 0001
B	100 0010	2	011 0010
C	100 0011	2	011 0010
D	100 0100	3	011 0011
E	100 0101	5	011 0101
F	100 0110	6	011 0110
G	100 0111	7	011 0111
H	100 1000	8	011 1000
I	100 1001	9	011 1001
J	100 1010	:	011 1010
K	100 1011	;	011 1011
L	100 1100	<	011 1100
M	100 1101	=	011 1101
N	100 1110	>	011 1110
O	100 1111	?	011 1111
P	101 0000	SP	010 0000
Q	101 0001	!	010 0001
R	101 0010	"	010 0010
S	101 0011	#	010 0011
T	101 0100	$	010 0100
U	101 0101	%	010 0101
V	101 0110	&	010 0110
W	101 0111	'	010 0111
X	101 1000	(	010 1000
Y	101 1001	)	010 1001
Z	101 1010	*	010 1010
[	101 1011	+	010 1011
\	101 1100	,	010 1100
]	101 1101	–	010 1101
^	101 1110	.	010 1110
_	101 1111	/	010 1111

TABLE 8.5
Gray Code Numbers with Equivalent Binary Numbers

Gray Code	Binary
0000	0000
0001	0001
0011	0010
0010	0011
0110	0100
0111	0101
0101	0110
0100	0111
1100	1000
1101	1001
1111	1010
1110	1011
1010	1100
1011	1101
1001	1110
1000	1111

Gray Code

Gray code is useful in mechanical encoders because a slight change in location only affects one bit in the code. Thus, it is often used to transmit binary code from rotating equipment when tracking the position of the shaft. As the shaft rotates from one position to another, only one bit changes in the code to permit a more stable and precise monitoring of its position. Table 8.5 shows the Gray code number next to the equivalent binary number.

Octal and Hexadecimal Numbers

Some special binary systems are used to simplify the usage of equipment, to aid in identification and location of equipment, or for ease in computations. Two special binary systems are the octal number and the hexadecimal number.

Octal Numbers

Programmable logic controllers (PLCs) use not only ASCII, but also octal numbers. The PLC uses octal numbers to address word and bit locations in its data tables or to address input and outpoint points. As the name implies, octal numbers are a base 8 system—that is, the system consists of eight numbers: 0, 1, 2, 3, 4, 5, 6, and 7. A subscript of 8 following an octal number designates it as such. So, a number such as 571_8 indicates that the number is octal and not decimal.

Binary to octal numbers are converted in two steps. First, combine the binary number in groups of three bits. For example, the binary number 110111101001001111 is regrouped as 110 111 101 001 001 111. Second, the equivalent decimal number for each group is determined. The easiest way to make the conversion is to use a table, such as table 8.6. The first row has the grouped binary numbers; the second row shows the decimal equivalents.

TABLE 8.6

**Binary Number in Groups of Three (top) and
Decimal Equivalents (bottom)**

110	111	101	001	001	111
6	7	5	1	1	7

Table 8.6 shows that the octal number for the binary number 11011110100100111 is 675117_8.

Hexadecimal Numbers

Many electronic display systems, such as computer screens, use hexadecimal numbers to electronically convert binary numbers for displaying characters and objects. The hexadecimal system is a base 16 system—that is, it uses 16 symbols, which are 0, 1, 2, 3, 4, 5, 6, 7, 8, 9, A, B, C, D, E, and F. A hexadecimal (hex, for short) number is indicated with a subscript of 16. For example, the number $42A9D_{16}$ is a hex number. Combining the binary in groups of 4 bits as shown in table 8.7 facilitates converting binary numbers to hex numbers.

As an example of converting a binary number to a hex number, suppose the binary number is 110111101001001111. First, arrange the binary numbers in groups of four, then find the equivalent alphanumeric numbers in the table and write them under the groups, as—

0011	0111	1010	0100	1111
3	7	A	4	F.

So, the hexadecimal equivalent for the binary number 00110111101001001111 is $37A4F_{16}$.

TABLE 8.7
Hexadecimal and Binary Equivalent Numbers and Letters

2^3	2^2	2^1	2^0	
8	4	2	1	Hex Value
0	0	0	0	0
0	0	0	1	1
0	0	1	0	2
0	0	1	1	3
0	1	0	0	4
0	1	0	1	5
0	1	1	0	6
0	1	1	1	7
1	0	0	0	8
1	0	0	1	9
1	0	1	0	A
1	0	1	1	B
1	1	0	0	C
1	1	0	1	D
1	1	1	0	E
1	1	1	1	F

BOOLEAN CONCEPTS

Digital logic involves statements or conditions that are true or false, yes or no, and energized or deenergized. Since digital implies either one state or the other, binary conditions can be established and results obtained with signals, components or other binary devices. Although digital logic is modern, the philosophical concepts behind it have been around a long time.

Classical Logic

Over 2,000 years ago, Greek philosophers studied the concepts of logical reasoning, or logic. One concept that the Greeks arrived at was that all observations could be simplified until a statement about an observation was either true or false. While this concept holds in certain instances, many observations are both true and false and many statements are either true or false.

They also came up with the idea of premises and conclusions—that is, they learned that if a person said or wrote two or more logical statements, or premises, then the person could reach a correct conclusion. For example, consider these statements: if my gas gauge reads empty, and my gas gauge is accurate, then I am out of gas. The statements can be broken down according to classical logic—

> Major premise: My gas gauge reads empty.
> Minor premise: My gas gauge is accurate, and
> Conclusion: I am out of gas.

Note that the conclusion is correct or true only if two conditions are met—that is, both statements must be true.

Logical AND Operation

Digital logic depends on three basic logic operations, or relationships. One logic operation is the logical AND operation, which the example about the gas gauge illustrates. In this case, both the statement regarding the empty reading AND the statement regarding the accuracy of the gauge must be true before the conclusion is true. (Writing the word "and" in all capital letters is the notation used for this type of logic.)

Another example of the logical AND operation is—

If Tom is taller than Bill and
Bill is taller than Dick, then
Tom is taller than Dick.

This statement is true, because if Tom is taller than Bill AND Bill is taller than Dick, then Tom must be taller than Dick.

Logical OR Operation

The logical OR operation also makes it possible to draw a conclusion from two or more statements. In this case, however, if either or both of the statements are true, the conclusion is true. This differs from the AND operation where both statements had to be true for the conclusions to be true. Following is an example of a logical OR operation.

If the sun is shining OR if artificial lights are on, then it must be light.

In this case, the logical OR operation holds that if the sun is shining, it is light OR if artificial lights are on, it is light. In this case, either the sun has to be shining OR the artificial lights have to be on for it to be light. On the other hand, both the sun can be shining and the artificial lights can be on for it also to be light.

Logical NOT Operation

Still another logical operation is the logical NOT operation. This operation means that a true statement is made false by the use of NOT. For example, consider the statement, if it is light, and someone says that it is NOT light, then the second statement is false. In other words, if it is true that it is light, the statement that it is NOT light is false.

Boolean Logic

In 1850, the English philosopher and mathematician George Boole developed a technique of using numerical values to relate to the truth or falsehood of a logical statement. He assigned a mathematical symbol to logical operations. Using this technique, logical statements can be expressed mathematically. For example, Boole's system can show so-called if-then operations as algebraic equations.

An if-then operation (also called an implication) is a statement that asserts that if a certain condition is true, then a certain other condition is also true. Stated algebraically, an if-then operation reads, "If p then q," which means that if p is true then q is also true. Boolean math makes it possible to manipulate equations just as any algebraic expression, with only a few exceptions, can be manipulated. The method of expressing logic as equations is called Boolean logic.

Boolean logic designates or labels each logical statement or premise. For example, the letter A can replace the premise, "Tom is taller than Bill." Similarly, B replaces, "Bill is taller than Dick," and C replaces, "Tom is taller than Dick." Now, the letters can replace the statements as, "if A AND B, then C." This shorthand notation greatly simplifies logical expressions.

The following symbols replace the three basic logic operations—

AND = $\cdot$ or $\times$, multiplication symbols;

OR = +, the addition symbol; and

NOT = a bar placed across the top of the symbol—for example, $\overline{A}$.

In many cases, as with conventional algebra, the AND symbol ($\cdot$ or $\times$) is left out, and two characters are simply written side by side. Using symbols and equations, the example, "If A AND B, then C," can be written in Boolean equation form as $A \cdot B = C$ or $AB = C$. This expression reads as, "A and B equal C."

In the same way, logical OR operations can also be written as a Boolean algebraic equation. For example, the previous logical OR operation, which stated that, if the light is on OR the sun is shining, then it must be light, can broken down as—

Statement A = the sun is shining,

Statement B = the artificial lights are on, and

Statement C = it is light.

The logical expression is written as $A + B = C$ and it reads as, "A or B equal C."

An example of a logical NOT operation is—

Statement A = it is light, and

Statement B = it is not light;

then, $\overline{A}$ = B.

This expression reads as, "not A equals B."

Boole also assigned the numerical symbol 1 to a true statement and 0 to a false statement. These numbered symbols are states or conditions of logical statements and, in themselves, have no numerical value. Thus, to express a true statement in Boolean logic one simply writes $A = 1$. This statement can be spoken as, "A is true." Similarly, to express a false statement, one simply writes $B = 0$. This statement reads, "B is false."

Boolean logic contains many definitions and expressions. This manual does not list them all, but a few include—

$0 \cdot 0 = 0$, which reads, a false statement and false statement are false;

$1 + 1 = 1$, which reads, a true statement or a true statement is a true statement;

$0 \cdot 1 = 1 \cdot 0 = 0$, which reads, a false statement and true statement are the same as a true statement and a false statement and both are false;

$0 + 1 = 1 + 0 = 1$, which reads, a false statement or a true statement is the same as a true statement or a false statement and both are true;

$1 \cdot 1 = 1$, which reads, a true statement and a true statement are true;

$0 + 0 = 0$, which reads, a false statement and a false statement are false;

$\overline{0} = 1$, which reads, a false statement is not a true statement;

$\overline{1} = 0$, which reads, a true statement is not a false statement;

$A \cdot 1 = A$,   which reads, A and a true statement is A;

$A + 0 = 0$,   which reads, A or a false statement is a false statement;

$A \cdot 0 = 0$,   which reads, A and a false statement is a false statement; and

$A + 1 = 1$,   which reads, A or a true statement is a true statement.

At the time Boolean logic was developed, few philosophers were interested in it. However, when the telephone was developed, engineers found that Boolean logic lent itself well to describing the operation of the complex relay switching gear that was necessary. A set of open relay contacts could be described by a logical 0 and closed contacts by a logical 1. Thus, racks of switches could be described by a few equations instead of several pages of electrical drawings. From that time, Boolean logic has been used mathematically to describe everything from simple relay systems to complex computer systems.

Truth Tables

In many cases, it is desirable to know the effect of all possible conditions on a logical statement. For example, consider the statement: If Tom is taller than Bill and Bill is taller than Dick, then Tom is taller than Dick. However, what if the part of the statement that says, "Bill is taller than Dick," is false? To be absolutely sure of the conclusion, it is necessary to list all possible conditions. Other possibilities include—

1. If Tom is taller than Bill and Bill is not taller than Dick, then Tom is not taller than Dick.

2. If Tom is not taller than Bill and Bill is taller than Dick, then Tom is not taller than Dick.

3. If Tom is not taller than Bill and Bill is not taller than Dick, then Tom is not taller than Dick.

Now, label the first statement A, the second B, and the third C. With these labels, the following equations are possible.

1. $A = 0$, $B = 0$

   $C = AB = 0$

This Boolean equation reads, "If A is false and B is false, then C is A and B, which is false."

2. $A = 0$, $B = 1$

   $C = AB = 0$

This equation reads, "If A is false and B is true, then C is A and B, which is false." (If A is false and B is true, C is false, because in a logical AND operation, both premises must be true for the conclusion to be true.)

3. $A = 1$, $B = 0$

   $C = AB = 0$

This equation reads, "If A is true and B is false, then C is A and B, which is false." (Both premises must be true for the conclusion to be true.)

4. $A = 1$, $B = 1$

   $C = AB = 1$

This equation reads, "If A is true and B is true, then C is A and B, which is true. (In this case, both premises in the logical AND operation are true, so the conclusion is true.)

A tabular format called a truth table is a simple method of expressing a logic equation. Table 8.8 is a truth table for the previous example.

Table 8.8 shows four possible combinations for consideration. Since two possible conditions exist for each input statement, the number of possible combinations or considerations for a possible decision or conclusion is 2^N, where N is the number of input conditions. If $C = AB$, the number of possibilities is 2^2, or 4, since A and B are two inputs. The truth table shows these four conditions, which are: (1) if A is false and B false, then C is false; (2) if A is false and B is true, then C is false; (3) if A is true and B is false, then C is false; and (4) is A is true and B is true, then C is true.

If the logic equation is $D = ABC$, which reads, D is either true or false depending on the truth values of A and B and C. (That is, whether D is true or false depends on whether A, B, and C are true or false.) In this case, the possibilities are 2^3, or 8 since A, B, and C are three inputs. Table 8.9 shows the possible combinations. Note that in only one case can D be true.

TABLE 8.8
Truth Table for C = AB

A	B	C
0	0	0
0	1	0
1	0	0
1	1	1

TABLE 8.9
Truth Table for D = ABC

A	B	C	D
0	0	0	0
0	0	1	0
0	1	0	0
0	1	1	0
1	0	0	0
1	1	0	0
1	1	1	0
1	1	1	1

Another truth table (table 8.10) can be drawn from the equation $D = A + B + C$, which reads, D is either true or false depending on the truth values of A or B or C. In this case, because it is a logical OR operation, D can be true in six out of eight possibilities.

Boolean logic has many applications but especially in computer use. And, in every hand-held calculator, digital computer, or digital controller, electronic binary devices produce the equivalent of 1 or 0 signals. By proper manipulation of these signals, engineers can create mathematical computations, display images, and controls.

TABLE 8.10
Truth Table for D = A + B + C

A	B	C	D
0	0	0	0
0	0	1	1
0	1	0	1
0	1	1	1
1	0	0	0
1	0	1	1
1	1	0	1
1	1	1	1

SELF-TEST

8. Advanced Math Concepts

*Multiply each question or problem answered correctly by ten
to arrive at your percentage of competency.*

1. Find the equivalent decimal numbers from the given binary numbers.

   a. 1011 _____

   b. 110111 _____

   c. 001110 _____

   d. 1011011011 _____

   e. 11111111 _____

2. Find the equivalent binary numbers from the given decimal numbers.

   a. 10 _____

   b. 100 _____

   c. 1000 _____

   d. 2456 _____

   e. 256 _____

3. Add the following binary numbers and check your answer with a decimal calculation.

   a.   11011

      + 1011

      _____

   b.   101010

      + 100010

      _____

   c.   10111011

      + 10101010

      _____

4. Subtract the following binary numbers and check your answer with a decimal calculation.

   a.   11011

      −1011

      _____

   b.   101010

      −100010

      _____

223

   c.   10111011

       $\underline{-10101010}$

5. Multiply the following binary numbers and check your answer with a decimal calculation.

   a.   11011

       $\underline{\times 1011}$

   b.   101010

        $\underline{\times 100010}$

   c.   10111011

        $\underline{\times 10101010}$

6. Divide the following binary numbers and check your answer with a decimal calculation.

   a. $1010 \div 101$ _____

   b. $101000 \div 1010$ _____

   c. $10001100 \div 1110$ _____

7. What are the equivalent octal numbers of the following binary numbers?

   a.   1000011110101010 _____

   b.   111011100001100 _____

   c.   1010101010101 _____

   d.   1011000110101110011 _____

8. What are the equivalent hexadecimal numbers of the following binary numbers?

   a.   1000011110101010 _____

   b.   111011100001100 _____

   c.   1010101010101 _____

   d.   1011000110101110011 _____

9. From the ASCII code table (table 8.4), determine the equivalent binary numbers of the following letters and characters.

   a. A _____

   b. Y _____

   c. \ _____

   d. ^ _____

e. ? _____ _____

f. / _____

g. & _____

10. Express the following statements in Boolean format.

a. A and B equals C

_____

b. A or B or C or D equals E

_____

c. Not A and B or C = D

_____

9

Advanced Oil Industry Applications

OBJECTIVES

Upon completion of chapter nine, the student will be able to—

1. Perform an electrical loading analysis of drilling, production, pipeline, and refining facilities.

2. Calculate electrical power factor of a system.

3. Analyze PLC applications using logic numbering systems.

4. Understand how electrical power influences diesel engine power loading.

5. Calculate mud control problems involving mud in the system, mud weighting, cycle time, and annular volume and velocity of mud.

6. Solve well-control problems dealing with hydrostatic pressure, circulating pressure, bottomhole pressure, shut-in drill pipe pressure, maximum allowable surface pressure, and gradients of mud and influx.

7. Determine the ton-miles of service required by a drilling line while making a round trip, making hole, coring, and setting casing.

8. Determine the amount of emulsion-treating chemical needed to use on a lease.

9. Calculate the hourly gas flow through an orifice meter.

10. Determine how to calibrate electronic instruments using basic math principles.

11. Calculate negative buoyancy involved in pipeline construction under water.

12. Convert hydrostatic head to pressure and vice versa in dealing with pumps.

13. Find the pump horsepower needed for an oil pipeline.

14. Determine the locations of pipeline pumping stations and plot a profile of ground elevations along a pipeline route.

15. Find the amount of heat required to raise the temperature of a volume of liquid.

16. Calculate the amount of product from a particular refining process, using known data on feeds and other products.

INTRODUCTION

This chapter gives several equations that are useful to those who work in the petroleum industry. However, those who work in other industries may also find them helpful. Further, be aware that Appendix A lists other industry-related formulas that are used in this text.

For more detailed explanations of electrical power and the power factors, see the Petroleum Extension Service (PETEX) publication entitled *Basic Electronics for the Petroleum Industry*, fourth edition. Other PETEX publications that readers should find helpful concerning the subjects covered in the chapter include *Basic Instrumentation*, fourth edition; *Diesel Engines and Electric Power*, third edition; *Drilling Fluids, Mud Pumps, and Conditioning Equipment*; *Practical Well Control*, fourth edition; *The Blocks and Drilling Line*, third edition; *Treating Oilfield Emulsions*, fourth edition; and *Gas and Liquid Measurement*.

ELECTRICAL POWER AND POWER FACTOR EQUATIONS

In alternating current (AC) electrical systems, the equipment being powered influences the form of power a transformer or generator delivers to the equipment. Also, whether the system is single-phase or three-phase influences wire and equipment sizing.

Three forms of electrical power exist in AC systems. They include—

1. apparent power, in kilovolt amperes (*kVA*);
2. real power, in kilowatts (*kW*); and
3. reactive power, in kilovolt-amperes-reactive (*kVARs*).

Apparent power is the vector sum of the power in watts plus the reactive power in volt-ampere reactive (*VAR*) in a circuit.

Real power is the component of apparent power that represents true work. Real power is expressed in watts and equals volt-amperes multiplied by the power factor.

Reactive power is the value of the power in an electric circuit obtained by multiplying the effective value of the current in amperes, the effective value of the voltage in volts, and the sine of the angular phase difference between current and voltage.

In addition, a ratio of real power to apparent power is a measure of how effective power is being delivered and used. This ratio is referred to as the power factor and is expressed in equation form as:

$$pf = \frac{kW}{kVA} = \frac{W}{VA}$$

where

$$
\begin{aligned}
pf &= \text{power factor} \\
kW &= \text{kilowatts} \\
kVA &= \text{kilovolt amperes} \\
W &= \text{watts} \\
VA &= \text{volt-amperes.}
\end{aligned}
$$

Apparent power is the result of multiplying voltage by current. Apparent power is also considered to be the total power. On the other hand, the capacitors and inductors in the load determine reactive power. Therefore, the power actually doing work is the real power, in *kW*, which is subtracted from the apparent power by the amount of reactive power.

A right triangle—a power triangle—shows the three forms of power (fig. 9.1). Side *a* of the triangle represents the amount of *kW*, side *b* represents the amount of *kVARs* and the hypotenuse (side *c*) represents total power, or *kVA*.

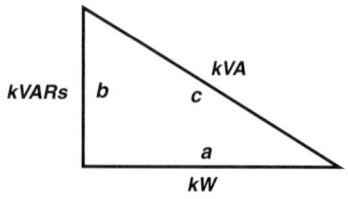

Figure 9.1   Power triangle

Recall the Pythagorean theorem concerning right triangles, which states that the sum of the square of the sides (a and b) is equal to the hypotenuse (c) squared. The equation is—

$$a^2 + b^2 = c^2. \tag{Eq. 1}$$

Because power calculations often involve the value of the hypotenuse (c), it is desirable to solve for c. The equation is—

$$c = \overline{a^2 + b^2}. \tag{Eq. 2}$$

Now, refer to the power triangle in figure 9.1. The power components can also be put in equation form as—

$$kVA = \overline{kW^2 + kVARs^2}. \tag{Eq. 3}$$

Several equations are available for solving problems involving the three forms of power. They are termed power equations and, for single-phase AC power, include—

$$VA_1 = V \times I \tag{Eq. 4}$$

$$kVA_1 = \frac{V \times I}{1,000} \tag{Eq. 5}$$

$$W_1 = VA_1 \times pf \tag{Eq. 6}$$

$$kW_1 = kVA_1 \times pf \tag{Eq. 7}$$

$$kVA = \overline{kW^2 + kVARs^2} \tag{Eq. 8}$$

$$kW_1 = \overline{kVA^2 - kVARs^2} \tag{Eq. 9}$$

$$kVARs_1 = \overline{kVA^2 - kW^2} \tag{Eq. 10}$$

where

V = line-line rms AC voltage, volts
I = line current, amperes
kVA_1 = single-phase apparent power, kilovolt amperes
W_1 = single-phase real power, watts
VA_1 = single-phase apparent power, volt-amperes
pf = power factor
kW_1 = single-phase real power, kilowatts
$VARs_1$ = single-phase reactive power, volt-amperes-reactive, or vars
$kVARs_1$ = single-phase reactive power, kilovolt-amperes-reactive, or kilovars.

The power equations for three-phase power include—

$$VA_3 = V \times I \times \overline{3} \tag{Eq. 11}$$

$$kVA_3 = \frac{V \times I \times \overline{3}}{1,000} \tag{Eq. 12}$$

$$W_3 = VA_3 \times pf \tag{Eq. 13}$$

$$kW_3 = kVA_3 \times pf \tag{Eq. 14}$$

$$kVA_3 = \overline{kW^2 + kVARs^2} \tag{Eq. 15}$$

$$kW_3 = \overline{kVA^2 - kVARs^2} \tag{Eq. 16}$$

$$kVARs_3 = \overline{kVA^2 - kW^2} \tag{Eq. 17}$$

where

VA_3 = three-phase apparent power, volt-amperes
kVA_3 = three-phase apparent power, kilovolt amperes
W_3 = three-phase real power, watts
kW_3 = three-phase real power, kilowatts
$kVARs_3$ = three-phase reactive power, kilovolt-amperes-reactive, or kilovars.

Electrical Loading Analysis

To successfully select the correct sizes and capacities of components in an electrical system, an accurate electrical loading analysis must be made. Such an analysis not only permits proper sizing of equipment, but also anticipates future additions to the system.

When the voltage is constant throughout the system, it simplifies the analysis because only current values in amperes are required to size and select the components. While voltage plays a role in equipment sizing and selection, the amount of current dictates the physical size of wiring, circuit breakers, and other equipment.

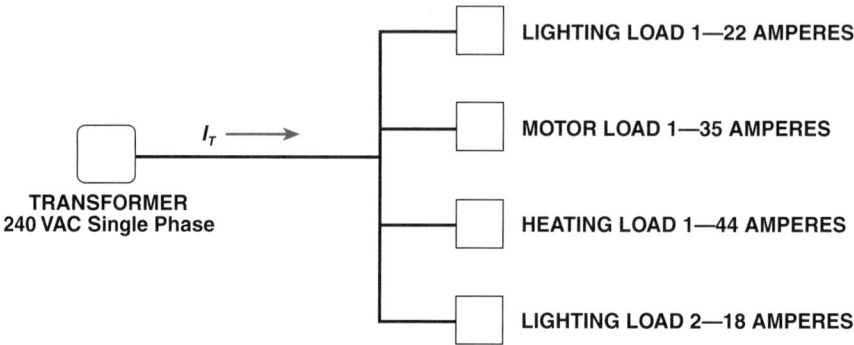

Figure 9.2   Diagram of a simple circuit

Example Problem: Figure 9.2 diagrams a simple circuit with lights, a motor, and a heater. A 240-volt, alternating current (VAC), single-phase transformer sends current to each of the components in the system. The figure gives the amount of load, in amperes, that each component carries. Determine the total amperes and *kVA* that the transformer must deliver to the indicated loads (all loads have a *pf* equal to 1.0).

Solution:

First, add the amperes at each load to arrive at a total amperage (I_T). In this case—

$$I_T = 22 + 35 + 44 + 18 = 119 \text{ amperes.}$$

Next, calculate the total load *kVA* with the equation—

$$kVA_1 = \frac{V \times I}{1,000}.$$

Thus,

$$kVA_1 = \frac{240 \times 119}{1,000}$$

$$kVA_1 = 28.56.$$

Analyzing three-phase power systems is similar to analyzing single-phase systems; however, the square root of three is factored in the equations. The square root of 3 accounts for the three voltages and currents and the phase displaced by 120 electrical degrees from each other in developing power.

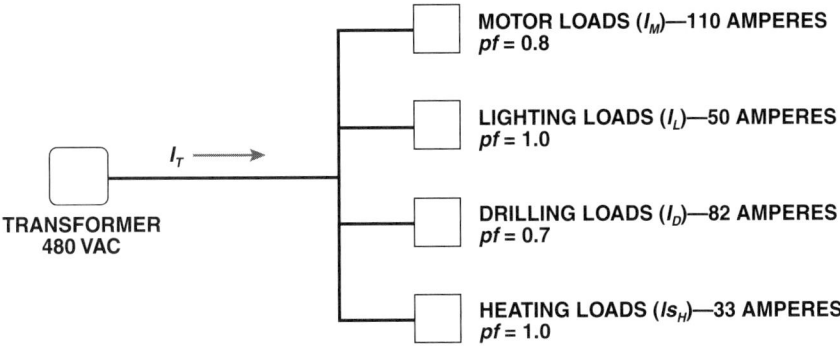

Figure 9.3   Diagram of a three-phase system

Example Problem: Figure 9.3 diagrams a three-phase, 480-VAC system with a transformer and four loads. The loads are motor loads (I_M), lighting loads (I_L), drilling loads (I_D), and heating loads (I_H). The power factor (*pf*) for the motor loads is 0.8, for the lighting loads 1.0, for the drilling loads 0.7, and for the heating loads 1.0. Using the diagram, calculate the total amperes, *kVA*, and *kW* of the system.

Solution:

Determine the amperes (*I*), *kVA*, and *kW* of each of the loads. For the motor loads—

$$I_M = 110 \text{ amperes}$$
$$kVA_M = \frac{480 \times 110 \times \overline{3}}{1,000}$$
$$kVA_M = 91.45$$
$$kW_M = kVA_M \times pf$$
$$= 91.45 \times 0.8$$
$$kW_M = 73.16.$$

For the lighting loads—

$$I_L = 50 \text{ amperes}$$
$$kVA_L = \frac{480 \times 50 \times \overline{3}}{1,000}$$
$$kVA_L = 41.57$$
$$kW_L = kVA_L \times pf$$
$$= 41.57 \times 1.0$$
$$kW_L = 41.57.$$

For the drilling loads—

$$I_D = 82 \text{ amperes}$$
$$kVA_D = \frac{480 \times 82 \times \overline{3}}{1,000}$$
$$kW_D = kVA \times pf$$
$$= 68.17 \times 0.7$$
$$kW_D = 47.72.$$

For the heating loads—

$$I_H = 33 \text{ amperes}$$

$$kVA_H = \frac{480 \times 33 \times \overline{3}}{1,000}$$

$$kW_H = kVA_H \times pf$$
$$= 27.43 \times 1.0$$
$$kW_H = 27.43.$$

Since the power factor (pf) varies between loads, the amperes cannot be added together directly. Instead, add all of the loads in kW together for total loads in kW (kW_T) and then add all of the kVA loads together for a total kVA (kVA_T). The system power factor (pf) and total amperes (I_T) can then be calculated with the following operations.

$$kW_T = kW_M + kW_L + kW_D + kW_D$$
$$kW_T = 73.16 + 41.57 + 47.42 + 27.43$$
$$kW_T = 189.58$$
$$kVA_T = kVA_M + kVA_L + kVA_D + kVA_H$$
$$= 91.45 + 41.57 + 68.17 + 27.43$$
$$kVA_T = 228.62$$

$$pf = \frac{kW_T}{kVA_T}$$

$$= \frac{189.58}{228.62}$$

$$pf = 0.83$$

By rearranging the kVA_3 equation (eq. 12), the total current (I_T) can be obtained—

$$I_T = \frac{kVA \times 1,000}{V \times \overline{3}}$$

$$= \frac{228.62 \times 1,000}{480 \times 1.732}$$

$$I_T = 275.$$

Electrical Power Factor Concepts

The concept of an electrical power factor developed from the fact that the load affects the amount of current required to accomplish a certain task. Remember: (1) real power in kW, accomplishes the work and (2) real power (kW) is a component of the total power, kVA.

The power factor can be obtained mathematically from the power triangle (see fig. 9.1) with one of two equations.

$$pf = \frac{kW}{kVA} \qquad \qquad \text{(Eq. 18)}$$

$$pf = \cos \theta = \frac{kW}{kVA} \qquad \qquad \text{(Eq. 19)}$$

where

$$pf = \text{power factor}$$
$$kW = \text{kilowatts}$$
$$kVA = \text{kilovolt amperes}$$

θ = the angle between the *kW* vector and the *kVA* vector in the power triangle. It is also the same angle between the current and voltage sinusoidal waveforms to a particular electrical load.

Resistive components such as lamps and heaters result in the current being in phase with, or in harmony with, the voltage as it progresses through its cycle. The voltage's effort produces a corresponding current effort—that is, *kVA* is the same as the *kW*. Therefore, for loads that are in phase, the power factor is 1.0.

Other electrical components do not produce equal effort. For example, a pure coil of wire (with theoretically no resistance) has the property of inductance, causing the current to lag the voltage by 90°. Consequently, no real power is produced from the source to the load and the power circulates between the source and the load. In this case, the power factor is zero.

Another electrical component in a circuit, a capacitor, causes the current to lead the voltage by 90°. Circulating power between the source and load results in no effective net power being delivered and the power factor is also zero.

In actual circuits, combinations of these three components (resistive, capacitive, and inductive) occur. For example, three-phase motor loads consisting of coils and wire resistance produce a lagging power factor in the neighborhood of 0.8. Fluorescent lighting usually contains capacitive ballasts that create a leading power factor of about 0.9.

In large systems, a large number of AC motors create significant current levels because of a poor power factor. Power utilities usually penalize companies if a company's power factor is less than 0.9. Consequently, the company may sometimes need to make a correction to reduce a power factor penalty. They can improve the power factor by using leading power factor components such as capacitors to offset the lagging power factor of inductive motor loads.

Example Problem: A three-phase power system has several AC motors that produce an apparent power of 1,000 *kVA* at 480 volts and 1,203 amperes. The 0.8 power factor of the motor creates a penalty on each monthly bill. What size capacitor, in *kVA*, is required to produce a 0.9 power factor and eliminate the penalty?

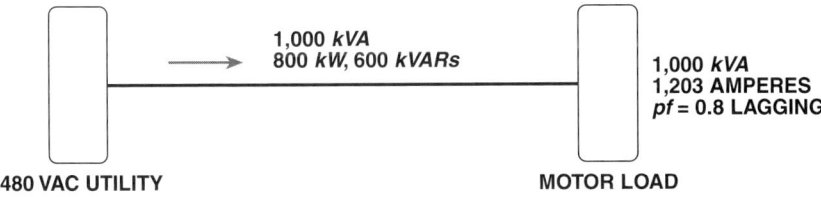

Figure 9.4   Diagram of a 480-VAC facility and load

Solution: Figure 9.4 diagrams the system. A 480-volt AC (VAC) utility supplies power at 1,000 *kVA*, 800 *kW*, and 600 *kVARs* to a motor load of 1,000 *kVA* at 1,203 amperes that has a lagging *pf* of 0.8. First, calculate *kW*—

$$kW = kVA \times pf$$
$$= 1,000 \times 0.8$$
$$kW = 800$$

Next, calculate *kVARs*—

$$kVARs = \overline{kVA^2 - kW^2}$$
$$= \overline{1,000^2 - 800^2}$$
$$kVARs = 600.$$

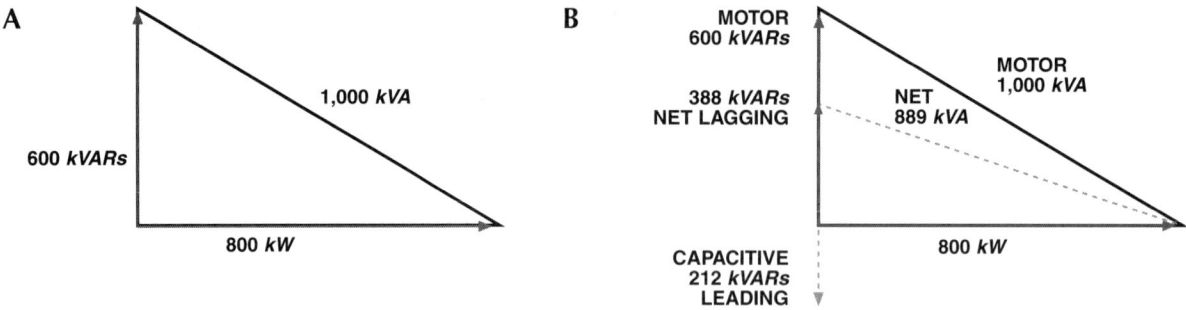

Figure 9.5   Power triangles: (A) system values and (B) calculated values

Next, refer to figure 9.5, which shows the power triangle and the example system values.

To produce a power factor of 0.9, calculate how much *kVA* must be reduced by using the formula—

$$kVA = \frac{kW}{pf}.$$

Plug in the values, which, in this example, are 800 *kW* and the desired power factor (*pf*) of 0.9. The result is—

$$= \frac{800}{0.9}$$

$$kVA = 889.$$

kVA must be reduced to 889.

To achieve this reduction, the net *kVARs* must also be reduced. To determine how much, use the formula—

$$kVARs = \overline{kVA^2 - kW^2}.$$

Then, plug in the values—

$$kVARs = \overline{889^2 - 800^2}$$
$$kVARs = 388.$$

Finally, subtract 388 from 600 to determine the reduction in *kVARs* to 212. Two hundred-twelve *kVARs* represent the amount of capacitive *kVARs* required to reduce the net *kVARs* in the system. The power triangle in figure 9.6 shows the concept. Figure 9.6 also diagrams the modified system that corrects the power factor to 0.9.

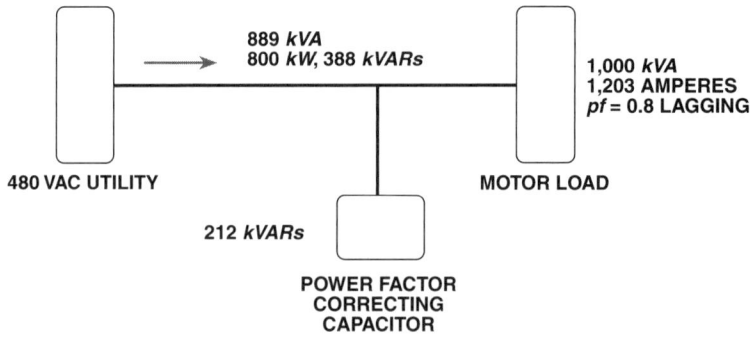

Figure 9.6   Diagram of a modified system

Diesel Engine-Gas Turbine Power Analysis

Diesel engines and gas turbines often provide mechanical power to electrical AC generators. These AC generators, in turn, supply electrical power to other loads such as motors, lighting, heaters, and production equipment.

Several important concepts relate electrical power to mechanical power and these concepts help us understand the effects electrical loads have on the prime mover, which may be a diesel engine or a gas turbine. One concept is: as electrical loads increase in a facility, the facility applies greater loads to the prime mover, even if the speed is constant. This additional loading is torque.

Both mechanical and electrical power can be expressed in kilowatts (kW). In the U.S., however, the mechanical power that motors produce is expressed in horsepower. Two equations express the relationship between the two power measurements—

$$hp = \frac{kW}{0.746} \qquad\qquad \text{(Eq. 20)}$$

or

$$kW = hp \times 0.746. \qquad\qquad \text{(Eq. 21)}$$

Example Problem: An electric motor is rated at 3.5 kW; what is its horsepower (hp) rating?

Solution: Plug in the values using equation 20—

$$hp = \frac{3.5}{0.746}$$
$$hp = 4.69.$$

Example Problem: A diesel engine is rated at 500 hp; what is its power rating in kW?

Solution: Plug in the values using equation 21—

$$kW = 500 \times 0.746$$
$$kW = 373.$$

When analyzing loads on a prime mover, an important consideration is the efficiency of the equipment in the system that lies between the prime mover and the driven equipment. Such equipment includes generators and transformers. The equation that defines efficiency is—

$$\text{Efficiency} = \frac{\text{Power Out}}{\text{Power In}} \times 100\%. \qquad \text{(Eq. 22)}$$

Example Problem: What is the efficiency of an AC generator if the electrical power out is 700 kW and the mechanical power in is 1,000 hp?

Solution: First, convert 1,000 hp to kW. In this case, it is—

$$kW = 1,000 \times 0.746$$
$$kW = 746.$$

Next, determine efficiency with equation 22, where power in is 746 kW and power out is 700 kW. Thus—

$$\text{Efficiency} = 700/746 \times 100\%$$
$$\text{Efficiency} = 93.8\%.$$

Example Problem: A power system using a turbine as its prime mover drives an alternator. What is the power required from the turbine if the electrical load is three-phase, 4,160 VAC, 2,000 *kVA* at 0.85 *pf* and the efficiency of the alternator is 95%?

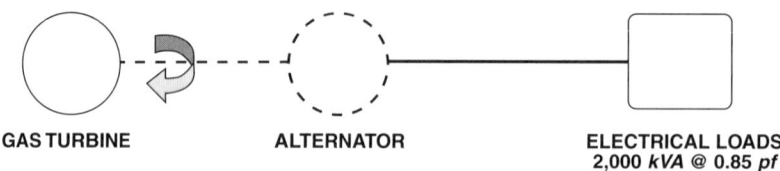

GAS TURBINE   ALTERNATOR   ELECTRICAL LOADS
                                                       2,000 *kVA* @ 0.85 *pf*

Figure 9.7   Diagram of example system

Solution: First, refer to figure 9.7, which sketches the system in the problem. Then, determine the real power, in *kW*, at the electrical load, using the equation—

$$kW = kVA \times pf$$
$$= 2,000 \times 0.85$$
$$kW = 1,700.$$

The amount of load on the turbine's output is 2,000 *kVA* and 1,700 *kW*.

Bear in mind, however, that the turbine produces only real power, or *kW*, not *kVA*. The component of power that is of value in determining loading of the turbine is the *kW*. Therefore, determine the power input to the generator by taking its output in *kW* and dividing it by the alternator's efficiency—

$$\text{Alternator Power Input} = \frac{1,700}{0.95} = 1,789 \; kW.$$

The turbine's output loading is the same value as the alternator's input requirements and is 1,789 *kW*. To express this value in horsepower—

$$hp = \frac{kW}{0.746}$$
$$= \frac{1,789}{0.746}$$
$$hp = 2,398.$$

Programmable Logic Controllers

Programmable logic controllers (PLCs) are electronic devices that control, measure, compare, switch, and communicate with a variety of human-machine interfaces. PLCs are made up of digital devices such as microprocessors, memory chips, and comparators. PLCs perform multiple functions, including math.

PLCs are rugged and flexible and do not require a great deal of maintenance. Moreover, they are relatively small and can perform small and large tasks. A PLC interfaces readily with field devices such as switches, controllers, positioners, process transmitters, solenoid valves, and lamps. PLCs can add, subtract, multiply, divide, derive square roots, and make comparisons.

One industrial use of PLCs is in process control, where PLCs control such process variables as liquid level, temperature, and pressure. In such control systems, process transmitters measure a process over a range from a low value to an upper value. A process transmitter produces a linear, or analog, 4-to-20 milliamp (mA), electronic signal. However, if the transmitter is hooked up to a PLC input, a special electronic device is required to convert this analog signal from the transmitter to digital form. Conversion from analog to digital is necessary because the microprocessor within the PLC can only sense and process digital signals. So, an analog to digital converter (ADC, or A/D converter) is required.

Process transmitters are analog devices because their signals are continuous with an infinite number of points between any two limits. When calibrated, for example, a pressure transmitter operating between 0 and 200 psi produces a current output signal of 4 to 20 mA. The input pressure and the output current are analog (or analogous) signals of an event that is taking place.

As stated earlier, because the PLC processor is a digital computer that can only recognize 1s and 0s, an A/D converter is required to convert the analog mA signal into a group of 1s and 0s that correspond to the pressure applied. This interface between analog current and digital processor must be capable of translating the analog signal into discrete, or digital, values that the processor can interpret.

Analog input modules are used in PLC systems to perform the conversion from analog signals to digital values (fig. 9.8). Circuitry in the module senses the range of analog signals (typically, 4 to 20 mA), converts it to an analog voltage of 1 to 5 volts DC, and produces a multiple-bit digital signal whose digital value corresponds to the analog input level. The A/D converts the analog signal to a digital signal.

In figure 9.8, note that the circuit converts the 4-to-20 mA current signal into analog voltage through a 250-ohm resistor. This analog voltage has a normal range of 1 to 5 volts DC and 0 to 5 volts DC if the signal wire breaks or becomes disconnected. Usually, a 4-, 8-, or 12-bit converter changes the analog

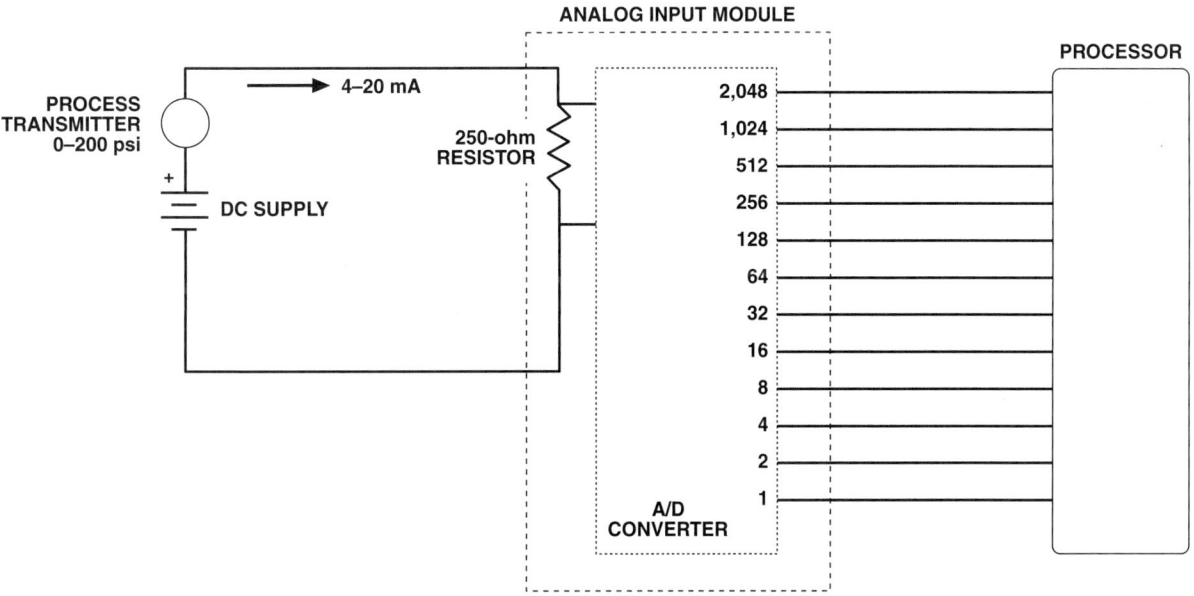

Figure 9.8   A/D converter in a PLC

signal voltage to digital signal voltage. The number of bits in the converter depends on the resolution and accuracy desired.

For example, a 12-bit A/D converter changes the analog signal range into 2^{12}, or 4,096, increments. In this case, the resolution is 1 part in 4,096 parts. Stated as a percentage, the 12-bit A/D converter reproduces the analog signal in digital form within 0.025% of its value. When overall accuracy of a measuring system is considered, a transmitter with 1% accuracy has only a 0.025% error added to its value through the PLC. This percentage is so small as to be negligible.

If we assume that the input signal covers the range of 0 to 20 mA (0 to 5 volts DC), the A/D converter begins with a decimal value of 0 and ends with a value of 4,095 by producing a combination of twelve 1s and 0s. The processor stores the digital values in a register for further operations within the PLC's circuitry.

Figure 9.8 diagrams a typical process pressure transmitter that has been calibrated over the range of 0 to 200 psi to produce an output signal of 4 to 20 mA. Its output signal is connected to an analog input module of the PLC where the signal is converted to a voltage of 1 to 5 volts DC.

Using a 12-bit analog input module, the digital output signal is produced prior to being delivered to the processor. Each bit output of the A/D converter is weighted in binary format beginning with 1 at the first bit followed by 2, 4, 8, 16, etc. and ending with 2,048 at the twelfth bit. Since a linear relationship exists at each point in the system, a table (table 9.1) can be developed that scales the system at different points in the process using selected values of pressure.

Note that the pressure of –50 in table 9.1 does not really exist; instead, it represents a value of 0 mA, which indicates a broken signal wire. The binary values are sent to the processor in block format, used by the software in the processor as a value, and inserted in the ladder logic program.

On the right side of figure 9.9 are typical binary and decimal values for a pressure of 150 psi.

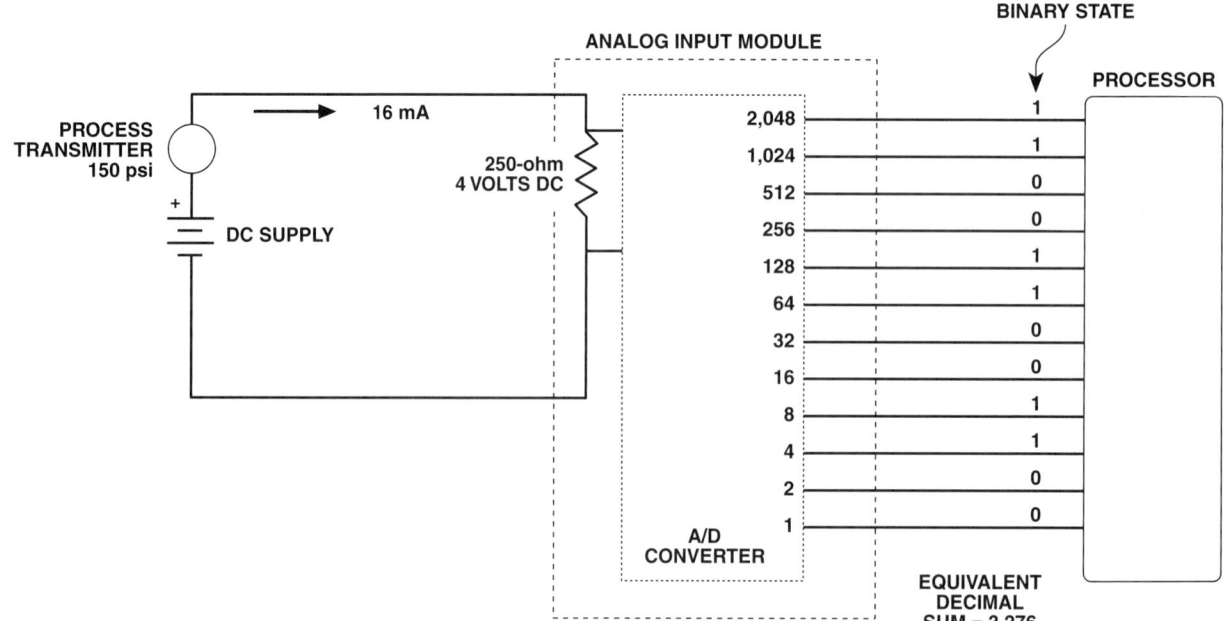

Figure 9.9   A/D converter example

TABLE 9.1
Scale of Values

Pressure	−50	0	100	150	200
mA	0	4	12	16	20
Volts DC	0	1	3	4	5
Decimal Value	0	819	2,457	3,276	4,095
Binary Value	000000000000	001100110011	100110011001	110011001100	111111111111

MATHEMATICS IN DRILLING OPERATIONS

Mathematics is important to drilling operations because many routine operations require working with equations. This section is designed to assist rig crewmembers by discussing several equations that cover calculations they will encounter on the job. The section also solves example problems to improve comprehension.

Torque, Speed, Power

Drilling rigs require prime movers—engines that produce power for all the rig's needs. While turbine engines power a few rigs, diesel engines power most of them. Whether diesel or turbine, engines provide mechanical power. Mechanical power either directly drives equipment or it drives electrical generators, which produce electrical power. In either case, the output shaft of the engine turns, or rotates, at a given speed and produces torque. Engine power is usually expressed in horsepower (*hp*) or kilowatts (*kW*).

The speed with which the engine's drive shaft turns is rotational motion and this motion is usually measured in revolutions per minute (rpm). That is, rpm is the rate of shaft rotation. The symbol for speed in mathematical equations is the letter N, which is also used to represent turns, or revolutions per unit of time. Engine speed varies from hundreds of rpm in diesel engines to thousands of rpm in gas turbines.

The rotational power of the engine shaft is called torque because torque is the force that causes rotation, twisting, or turning. Torque (T) is often expressed as a force-distance, such a foot•pounds (ft•lb). In this case, the distance is in feet and the force is in pounds. Figure 9.10 sketches an engine, a rod, and a scale that measures force. The large arrow around the engine's shaft represents rotation, or torque. The iron rod has a length (l) and the engine produces a force (f) in lb. Multiplying length in ft times force in lb results in torque. The equation is—

$$T = l \times f \qquad \text{(Eq. 23)}$$

where

T = torque, ft•lb

l = length, ft

f = force, lb.

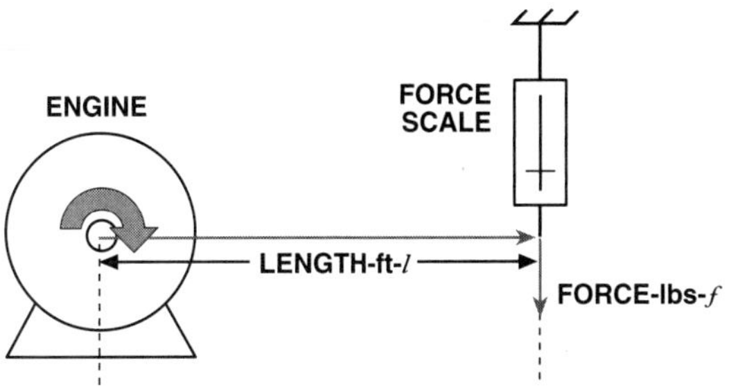

Figure 9.10 *Diagram demonstrating engine torque*

Example Problem: Referring to figure 9.10, what is the torque, in ft•lb, the engine's shaft produces if the length of the rod is 3 ft and the force scale displays a force of 1,000 lb?

Solution: Use equation 23 ($T = l \times f$) and plug the given values into the equation. Thus—

$$T = 3 \times 1,000$$
$$T = 3,000 \text{ ft•lb.}$$

Example Problem: Refer to figure 9.10 again. Assume the rod is ½- (0.5-) ft long and the force is 6,000 lb. How much torque is the engine producing?

Solution:

$$T = 0.5 \times 6,000$$
$$T = 3,000 \text{ ft•lb.}$$

Example Problem: How much ft•lb of torque is the engine producing if the rod's length is 10 ft and the force is 300 lb?

Solution:

$$T = 10 \times 300$$
$$T = 3,000 \text{ ft•lb.}$$

In the last three examples, note that the length of the rod and the force at the end of the rod determine the torque. That is, when the rod is 3 ft long and the weight-force is 1,000 lb, the engine produces 3,000 ft•lb of torque. Also, when the rod is 0.5 ft long and the weight-force is 6,000 lb, the torque is 3,000 ft•lb. And, when the rod is 10 ft long and the force is 300 lb, the torque is also 3,000 ft•lb. The point is that the length of the rod and the weight-force determine the torque.

When an engine rotates at a certain speed, or rpm, it also delivers rotational torque at the shaft. Put another way, the engine delivers rpm and torque at the same time. The combination means that the engine delivers work at a certain rate. This rate can be expressed in ft•lb/second (s), *hp*, or *kW*. When an engine produces 550 ft•lb/s it also produces 1 *hp*.

The equation to express engine *hp* is—

$$hp = \frac{T \times N}{k} \qquad \text{(Eq. 24)}$$

where

T = torque, ft•lb, newton•meters (N•m), etc.
N = speed, rpm
k = constant (its value depends on units used in equation); for ft•lb, it is 5,252.

Example Problem: An engine rotates at 1,800 rpm and produces 5,000 ft•lb of torque. How much horsepower does it produce?

Solution: For the conventional units given, the constant, k, is 5,252. Plug the values into equation 24, which is—

$$hp = \frac{T \times N}{k}$$

$$= \frac{5,000 \times 1,800}{5,252}$$

$$= \frac{9,000,000}{5,252}$$

$$hp = 1,714.$$

Where kilowatts (*kW*) measure rotational power, the equation—

$$kW = hp \times 0.746 \qquad \text{(Eq. 25)}$$

converts *hp* to *kW*. Where you need to convert *kW* to *hp*, use the equation—

$$hp = \frac{kW}{0.746}. \qquad \text{(Eq. 26)}$$

Example Problem: How many *kW* does a 1,000-*hp* engine develop?

Solution: Plug the value into equation 25. Thus—

$$kW = 1,000 \times 0.746$$
$$kW = 746.$$

Example Problem: An engine develops 926 *kW*. How much *hp* does it develop?

Solution: Plug the value into equation 26. Thus—

$$hp = \frac{926}{0.746}$$

$$hp = 1,241.$$

Optimization

Most rigs today use diesel engines to drive alternators, which produce alternating current. (Generators produce direct current.) Large electrical cables send this current to special electronic gear that, in some cases, converts (rectifies) the AC to DC. DC is needed because many rigs use DC motors. In other cases, the gear simply transmits AC current to AC motors. Whether AC or DC, these motors are located on or near the rig component that requires power, such as the large hoist (the drawworks) that raises and lowers the drill stem and other components into and out of the hole.

When a mechanical engine, such as a diesel engine, drives an electrical alternator to produce electricity, power is being converted in several different forms. For example, chemical energy (fuel) is ignited in the engine and converted to thermal energy, which, in turn, expands to create forces within the engine to produce rotation and torque at its shaft. This mechanical power, expressed in *hp* or *kW*, drives the alternator to produce electrical power expressed as *kW*. Figure 9.11 diagrams the conversion of mechanical power in *hp* or *kW* into electrical power in *kW*, *kVA*, or *kVARs* by an alternator. The electrical energy, in turn, drives electrical loads.

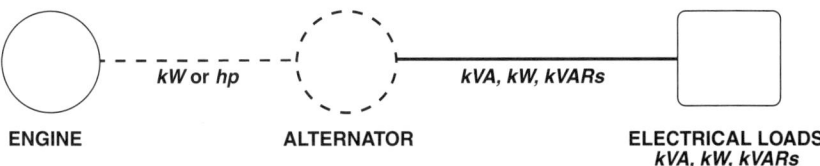

ENGINE   **ALTERNATOR**   **ELECTRICAL LOADS**
                                                                kVA, kW, kVARs

Figure 9.11   An engine produces mechanical power, which an alternator converts to electrical power.

Example Problem: Figure 9.12 diagrams an engine-alternator set whose electrical loads combine to produce 1,000 *kVA*, 800 *kW*, and 600 *kVARs* at a power factor of 0.8. The alternator is 95% efficient. How much power is the engine required to deliver?

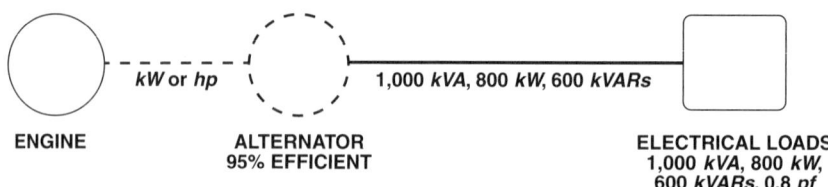

Figure 9.12   Power loading example

Solution: While the alternator can deliver three forms of electrical power in terms of *kVA*, *kW*, or *kVARs*, the engine can only deliver one type of power—*kW* or *hp*. Thus, the *kW* required by the electrical load can be transferred to the engine shaft by considering the efficiency of the alternator. Put another way, *kW*-mechanical is the same as *kW*-electrical because both are the same units of power, or the rate of doing work. So, to convert the 800 *kW* put out by the alternator into *kW* at the engine shaft, alternator efficiency must be considered. The equation is—

$$kW_{\text{engine}} = \frac{kW_{\text{alternator}}}{\text{efficiency}}. \qquad \text{(Eq. 27)}$$

In this case,

$$kW_{\text{engine}} = \frac{800}{0.95}$$

$$kW_{\text{engine}} = 841 \ kW.$$

To convert 841 *kW* to *hp*, simply divide 841 by 0.746 to obtain 1,100 *hp*. So, the engine must develop 841 *kW* or 1,100 *hp* to handle the loads.

Mud Control

Two concerns of rig crewmembers are the volume of drilling mud in the entire system and the time it takes for the mud to make an entire cycle from the pump's suction, to the bottom of the hole, and back to the pump's suction. Usually, crewmembers must add weighting material and treatment materials to the mud at a rate that allows the mud to make at least one complete cycle during the treatment. In this manner, the materials are evenly distributed throughout the system. To determine the mud's cycling time, crewmembers must determine the volume of mud in the hole and in the active pits, or tanks, on the surface and the output rate of the mud pump.

Amount of Mud in the System

To determine the amount of mud in the system, use the equation—

$$V = V_p + V_h \qquad \text{(Eq. 28)}$$

where

V = volume of mud in system, bbl

V_p = volume of mud in pits, bbl

V_h = volume of mud in hole, bbl.

To determine the volume of mud in the hole use the equation—

$$V_h = d^2 \qquad \text{(Eq. 29)}$$

where

V_h = volume of mud in hole, bbl/1,000 ft
d = average diameter of hole, in.

Note that equation 29 gives an approximate amount of volume. Because a hole is a cylinder, for an exact volume, the formula for determining the volume of a cylinder ($V = \pi r^2 h$) should be used. However, for field purposes, equation 29 is adequate.

To determine the volume of mud in the active pits, use the equation—

$$V_p = \frac{l \times w \times h}{5.6} \qquad \text{(Eq. 30)}$$

where

V_p = volume of mud in pit, bbl
l = length of pit, ft
w = width of pit, ft
h = height, or depth, of mud in pit, ft
5.6 = number of ft^3/bbl.

Example Problem: A rig has five active pits in the mud system. Three of the pits are 22 ft long by 15 ft wide and the depth of mud in each is 10 ft. One of the pits is 15 ft long by 12 ft wide, and the mud's depth is 10 ft. Finally, the last pit is 18 ft long by 15 ft wide and the mud's depth is 10 ft. How many bbl of mud are in the pits?

Solution: First, determine the volume of mud in ft^3 for each pit and then the volume in bbl, using equation 30. In this case, the volume of mud in the three largest pits is 22 × 15 × 10 ÷ 5.6 = 589 bbl × 3 = 1,767 bbl total. The volume of the 15 × 12 pit is 15 × 12 × 10 ÷ 5.6 = 321 bbl. The volume of the 18 × 15 pit is 18 × 15 × 10 ÷ 5.6 = 482 bbl. Finally, add all the volumes to determine that the total volume of mud in all the pits is 2,570 bbl.

Example Problem: A hole is 4,000 ft deep. It is lined from the surface to a depth of 500 ft with 17½-in. casing. Below the surface casing, 2,500 ft of 12¼-in. intermediate casing lines the hole. The remainder of the hole is being drilled with a 9⅝-in. bit. The rig is using four active mud pits, three of which hold 600 bbl each. The remaining pit holds 350 bbl. Find the approximate amount of mud in the system.

Solution: First, determine the average diameter of the hole. In this case, it is 17½ + 12¼ + 9⅝ = 17.5 + 12.25 + 9.625 = 39.375 ÷ 3 = 13.125, which rounds to 13. Now use equation 29, which is—

$V_h = d^2$
$V_h = 13^2$
$ = 169$ bbl/1,000 ft
$ = 169 \times 4$
$V_h = 676$ bbl of mud in hole.

Then, add the volume of mud in the pits to the volume of mud in the hole. In this case, it is $3 \times 600 + 350 = 2{,}150$ bbl. So, the total volume in the system is $2{,}150 + 676 = 2{,}826$, or about 2,825 total bbl of mud in the system.

Mud Cycling Time

The equation for determining the mud's cycling time is—

$$t = \frac{V}{pS} \qquad \text{(Eq. 31)}$$

where

t = cycling time, min
V = volume of mud in the system, bbl
p = volume of mud per pump stroke, bbl
S = pumping rate, strokes per minute (spm).

Example Problem: The total volume of mud in a system is 2,825 bbl and the pump moves 0.5 bbl of mud with each stroke. The pumping rate is 120 spm. How long does it take for the mud to complete one cycle?

Solution: Plug the values in to equation 31—

$$t = \frac{2{,}825}{0.5 \times 120}$$

$$t = 47 \text{ min.}$$

Incidentally, you can find how much mud a pump puts out per stroke from tables provided by the pump manufacturer. Output depends on the type of pump, its liner and piston size, and the speed at which it is pumping mud.

Weighting Up Mud

Another important task rig crewmembers perform is weighting up the mud. Many minerals are available for this job; however, barite is one of the more common ones. An equation is available that tells crewmembers how many sacks of barite to add to the system per 100 bbl of mud. Note that the equation is for barite and mud weight in pounds per gallon (ppg). The equation is—

$$B = \frac{1{,}470\,(W_2 - W_1)}{35 - W_2} \qquad \text{(Eq. 32)}$$

where

B = sacks of barite per 100 bbl of mud
W_1 = initial mud weight, ppg
W_2 = desired mud weight, ppg.

Table 9.2 is also a handy way of determining values of B. The table not only gives the number of sacks of barite to weight up the mud, but also shows how many barrels of water to add to 100 barrels of mud to reduce its weight. The upper right part of the table is for mud-weight increases and the lower right is for mud-weight reductions.

Example Problem: Suppose 900 bbl of mud are in the system and that the mud weighs 12-ppg. How many sacks of barite are necessary to add to increase the mud weight to 14.5 ppg?

TABLE 9.2
Lightening and Weighting Drilling Mud

Initial Mud Weight (ppg)	Desired Mud Weight (ppg)																	
	9.5	10.0	10.5	11.0	11.5	12.0	12.5	13.0	13.5	14.0	14.5	15.0	15.5	16.0	16.5	17.0	17.5	18.0
9	29	59	90	123	156	192	229	268	308	350	395	442	490	542	596	653	714	778
9.5		29	60	92	125	160	196	234	273	315	359	405	452	503	557	612	672	735
10	43	29	30	61	93	128	164	201	239	280	323	368	414	464	516	571	630	691
10.5	85	30		31	62	96	131	167	205	245	287	331	376	426	479	531	588	648
11	128	60	23		31	64	98	134	171	210	251	294	339	387	437	490	546	605
11.5	171	90	46	19		32	66	101	137	175	215	258	301	348	397	449	504	562
12	214	120	69	37	16		33	67	103	140	179	221	263	310	357	408	462	518
12.5	256	150	92	56	32	14		34	68	105	144	184	226	271	318	367	420	475
13	299	180	115	75	48	27	12		34	70	108	147	188	232	278	327	378	432
13.5	342	210	138	94	63	41	24	11		35	72	111	150	194	238	286	336	389
14	385	240	161	112	76	54	36	21	10		36	74	113	155	199	245	294	345
14.5	427	270	185	131	95	68	48	32	19	9		37	75	116	159	204	252	302
15	470	300	208	150	110	82	60	43	29	18	8		37	77	119	163	210	259
15.5	513	330	231	169	126	95	72	54	39	26	16	8		39	79	122	168	216
16	556	360	254	187	142	109	84	64	48	35	24	15	7		40	81	126	172
16.5	598	390	277	206	158	123	96	75	58	44	32	23	14	7		41	84	129
17	641	420	300	225	174	136	108	86	68	53	40	30	21	13	6		42	86
17.5	684	450	323	244	189	150	120	96	77	62	49	38	28	20	12	6		43
18	726	480	346	262	205	163	132	107	87	71	57	45	35	26	18	12	5	

The lower half of the table shows the number of barrels of water that must be added to 100 bbl of mud to produce desired weight reductions. To use this portion of the table, locate the initial mud weight in the vertical column at the left, then locate the desired mud weight in the upper horizontal row. The number of barrels of water to be added per 100 bbl of mud is read directly across from the initial weight and directly below the desired mud weight. For example, to reduce an 11 ppg mud to a 9.5 ppg mud, 128 bbl of water must be added for every 100 bbl of mud in the system.

The upper right half of this table shows the number of sacks of barite that must be added to 100 bbl of mud to produce desired weight increases. To use this portion of the table, locate the initial mud weight in the vertical column to the left, then locate the desired mud weight in the upper horizontal row. The number of sacks of barite to be added per 100 bbl of mud is read directly across from the initial weight and directly below the desired mud weight. For example, to raise an 11 ppg mud to 14.5 ppg, 251 sacks of barite must be added per 100 bbl of mud in the system.

Solution: Solve this problem using equation 32 or table 9.2. From table 9.2, note that 179 sacks are needed to increase 100 bbl of mud from 12 ppg to 14.5 ppg. So, for a 900-bbl system, $B = 179 \times 9 = 1,611$ sacks of barite to be added. Using equation 32, plug the values into it and solve it. Thus,

$$B = \frac{1,470\,(14.5 - 12.0)}{35 - 14.5}$$

$$= \frac{1,470 \times 2.5}{20.5}$$

$$= \frac{3,675}{20.5}$$

$$B = 179.$$

Again, the equation shows that for a 900-bbl system, 1,611 sacks of barite are needed because 179 sacks per 100 bbl in a 900-bbl system is $179 \times 9 = 1,611$ sacks.

Reducing Mud Weight

Besides increasing the weight of mud, crewmembers may also be required to reduce the mud weight by adding water to the system. As mentioned earlier, table 9.2 can be used to determine the bbl of water to add to reduce the mud weight by a given amount. For example, if the initial mud weight is 12.5 ppg and the desired mud weight is 11.0 ppg, then 56 bbl of water should be added per 100 bbl of mud. Thus, in a 900 bbl system, 9×56 bbl, or 504 bbl of water should be added. An equation is also available. It is—

$$x = \frac{100\,(W_1 - W_2)}{W_2 - 8.33} \tag{Eq. 33}$$

where

x = water to be added, bbl/100 bbl of mud
W_1 = initial mud weight, ppg
W_2 = desired mud weight, ppg.

Example Problem: The initial mud weight is 12.5 ppg and its weight should be reduced to 12.2 ppg. The system contains 1,475 bbl of mud. How much water should be added to obtain the desired weight?

Solution: First, plug the values into equation 33. Thus—

$$x = \frac{100\,(12.5 - 12.2)}{12.1 - 8.33}$$

$$= \frac{100\,(0.3)}{3.77}$$

$$= \frac{30}{3.77}$$

$$x = 7.9 \text{ or } 8 \text{ bbl/100 bbl mud.}$$

Finally, $8 \times 14.75 = 118$ bbl of water.

Weight Increase with Bentonite

Although barite is often used as a weighting material, virtually any solid substance added to the mud increases its weight, or density. One material commonly used with fresh water drilling fluids it's bentonite. Bentonite gives mud many desirable properties—for example, it provides viscosity, stabilizes the hole (keeps it from caving in), and provides gel strength, which gives mud the ability to carry cuttings up the hole and suspend them when pumping stops. Because it is important for crewmembers to be aware of the mud's weight and because bentonite is commonly added to the drilling fluid, the following equation is useful. It is—

$$B = \frac{874\,(W_2 - W_1)}{20.8 - W_2} \qquad \text{(Eq. 34)}$$

where

B = sacks of bentonite per 100 bbl of mud
W_1 = initial mud weight, ppg
W_2 = desired mud weight, ppg.

Example Problem: A freshwater drilling fluid weighs 8.33 ppg. The total amount of mud in the system is 1,877 bbl. Bentonite is to be added to the system to give the water desirable drilling properties. However, the mud's weight should not exceed 8.5 ppg. What is the maximum number of sacks of bentonite that can be added to this system to achieve 8.5-ppg mud?

Solution: First, use equation 34 and plug in the values. Thus—

$$B = \frac{874\,(8.5 - 8.33)}{20.8 - 8.5}$$

$$= \frac{874\,(0.17)}{12.3}$$

$$= \frac{148.58}{12.3}$$

$$B = 12.07, \text{ or } 12.$$

Then, because the system contains 1,877 bbl, multiply 12 by 18.77 to determine that 225 sacks of bentonite can be added without exceeding a mud weight of 8.5 ppg.

Mud Volume Increase

Rig crewmembers also need to be aware that when they add solid materials such as bentonite and barite to the mud, the mud volume increases. To understand this phenomenon, imagine a glass half full of water. Put a solid object in the glass, such as a steel block, and the level of water rises. Put another way, the solid object displaces an amount of water equal to its weight, or density. Equation 35 states this phenomenon mathematically and can be used to determine the amount that the mud volume increases when solid material is added to the system.

$$V = \frac{100\,(W_2 - W_1)}{N - W_2} \qquad \text{(Eq. 35)}$$

where

V = volume increase, bbl/100 bbl of mud
W_1 = initial mud weight, ppg
W_2 = final mud weight, ppg
N = weight of solid material, ppg. (For barite, N = 35.0; for bento-nite, N = 20.8.)

Example Problem: Suppose that a 12.0-ppg mud is weighted up to 14.5 ppg with barite. The current volume of the mud is 1,500 bbl. How much does the mud volume increase by the addition of barite?

Solution: Use equation 35 and plug in the values. Thus—

$$V = \frac{100\ (14.5 - 12.0)}{35.0 - 14.5}$$

$$= \frac{100 \times 2.5}{20.5}$$

$$= \frac{250}{20.5}$$

$$V = 12.2 \text{ bbl/100 bbl of mud.}$$

Then, because the system contains 1,500 bbl of mud, multiply 12.2 × 15 = 183 bbl. The weighting material (barite in this case) increases the mud volume from 1,500 bbl to 1,500 + 183 = 1,683 bbl.

Annular Volume

To determine the amount of mud in the hole, the volume of mud in the annulus must be known. Annular volume depends on the size and displacement of the drill stem. That is, the drill stem elements, such as the drill pipe and drill collars, take up space in the hole and must be taken into account when determining annular volume. The equation is—

$$V_a = V_h - D_{dp} \qquad\qquad \text{(Eq. 36)}$$

where

V_a = annular volume, bbl/ft
V_h = volume of mud in hole (open hole capacity), bbl/ft
D_{dp} = capacity of drill stem and displacement of drill stem, bbl/ft.

To find the capacity and displacement of the drill stem elements, several sources are available. One is Halliburton Energy Service's *Cementing Tables* (also known as "The Red Book"), which is a pocket-size manual that lists capacities and displacements for virtually all oilfield tubular goods. Contact Halliburton at www.halliburton.com. Another source is Éditions Technip's *Drilling Data Handbook*, which is available in English and French. Contact Éditions Technip at www.editions technip.com.

Annular Velocity

Another important factor when dealing with the mud system is the speed, or velocity, of the mud as it flows up the annulus after leaving the drill bit. Annular velocity plays a role in well control because velocity affects bottomhole pressure, which is of paramount importance in well control. The equation for annular velocity is—

$$v_a = \frac{P_o}{V_a} \qquad \text{(Eq. 37)}$$

where

v_a = annular velocity, feet per minute (ft/min)
P_o = pump output, bbl/min
V_a = annular volume of mud, bbl/ft.

Remember that you can find pump output by referring to the manufacturer's tables that accompany the pumps on your rig. Because many tables refer to pump output in gallons per minute (gpm), rather than bbl/min, an equation is available for determining annular velocity gpm. It is—

$$v_a = \frac{P_o \times 0.024}{V_a} \qquad \text{(Eq. 38)}$$

where

v_a = annular velocity, ft/min
P_o = pump output, gpm
V_a = annular volume of mud, bbl/ft.

Well Control

Well control and mud control are closely related because the hydrostatic pressure of the drilling mud counteracts the pressure of formation fluids. When the formation pressure increases faster than the hydrostatic pressure or the hydrostatic pressure is reduced, a kick occurs. If prompt action is not taken to control the kick, or kill the well, a blowout may occur. For more details about well control, see the Petroleum Extension Service (PETEX) publications entitled *Introduction to Well Control* and *Practical Well Control* (www.utexas.edu/cee/petex).

Hydrostatic Pressure

As mentioned earlier, the equation for finding hydrostatic pressure is—

$$hp = MW \times TVD \times C \qquad \text{(Eq. 39)}$$

where

hp = hydrostatic pressure; psig or kPa
MW = mud weight; ppg, pcf, or kg/m^3
TVD = true vertical depth of hole, ft or m
C = a constant.

Note that the value of the constant C is 0.052 for ppg, 0.00695 for pcf, and 0.009806 for kg/m^3).

Example Problem: A well has a TVD of 3,426.3 m and the hole is full of mud that weighs 1,332.8 kg/m^3. What is the hydrostatic pressure at the bottom of the hole?

Solution: Solve for hydrostatic pressure, using equation 39.

$$
\begin{aligned}
hp &= MW \times TVD \times 0.009806 \\
&= 1{,}332.8 \times 3{,}426.3 \times 0.009806 \\
hp &= 47{,}779.8 \text{ kPa.}
\end{aligned}
$$

Initial Circulating Pressure

Part of a well-killing operation is removing the intruded fluids from the well and pumping kill-weight mud into the well. In one method, kill-weight mud is pumped down the drill stem while the intruded kick fluids in the annulus are pumped out. In this case, the circulating pressure that occurs when pumping first starts is different from the circulating pressure that occurs when the drill stem is full of kill-weight mud. That is, initial circulating pressure is different from final circulating pressure. The equation for determining the initial circulating pressure as kill-weight mud is pumped down the drill stem is—

$$ICP = KRP + SIDPP \qquad \text{(Eq. 40)}$$

where

ICP = initial circulating pressure
KRP = kill-rate pressure
$SIDPP$ = shut-in drill pipe pressure.

Example Problem: A pump develops 900 psi kill-rate pressure (KRP) when it is run at the speed required to pump a kick from the well. The well kicked and the shut-in drill pipe pressure ($SIDPP$) is 250 psi. What is the initial circulating pressure (ICP)?

Solution: Plug the values into equation 40. Thus—

ICP = 900 + 250
ICP = 1,150 psi.

Example Problem: If KRP is 800 psi and ICP is 1,000 psi, what is $SIDPP$?

Solution: Use equation 40 and transpose it to solve for $SIDPP$.

$SIDPP$ = $ICP - KRP$
       = 1,000 − 800
$SIDPP$ = 200 psi.

Final Circulating Pressure

When killing a well using the wait-and-weight method of well control, the technique involves pumping kill-weight mud down the drill stem to the bottom of the well. Once the heavier kill-weight mud (the new mud) fills the drill stem and displaces the mud in use when the kick occurred (the old mud), the pump pressure drops from an initial circulating pressure (ICP) to a final circulating pressure (FCP). To determine FCP, use equation 41.

$$FCP = KRP \times W_2 \div W_1 \qquad \text{(Eq. 41)}$$

where

FCP = final circulating pressure
KRP = kill-rate pressure
W_2 = new mud weight
W_1 = old mud weight.

Bottomhole Pressure

When a well is completely shut in on a kick, pressure appears on the drill pipe pressure gauge. This pressure is shut-in drill pipe pressure ($SIDPP$). The drill stem can be thought of as a gauge with a long stem on it—that is, the gauge is on the surface, at the top of the drill stem and the drill stem extends to the bottom of the well. Consequently, the amount of pressure on the drill pipe pressure gauge, along with hydrostatic pressure (hp) indicates the amount of bottomhole pressure. The equation is—

$$BHP = HP + SIDPP \qquad \text{(Eq. 42)}$$

where

BHP = bottomhole pressure
HP = hydrostatic pressure
$SIDPP$ = shut-in drill pipe pressure.

Maximum Allowable Surface Pressure

One concern when killing a well is the pressure a gas kick develops as it is circulated up the hole and out of the well. As gas rises, it expands because the hydrostatic pressure on it lessens as the gas nears the surface. As the gas expands in the annulus, the pressure increases. This pressure increase is shown on the annular pressure gauge (the casing pressure gauge) at the surface. Consequently, shut-in casing pressure ($SICP$) is allowed to increase as the kick is circulated out of the well. However, if $SICP$ increases too much, it can fracture the formation at its weakest point, which is virtually always at the bottom of the shallowest casing shoe. Should the formation fracture at the bottom of the casing, kick fluids can flow into the formation and an underground blowout can occur. Thus, it is imperative to know the maximum amount of surface pressure that can be allowed. The equation is—

$$MASP = (FG - MG) \times D_{cs} \qquad \text{(Eq. 43)}$$

where

$MASP$ = maximum allowable surface pressure, psi
FG = fracture gradient, psi/ft
MG = mud gradient, psi/ft
D_{cs} = casing seat depth, ft.

Normally, a formation fracture gradient is determined by a leak-off test. A leak-off test involves pumping mud into the well after the casing string is cemented in place. Pressure is increased gradually until a sharp drop in pressure is noted, which indicates that the formation has fractured. This fracture pressure is recorded and used to determine the fracture gradient (FG).

Mud Weight Increase

When a well kicks, it is often because the mud weight in use when the well kicked was not heavy enough and therefore did not develop enough hydrostatic pressure on bottom to balance or overcome formation pressure. Thus, it is important to know how much the mud needs to be weighted up to balance formation pressure and kill the well. The equation is—

$$W_2 = W_1 + WI_{SIDPP} + WI_{OP} \qquad \text{(Eq. 44)}$$

where

W_2 = kill, or new, mud weight, ppg
W_1 = original, or old, mud weight, ppg
WI_{SIDPP} = mud weight increase required to equal formation pressure
WI_{OP} = mud weight increase required for overbalance pressure.

Note that the mud weight is usually increased a few points (decimal points) above that weight needed to merely balance formation pressure. For example, if the weight needed to equal formation pressure is 11.2 ppg, then the mud may actually be weighted up to 11.4 ppg.

To determine the mud weight increase required, two equations are available. They are—

$$WI_{SIDPP} = SIDPP \div (TVD \times C) \qquad \text{(Eq. 45)}$$
$$WI_{OP} = OP \div TVD \div C \qquad \text{(Eq. 46)}$$

where

WI_{SIDPP} = mud weight increase to balance formation pressure, ppg
WI_{OP} = mud weight increase required for overbalance pressure, ppg
$SIDPP$ = shut-in drill pipe pressure, psi
TVD = true vertical depth, ft
OP = overbalance pressure, psi
C = constant, 0.052.

Mud Pressure Gradient

The more a mud weighs—the denser it is—the more hydrostatic pressure it develops. It develops this pressure in terms of a gradient. A pressure gradient is a measure of the change in pressure as it relates to the depth. Thus, the pressure gradient of a mud increases with well depth. Because it is important to know the pressure gradient of a mud of a given weight, crewmembers can use the following equation to determine it—

$$MG = MW \times C \qquad \text{(Eq. 47)}$$

where

MG = mud pressure gradient, psi/ft or kPa/m
MW = mud weight, ppg, pcf, or kg/m^3
C = constant, which is 0.052 when using ft, ppg, and psi; 0.00695 when using ft, pcf, and psi; and 0.009806 when using m, kg/m^3, and kPa.

Pressure Gradient of Intruded Fluids

While a well can be successfully killed without knowing the pressure gradient of the intruded fluids (the influx), it is interesting to note that an equation exists for determining it. Critical to the accuracy of equation 48 is the height of the influx (IH), which can be difficult to determine. Therefore, use it with caution.

$$IG = MG - \frac{MASP - SIDPP}{IH} \qquad \text{(Eq. 48)}$$

where

IG = influx gradient, psi/ft
MG = mud gradient, psi/ft
$MASP$ = maximum allowable surface pressure, psi
$SIDPP$ = shut-in drill pipe pressure, psi
IH = influx height, ft.

Equation 49 determines influx height. However, it assumes that the influx consists of a single fluid, such as methane or salt water. In reality, kicks are usually a mixture of several formation fluids, whose composition is difficult or impractical to determine. Nevertheless, it is interesting to note.

$$IH = \frac{pg}{V_a}$$ (Eq. 49)

where

IH = influx height, ft
pg = pit gain, bbl
V_a = annular volume, or capacity, bbl/ft.

Note that annular volume, or capacity, can be found by using charts provided in Halliburton's *Cementing Tables* or Édition Technip's *Drilling Data Handbook*.

Pump Pressure

When killing kicks, crewmembers usually use a pump speed slower than normal to circulate the kick out of the well and pump heavier mud into it. As pump speed changes, so does pump pressure. Put another way, pump pressure is directly related to the pump rate. Equation 50 can be used to determine the pump's pressure when the old pump pressure and rate are known. The equation is—

$$P_2 = P_1(S_2 \div S_1)^2$$ (Eq. 50)

where

P_2 = new pump pressure, psi
P_1 = old pump pressure, psi
S_2 = new pump speed, strokes per minute (SPM)
S_1 = old pump speed, SPM.

Drilling Line Service

Drilling line is strung between the crown and traveling blocks. As the traveling block is raised and lowered, tools such as drill pipe, drill collars, and casing are lowered into the hole and raised from it. The drilling line moves through and over sheaves (pulleys) in the blocks and this movement creates wear on the line. This wear is measured in ton-miles or, in the case of the SI system, in megajoules. After the line works for a given number of ton-miles or megajoules, crewmembers must move, or slip, the line so that the wear points on the line are changed. Slipping the line extends the service life of the line and ensures that it performs as it should.

Because much of the work the line does involves the weight of the drill pipe and the drill string, it is important to know their weights. Moreover, because the drill stem elements are usually in a hole that is filled with mud, you must know the weight of the pipe and collars in mud. Just as a ship floats in water, so do drill pipe and drill collars float in a mud-filled hole. Put another way, drill pipe and drill collars weigh less when they are in a hole full of mud than they do when they are in the air.

Figures 9.13 and 9.14 plot the weight of pipe and collars against the weight of the mud in which they float. To use the charts, locate the line representing the drill pipe or collars being used and find the point on that line directly above the mud weight being used. Reading across to the left axis from this point gives the effective weight of the drill pipe (W_m) or collars (E_c) in the given mud. These

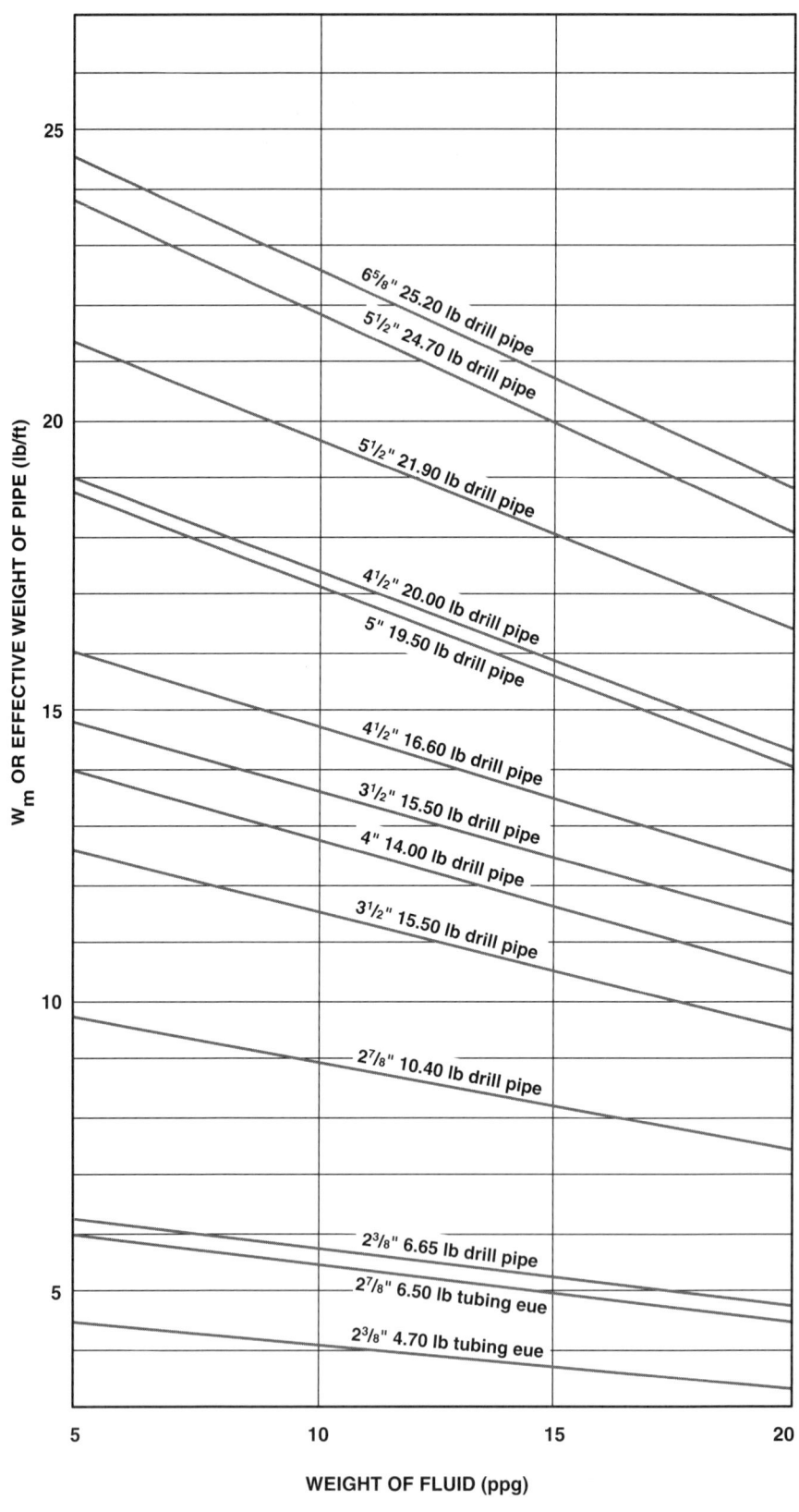

Figure 9.13   Effective weight of pipe in drilling mud

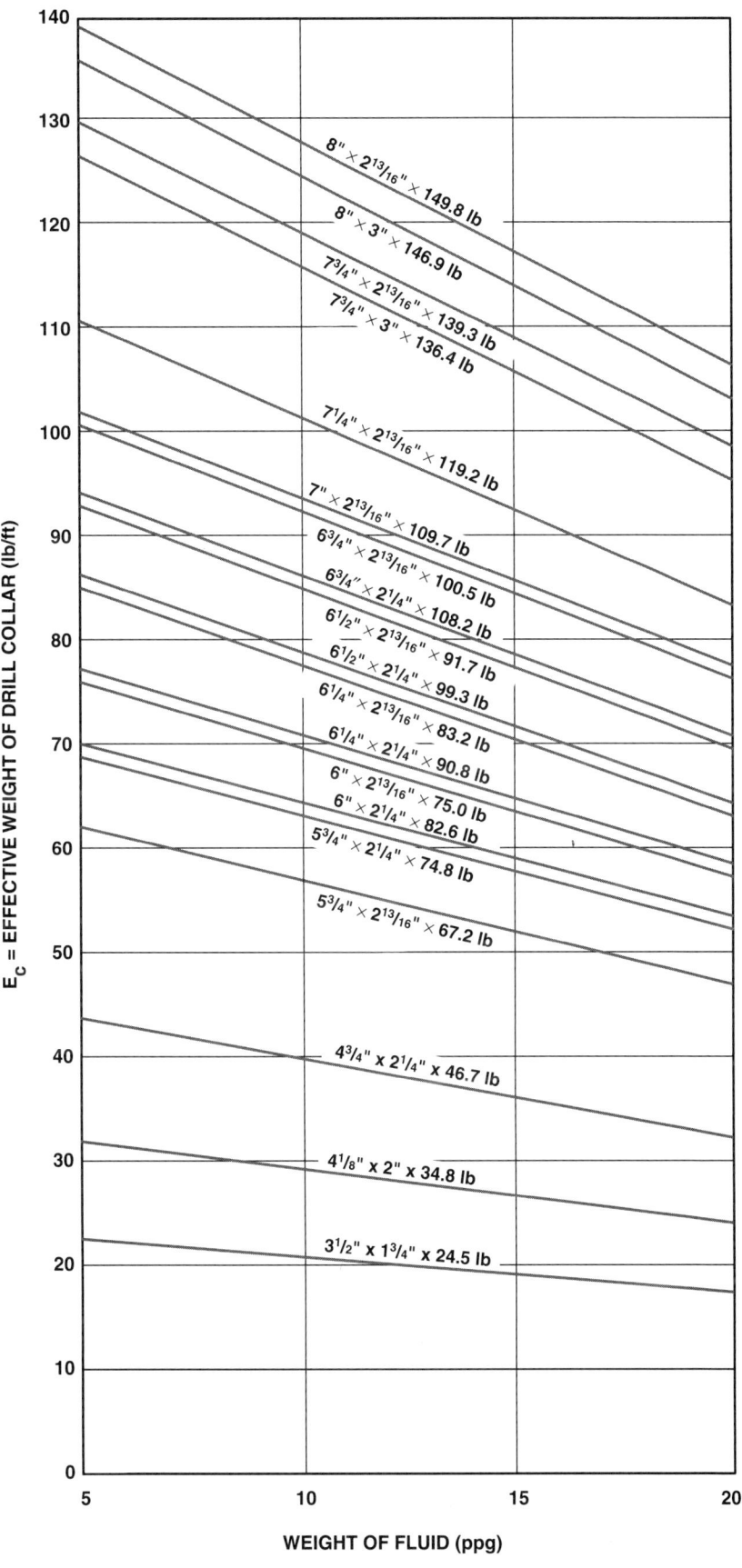

Figure 9.14   Effective weight of drill collars in drilling mud

values are used in equation 51 to determine the ton-miles of use on a drilling line. The equation is—

$$T_r = \frac{W_m D(L_s + D) + 2D(2M + E_c)}{10,560,000} \qquad \text{(Eq. 51)}$$

where

T_r = drilling line use, ton-miles, which is the weight in tons moved times the distance moved in miles

D = depth of hole, ft

L_s = average length of a drill pipe stand, ft

W_m = effective weight of drill pipe in mud, pounds per foot (lb/ft)

M = total weight of traveling block-elevator assembly, lb

E_c = effective weight of drill collar assembly in mud minus the effective weight of the same length of drill pipe in mud, lb.

Example Problem: A rig crew trips the drill stem out of the hole, puts on a new bit, and then trips the stem back in (they make a round trip). The total depth of hole is 5,000 ft. The length of the drill collar string is 200 ft and the collars have an outside diameter (OD) of 5¾ in. and an inside diameter (ID) of 2¼ in. (stated as 5¾" × 2¼" in fig. 9.14). The drill pipe string is made up of 4½-in. drill pipe that weighs 16.6 lb/ft. The average length of a stand of drill pipe (L_s) is 100 ft. The weight of the traveling block-elevator assembly (M) is 10,720 lb. The mud weight is 10.5 ppg. Find the ton-miles of wear on the drilling line.

Solution: First, use figures 9.13 and 9.14 to determine W_m and E_c. Then, write down the values for D, L_s, and M. Finally, plug the values into equation 51. Thus—

W_m = 14.5 lb/ft

D = 5,000 ft

L_s = 100 ft (average value)

M = 10,720 lb

E_c = (60.7 − 14.5)200

E_c = 9,240 lb

$$T_r = \frac{14.5 \times 5,000 \times (5,000 + 100) + (2 \times 5,000) \times (21,440 + 9,240)}{10,560,000}$$

T_r = 64 ton-miles.

Wear on the drilling line not only occurs while tripping, but also when drilling. Equation 52 determines wear in ton-miles in terms of the amount of hole drilled in ft. The equation is—

$$T_d = 3(T_2 - T_1) \qquad \text{(Eq. 52)}$$

where

T_d = ton-miles drilling

T_1 = ton-miles for one round trip at depth D_1 (where drilling began)

T_2 = ton-miles for one round trip at depth D_2 (when drilling stopped).

Example Problem: Find the number of ton-miles of service when drilling from 11,000 to 11,500 ft with the following data—

500 ft of 5¾ × 2¼ in. drill collars

5-in. 19.5 lb/ft drill pipe

M = 24,450 lb

mud weight, 12.2 ppg.

Solution: First, using equation 52, find T_2 (service from a round trip at 11,500 ft) and T_1 (service from a round trip at 11,000 ft).

$$T_2 = 328 \text{ ton-miles}$$
$$T_1 = 304 \text{ ton-miles.}$$

Then substitute these values in equation 51. Thus—

$$T_d = 3 \times (328 - 304)$$
$$T_d = 72 \text{ ton-miles.}$$

Coring operations require a different equation from that used for drilling operations. Core barrels do not put as much wear on the line as drilling operations. So, the equation for coring operations is—

$$T_c = 2(T_4 - T_3) \qquad\qquad \text{(Eq. 53)}$$

where

T_c = ton-miles coring

T_4 = ton-miles for one round trip at depth D_4 (where coring stopped)

T_3 = ton-miles for one round trip at depth D_3 (where coring began).

Example Problem: Given the same statistics as in the previous problem, find the number of ton-miles of service when coring from 11,000 ft to 11,500 ft instead of drilling.

Solution: Use equation 53. T_4 in this equation is the same at 11,500 ft as T_2 is in the previous problem, and T_3 at 11,000 ft is identical to T_1 in the previous problem. Thus—

$$T_c = 2 \times (328 - 304)$$
$$T_c = 48 \text{ ton-miles.}$$

Running casing puts more work on the drilling line because casing strings are usually much heavier than drill pipe and drill collars. The equation for determining wear when running casing is—

$$T_s = \frac{D\,(L_{cs} + D)\,W_{cm} + 4DM}{21,120,000} \qquad\qquad \text{(Eq. 54)}$$

where

T_s = ton-miles setting casing

L_{cs} = length of joint of casing, ft

W_{cm} = effective weight of casing in mud, lb/ft (may be estimated from drill pipe data given in figure 9.13).

Example Problem: Find the number of ton-miles of service on the line when setting 6-in. casing that weighs 20 lb/ft at 11,500 ft. M = 24,450 lb and the mud weight is 12.2 ppg.

Solution: Use equation 54 and rearrange it. Thus—

$$T_s = \frac{DW_{cm}(D + L_{cs}) + 4DM}{21,120,000}$$

Estimating from figure 9.13—

W_{cm} = 16.7 lb/ft

L_{cs} averages about 30 ft.

Then—

$$T_s = \frac{11,500 \times 16.7 \times (11,500 + 30) + (4 \times 11,500 \times 24,450)}{21,120,000}$$

T_s = 158 ton-miles.

Practice Problems

1. A well is 6,000 ft deep, 8½ in. in diameter, and is filled with mud. The mud tanks and surface lines contain 800 bbl of mud. How many 100-lb sacks of barite would be needed to raise the weight of the mud from 11 ppg to12.5 ppg?

2. In the well in problem 1, how much water would have to be added to bring the weight from 12.5 ppg down to 12 ppg? Use table 9.2 for this calculation.

3. A diesel engine is driving a 93% efficient alternator that is delivering 440 *kVA* to the rig's electrical loads. The load's power factor is 0.7. How much *hp* is required for the engine to deliver?

4. A 10,000-ft hole contains 800 ft of 6½- × 2¹³⁄₁₆-in. drill collars and a string of 5-in. OD, 19.5 lb/ft drill pipe. The total weight of the traveling block-elevator assembly is 15,000 lb. The average length of a stand of drill pipe is 100 ft, and the mud weight is 15 ppg. What is the ton-miles of wear put on the drilling line by a round trip?

5. Using the statistics in problem 4, find the ton-miles of wear on the drilling line when it drills from 10,000 ft to 12,000 ft.

6. Solve for hydrostatic pressure when the mud weight is 13.5 ppg and the true vertical depth is 6,000 ft.

7. Find the mud weight when the hydrostatic pressure is 3,245 psi and the true vertical depth is 5,340 ft.

8. Find the initial circulating pressure if the kill rate pressure is 1,125 psi and the shut-in drill pipe pressure is 200 psi.

_____

9. Find the kill mud weight for a well with the following data:

   Shut-in drill pipe pressure = 310 psi
   True vertical depth = 7,000 ft
   Old mud weight = 9.5 psi
   Overbalance pressure = 200 psi.

_____

10. What is the mud pressure gradient for 10-ppg mud?

_____

11. A pump has a pressure of 2,500 psi when its speed is 50 strokes per minute. If the speed is reduced to 25 spm, what is its pressure?

_____

12. What is the electrical power factor if a drill rig generator delivers 200 amperes at 480 VAC, three phase, 130 kW?

_____

MATHEMATICS IN PRODUCTION OPERATIONS

Productivity Index

The ability of reservoir pressure to move fluids into the wellbore at a given flowing pressure depends on drawdown. Drawdown is the difference, or differential, between a well's shut-in and flowing pressure. That is, drawdown is the difference between shut-in pressure and flowing pressure. Drawdown data is used to plot a well's ability to produce—its deliverability. From a well's deliverability, engineers can determine its productivity index (PI) or its inflow performance ratio (IPR). The productivity index shows the relationship between the well's pressure drawdown and its daily production. The equation is—

$$PI = \frac{PR}{(SIBHP - FBHP)} \qquad \text{(Eq. 55)}$$

where

$\quad\quad PI$ = productivity index, in barrels per day (bpd) per psi
$\quad\quad PR$ = production, bpd
$\quad SIBHP$ = initial shut-in bottomhole pressure, psi
$\quad FBHP$ = flowing bottomhole pressure, psi.

Pump Horsepower

Oilfield gathering systems use many types of pumps to move produced fluids from one point to another on a lease. One common type of pump is a positive-displacement pump called a duplex, or double-acting, pump. A duplex pump has two pistons that move back and forth inside two cylinders. They deliver fluid on both sides of the stroke—that is, as the piston goes forward, it delivers fluid, and, as it goes back, it delivers fluid. Equation 56 provides a close estimate of the *hp* needed to drive a duplex pump.

$$hp = (R \times P) \div 48{,}000 \qquad \text{(Eq. 56)}$$

where

hp = horsepower required to drive pump
R = rate of flow, bpd
P = delivery pressure, psig
48,000 = a constant.

Example Problem: A duplex pump is designed to deliver 5,000 bpd of oil at a pressure of 120 psi. How much horsepower is required to drive the pump?

Solution: Use equation 56 and substitute known values.

$$hp = (5{,}000 \times 120) \div 48{,}000 = 12.5 \ hp.$$

Oil Recovery

Oil exists in the pores or fractures of buried formations, or reservoirs. Almost always, it exists in rock openings with other fluids, such as salt water and gas. The characteristics of the fluid in the reservoir, reservoir pressure, the total amount of fluid in the reservoir, and a host of other factors determine the amount of oil that a company can ultimately recover. Equation 57 gives a rough estimate of recoverable oil. It is—

$$R = FAtpsr\,(1 - c) \qquad \text{(Eq. 57)}$$

where

R = ultimate recovery, bbl
F = 7,758 barrels per acre-foot (bbl/ac•ft)
A = acres in the lease
t = average thickness of pay zone, ft
p = porosity of reservoir rock
s = shrinkage factor
r = recovery factor, depending on type of drive and other reservoir conditions
c = percentage of connate water in pore spaces.

Treating Emulsions

Many wells produce oil in the form of an emulsion. An emulsion is a special combination of oil and water wherein the water is usually spread out, or dispersed, as droplets within the oil. (In rare cases, oil is dispersed in the water.) Normally, oil and water do not mix; however, in the presence of an emulsifying substance, droplets of water can disperse within the oil. Once dispersed, these droplets can sometimes be difficult to remove. (For more information about emulsions, see the Petroleum Extension Service publication entitled *Treating Oilfield Emulsions*.)

Given enough time, gravity causes most or all of the water droplets to settle out. In many cases, however, too much time is required for the process to be practical. Consequently, companies not only place emulsions into settling tanks to give the water a chance to settle out of the oil, but they also add emulsion-treating chemicals to the emulsion. These chemicals, sometimes combined with heat to thin the emulsion, cause the water droplets to merge, form larger drops, and settle to the bottom of the vessel in which the emulsion is placed for treatment.

To determine the amount of emulsion-treating chemical to use on a lease, it is necessary to run a series of tests known as a bottle test. The major steps in this test are shown here.

1. Prepare a mixture of three parts xylol and one part methanol to use as a solvent for the chemicals.
2. Choose an emulsion-treating chemical to be tested and add 2 millilitres (mL) of the chemical to 98 mL of the xylol-methanol mixture. The resulting mixture is a 2% solution of the chemical.
3. Put 100 mL of the emulsion in each of three prescription bottles and add 0.2 mL of the 2% solution to one bottle, 0.4 mL to one bottle, and 0.8 mL to one bottle. Shake well, and if heaters are to be used in lease treating, heat the oils in the bottles to the same temperature to be used on the lease. If no bottle shows a clear separation of oil and water, choose another chemical and repeat the procedure. If one bottle clearly shows a better separation than the others, note the amount of solution used in this bottle and use that amount in the following tests.
4. Make up a 2% solution of each of the other chemicals that might be used for treating, and add an amount of solution of each chemical equal to that determined in step 3 to separate 100 mL fresh samples of emulsion. Choose the bottles that show the best results, reduce the amount of each chemical to 0.1 mL less than on the previous test, and repeat the test. Continue this process, reducing the amount of chemical used each time, until the one chemical that achieves good separation with the least amount of solution is found.
5. Then use the amount of chemical concentrate needed in the bottle test and the percent concentration of the solution to solve for the number of pints (pt) needed to treat each 100 bbl of emulsion. The general formula for this relationship is—

$$p = 336sc \qquad \text{(Eq. 58)}$$

where

p = pints of chemical concentrate (solution) per 100 bbl of emulsion

s = amount of solution needed in bottle test, in mL/100 mL of emulsion

c = concentration of solution expressed as a decimal fraction

336 = number of pt/bbl.

Example Problem: Suppose the amount of chemical needed in the bottle test is 0.2 mL/ 100 mL of emulsion, and the concentration is 2%. How many pints of treating fluid should you use per 100 bbl of emulsion?

Solution: Use equation 58 and substitute the values. Thus—

$p = 336sc$
  $= 336 \times 0.2 \times 0.02$
$p = 1.34.$

The amount of chemical needed for each 100 barrels of emulsion produced is 1.34 pt. Table 9.3 shows tabulated values of p for various values of s and c.

TABLE 9.3
Converting Bottle Test Results to Chemical Pump Rates per 100 bbl of Emulsion Produced

Compound per 100 cc Emulsion as Determined by Bottle Test (cc)			Concentrated Treating Compound per 100 bbl of Emulsion Produced (pints)
1% Solution	2% Solution	Concentrate	
0.025	0.0125	0.00025	0.084
0.05	0.025	0.0005	0.153
0.1	0.05	0.001	0.336
0.2	0.10	0.002	0.672
0.3	0.15	0.003	1.008
0.4	0.20	0.004	1.344
0.5	0.25	0.005	1.680
0.6	0.30	0.006	2.016
0.7	0.35	0.007	2.352
0.8	0.40	0.008	2.688
0.9	0.45	0.009	3.024
1.0	0.50	0.010	3.360
1.1	0.55	0.011	3.70
1.2	0.60	0.012	4.03
1.3	0.65	0.013	4.27
1.4	0.70	0.014	4.70
1.5	0.75	0.015	5.04
1.6	0.80	0.016	5.38
1.7	0.85	0.017	5.71
1.8	0.90	0.018	6.05
1.9	0.95	0.019	6.38
2.0	1.00	0.020	6.72
2.1	1.05	0.021	7.06
2.2	1.10	0.022	7.39
2.3	1.15	0.023	7.73
2.4	1.20	0.024	8.06
2.5	1.25	0.025	8.40
2.6	1.30	0.026	8.74
2.7	1.35	0.027	9.07
2.8	1.40	0.028	9.41
2.9	1.45	0.029	9.74
3.0	1.50	0.030	10.08
3.1	1.55	0.031	10.42
3.2	1.60	0.032	10.75
3.3	1.65	0.033	11.09
3.4	1.70	0.034	11.42
3.5	1.75	0.035	11.76
3.6	1.80	0.036	12.10
3.7	1.85	0.037	12.43
3.8	1.90	0.038	12.77
3.9	1.95	0.039	13.10
4.0	2.00	0.040	13.44

Meter Runs

While current technology is advancing the use of sophisticated methods to measure quantities of gas flowing in a line, an old standby method is still widely used. This method involves an orifice meter. In gas measurement, an *orifice* is a precisely drilled hole of a given size in a flat metal plate. This orifice plate is placed inside a special holder that is inserted into a pipe (a line) in which gas is flowing. The assembly is an orifice meter and such meters measure volume based on the pressure change that occurs in the gas as it flows through the orifice. Put another way, gas volume can be inferred by the change in pressure that occurs in the gas on either side of the orifice.

To measure gas volumes accurately with an orifice meter, a smooth flow of gas must be provided around the meter. Because bends, ells, and constrictions in pipe are sources of turbulence, straight lengths of pipe must be placed just upstream and downstream from the orifice plate.

If space considerations prevent long straight runs of pipe, straightening vanes inserted upstream from the orifice plate reduce the length of straight pipe necessary. Figure 9.15 and table 9.4 illustrate common sources of turbulence and prescribed minimum lengths of pipe necessary in each case. The boxheads across the top of table 9.4 represent the ratio of orifice diameter to nominal pipe diameter. The minimum lengths of straight pipe are stated in terms of nominal pipe diameters.

Example Problem: Two ells not in the same plane are upstream from a 4-in. orifice in a 10-in. pipe. What straight lengths of pipe must be placed upstream and downstream from the orifice? If straightening vanes are used, where should they be located? What lengths of pipe are necessary with straightening vanes?

Solution: The ratio of orifice diameter to nominal pipe diameter is 0.4. The two ells not in the same plane correspond to case 3 in figure 9.15. From table 9.4, $A = 18.0$ and $B = 3.3$ diameters, or 180 in. and 33 in., respectively. If straightening vanes are used, B remains the same, but the upstream length is given by $A' = 9.5$ diameters, or 95 in. The downstream end of the straightening vanes should be a distance C upstream from the orifice, and from table 9.4, $C = 5.0$ diameters, or 50 in.

Electronic Instrumentation

In the process industry, quantities of pressure, level, flow, temperature, and other factors to control quality, must be measured to document quantity and perform accounting functions. Mechanical, pneumatic, hydraulic, or electronic instruments achieve such measurements with varying accuracy, reliability, maintenance, documentation, and control. Electronic instruments or transmitters provide the greatest accuracy, highest reliability, and ease of calibration. They can also electrically transmit the measured results to a remote site.

Instruments used to measure process variables (PV) are linear analog devices. They are so called because the electrical output is directly proportional (linear) to the process input. Although several standards exist, the most common range of electrical output is 4 to 20 mA as the process changes from its minimum level to the maximum level. That is, output amperage is 4 mA at its minimum and 20 mA at its maximum. With most instruments, the input range of the process can be changed, or reranged, but the output current level is always within the range of 4 to 20 mA.

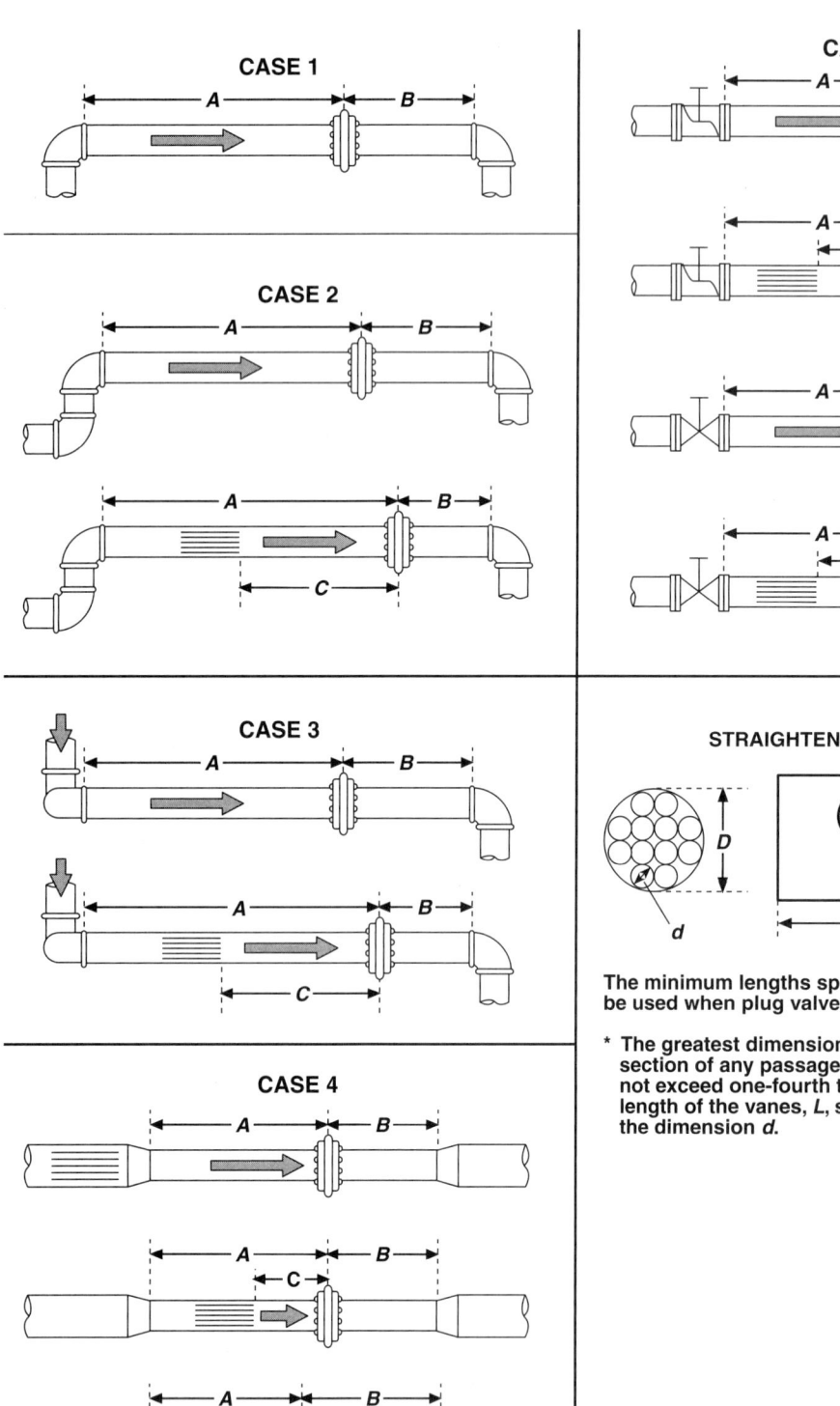

CASE 1

CASE 2

CASE 3

CASE 4

CASE 5

STRAIGHTENING VANES

The minimum lengths specified for Case 5 should be used when plug valves are involved.

* The greatest dimension *d* of the inside cross section of any passage through the vanes shall not exceed one-fourth the ID of the pipe *D*. The length of the vanes, *L*, shall be at least ten times the dimension *d*.

LEGEND

ORIFICE FLANGES

STRAIGHTENING VANES

Figure 9.15   Standard specifications for meter runs

TABLE 9.4
Minimum Length Requirements for Meter Runs

		Ratio of Diameters (orifice/pipe)									
	Sec.	0.1	0.2	0.3	0.4	0.5	0.6	0.67	0.70	0.75	0.80
Case 1. One ell or tee ahead of orifice run	A	6.0	6.0	6.0	6.0	6.8	9.0	12.5	14.0	4.4	4.6
	B	2.5	2.8	3.0	3.3	3.7	4.0	4.1	4.3	4.4	4.6
Case 2. Two ells or bends ahead of orifice run (bends in same plane)	A	8.7	8.7	8.7	8.7	10.0	13.5	17.0	19.0	22.0	25.0
	A'	8.7	8.7	8.7	8.7	9.0	10.4	11.6	12.1	14.0	15.5
	B	2.5	2.8	3.0	3.3	3.7	4.0	4.1	4.3	4.4	4.6
	C	5.0	5.0	5.0	5.0	5.1	5.1	6.4	6.7	7.3	8.0
Case 3. * Two ells or bends ahead of orifice run (bends not in same plane)	A	14.0	14.6	16.0	18.0	20.6	25.0	29.0	31.0	35.0	40.0
	A'	9.5	9.5	9.5	9.5	10.0	11.0	12.7	13.5	15.0	16.5
	B	2.5	2.8	3.0	3.3	3.7	4.0	4.1	4.3	4.4	4.6
	C	5.0	5.0	5.0	5.0	5.1	5.6	6.4	6.7	7.3	8.0
Case 4. Reducer ahead of orifice run	A	6.0	6.0	6.0	6.2	7.5	9.6	11.3	12.3	13.7	15.0
	A'	8.7	8.7	8.7	8.7	9.0	10.4	11.8	12.3	13.7	15.0
	B	2.5	2.8	3.0	3.3	3.7	4.0	4.1	4.3	4.4	4.6
	C	5.0	5.0	5.0	5.0	5.1	5.6	6.4	6.7	7.3	8.0
Case 5. Valve or regulator ahead of orifice run	A	16.0	17.8	19.5	21.7	25.0	30.0	35.7	39.0	44.5	50.0
	A'	9.3	9.3	9.3	9.5	10.3	12.0	14.6	15.4	17.0	19.5
	B	2.5	2.8	3.0	3.3	3.7	4.0	4.1	4.3	4.4	4.6
	C	5.0	5.0	5.0	5.0	5.1	5.6	6.4	6.7	7.3	8.0

Note 1. Dimensions given refer to meter-run installations shown in figure 9.9.

Note 2. When pipe taps are used, A, A', and C should be increased by two pipe diameters and B increased by eight diameters.

Note 3. When the diameter of the orifice may require changing to meet different conditions, the lengths of straight pipe should be those required for the maximum orifice-to-pipe-diameter ratio that may be used.

*Note 4. When in Case 3 the two ells are closely preceded by a third that is not in the same plane as the middle or second ell, the length of A should be doubled.

Figure 9.16 diagrams an electronic transmitter. A source of DC voltage powers the electronic transmitter. A sensor receives the process variable and puts it into the transmitter. Depending on the value of the variable, a 4 to 20 mA signal goes to a meter or a PLC, which processes the signal. The transmitter can be set up to send the process variable to a remote point.

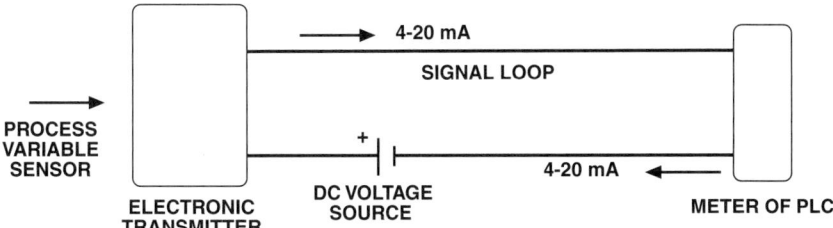

Figure 9.16   Block diagram of electronic transmitter

Many terms are used in electronic instrumentation and you should become familiar with them. Following are several terms commonly employed in instrumentation.

Process Variable (PV): The quantity of the process under consideration. This variable may be pressure, temperature, level, flow, density, or other characteristics of the process that can be measured.

Calibration: Sometimes called reranging, calibration is the process of establishing the lowest process variable to be measured along with the highest process variable to be measured through mechanical or electronic adjustment

Upper Range Value (URV): The maximum value of the process variable to be measured and calibrated. URV is sometimes referred to as span when adjusting the transmitter.

Lower Range Value (LRV): The lowest value of the process variable to be measured and calibrated.

Current Output: The range of electrical current, measured in milliamperes (mA), which the electronic instrument produces during operation. While current output is usually from 4 to 20 mA, some systems may use 1 to 5 mA, 10 to 50 mA, or a voltage range.

Calibration Procedure

Calibrating, or reranging, an electronic process instrument involves the use of precision reference sources that simulate the process variable or uses a smart calibration unit that duplicates the process variable.

With many electronic instruments, a precision reference source is needed to simulate the process variable and allow the operator to set the output from 4 to 20 mA with a screwdriver adjustment. For example, if the instrument is a pressure measuring device, a minimum or maximum pressure range exists over which the instrument is capable of being calibrated.

Example Problem: The nameplate on a pressure transmitting instrument indicates the transmitter's range is 0 to 2,000 psi. However, the transmitter needs to be reranged so that it can handle a range from 0 to 500 psi.

Solution: First, set the reference pressure source on the instrument to 0.00 psi. Next, set the zero screwdriver adjustment to the output current of 4.00 mA. Use an ammeter to make this adjustment. Now, elevate the reference pressure source to 500.00 psi and adjust the span, or range, screwdriver adjustment to an output current of 20.00 mA. Again, use an ammeter to make the adjustment. Make fine adjustments at the LRV and URV until the settings are established at 4 and 20 mA. Finally, perform a final check at the midpoint of 12.00 mA and 250.00 psi to assure transmitter linearity.

Since electronic transmitters are linear over the full range, equations can be established to verify operation or to determine the process variable by viewing the output current level. Pressure transmitter equations include—

$$p = \frac{mA - 4}{16} \times P \qquad \text{(Eq. 59)}$$

or

$$mA = \left(\frac{p}{P} \times 16 \right) + 4 \qquad \text{(Eq. 60)}$$

where

p = current pressure to transmitter
mA = transmitter output, mA
P = upper range value of pressure.

Temperature transmitter equations include—

$$t = \frac{mA - 4}{16} \times T \qquad \text{(Eq. 61)}$$

or

$$mA = \left(\frac{t}{T} \times 16 \right) + 4 \qquad \text{(Eq. 62)}$$

where

t = current temperature signal to the transmitter
mA = transmitter output, mA
T = upper range value of temperature.

Level transmitter equations include—

$$l = \frac{mA - 4}{16} \times L \qquad \text{(Eq. 63)}$$

or

$$mA = \left(\frac{l}{L} \times 16 \right) + 4 \qquad \text{(Eq. 64)}$$

where

l = current level signal to the transmitter
mA = transmitter output, mA
L = upper range value of level.

Example Problem: A level transmitter has been calibrated over the range of 0 to 100 in. of water. The transmitter is checked and its current output is 14.6 mA. What is the present level of liquid in the tank?

Solution: Use the level equation 63, and plug in the values. Thus—

$$l = \frac{14.6 - 4}{16} \times 100$$

$$l = 66.25 \text{ in.}$$

Gas Flow

As mentioned previously, orifice meters are often used to measure the volume of gas flowing through a line. Although nowadays computers connected to orifice meters usually sense and calculate gas volumes, you should know that calculations are based on a formula. The formula is—

$$Q_h = C' \ \overline{h_w P_f} \qquad \text{(Eq. 65)}$$

where

Q_h = volume of gas per hour flowing through the meter, ft^3
C' = orifice flow constant, calculated from several factors, usually stamped on the meter
h_w = differential pressure read from the meter chart, in. of water
P_f = static pressure, psia.

The static pressure is found by adding the atmospheric pressure at the meter location to the static pressure reading from the meter chart.

Example Problem: The chart in figure 9.17 is from a meter with an orifice flow constant of 800. Find the volume of gas that flowed through the meter between 5 A.M. and 6 A.M. if the atmospheric pressure is 14.5 psi.

Solution: The chart shows the average differential pressure to be about 33.5 in. of water from 5 A.M. to 6 A.M.. The static pressure averages about 32 psig during this period, or 32 + 14.5 = 46.5 psia.

Then, using equation 65, the volume of gas is—

$$Q_h = 800 \times \overline{33.5 \times 46.5}$$
$$Q_h = 31,500 \text{ ft}^3.$$

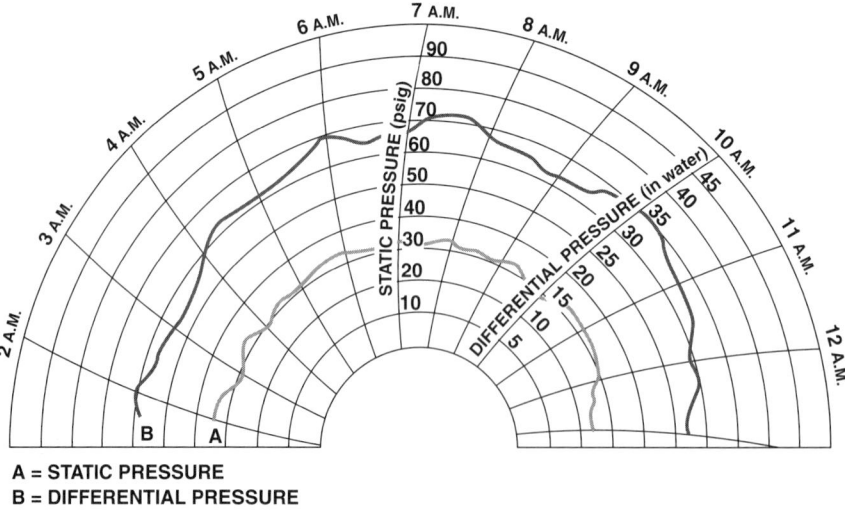

A = STATIC PRESSURE
B = DIFFERENTIAL PRESSURE

Figure 9.17   Gas measurement chart (linear)

Practice Problems

1. A well with an initial shut-in bottomhole pressure of 2,000 psi is tested and found to produce 185 bpd at a stabilized flowing bottomhole pressure of 1,750 psi. What is the productivity index (the barrels per day per psi of pressure drawdown) of the well?

_____

2. How much horsepower is required to drive a duplex double-acting pump that is designed to deliver 4,500 bbl of oil per day at a pressure of 150 psi?

_____

3. Suppose the amount of chemical needed in a bottle test is 0.6 mL per 100 mL of emulsion. If you have a 2% solution, how many pints of treating fluid should you use for each 100 barrels of raw crude?

_____

4. Using an orifice flow constant of 1,000, find the rate of flow (in Mcf/hr) at 2:00 A.M. on the chart shown in figure 9.17. Use 14.65 for atmospheric pressure.

_____

5. Find the volume of gas flowing through the meter between 8 A.M. and 8 P.M. if the differential pressure averages 100 in. of water and the static pressure averages 150 psig. (Graduations on the chart are from 0 to 500 psig.) The orifice flow constant is 1,000. Use 14.65 psi for atmospheric pressure.

_____

6. A temperature transmitter is measured to have 18 mA of current after having been calibrated from 0 to 200 deg. F. What is the current temperature being measured by the transmitter?

_____

MATHEMATICS IN PIPELINE OPERATIONS

Pipe Buoyancy

In constructing underwater pipeline crossings and in laying pipe offshore, it is often necessary to coat the line pipe with concrete to overcome the fact that the pipe floats in the water. Put another way, the pipe is buoyant. When line pipe is coated with concrete, the pipe gains negative buoyancy—that is, it becomes heavier than the water it displaces and submerges. The formula for calculating negative buoyancy is—

$$b = W + (-0.34d^2 + 2.78dt + 2.78t^2) \qquad \text{(Eq. 66)}$$

where

b = negative buoyancy, or effective weight of pipe in fresh water, lb/ft

W = weight of uncoated pipe, lb/ft

d = nominal diameter of pipe, in.

t = thickness of concrete coating, in.

Table 9.5 tabulates values for the expression $(-0.34d^2 + 2.78dt + 2.78t^2)$ in equation 66.

To find pipe buoyancy, the given pipe size, weight per ft, and concrete thickness, add the table value to W in equation 66. If the value of b is greater than 0, the pipe has negative buoyancy and submerges. If the value of b is less than 0 (it is a negative number), the pipe will float. In this case, b is a measure of the buoyant force of the water.

Example Problem: Find the negative buoyancy resulting from a 3-in. coating of concrete on an 18-in. pipe that weighs 82.06 lb/ft.

TABLE 9.5
Negative Pipe Buoyancy

Nominal Pipe Diameter (inches)	Thickness of Concrete Coating (inches, based on 190 lb/cu ft concrete)													
	0"	¼"	½"	¾"	1"	1¼"	1½"	1¾"	2"	2¼"	2½"	2¾"	3"
8⅝"	−25.3	−19.1	−12.6	−5.7	1.5	9.1	17.0	25.2	33.8	42.8	52.1	61.7	71.7
10¾"	−39.4	−31.7	−23.7	−15.3	−6.7	−2.3	11.7	21.2	31.5	42.0	52.7	63.9	75.3
12¾"	−55.2	−46.1	−36.8	−27.0	−17.0	−6.6	4.3	15.3	26.8	38.7	50.8	63.3	76.3
14"	−66.6	−56.7	−46.4	−35.8	−24.9	−12.7	−1.9	10.0	22.3	35.0	48.1	61.4	75.2
16"	−87.0	−75.7	−64.1	−52.1	−39.8	−27.1	−14.0	−0.8	13.0	27.1	41.6	56.2	71.3
18"	−110.0	−97.3	−84.3	−70.9	−57.2	−43.2	−28.7	−14.0	1.1	16.6	32.4	48.5	65.0
20"	−136.0	−121.9	−107.5	−92.8	−77.6	−62.3	−46.4	−30.3	−13.9	1.1	20.3	37.9	55.8
22"	−164.3	−148.8	−133.1	−116.9	−100.4	−83.7	−66.4	−49.0	−31.1	−12.7	5.9	24.7	43.8
24"	−195.8	−178.9	−161.7	−144.2	−126.3	−108.1	−89.5	−70.6	−51.3	−31.7	−11.7	8.6	29.2
26"	−229.5	−211.2	−192.6	−173.7	−154.5	−134.9	−114.8	−94.6	−73.9	−52.8	−31.4	−9.7	12.4
28"	−266.3	−246.7	−226.8	−206.5	−185.9	−165.0	−143.7	−121.9	−100.0	−77.4	−54.9	−32.0	−8.3
30"	−306.0	−285.0	−263.6	−241.9	−219.8	−197.6	−174.7	−151.6	−128.1	−104.1	−80.2	−55.8	−31.0
32"	−348.0	−325.6	−302.9	−279.7	−256.2	−232.6	−208.5	−184.0	−159.0	−133.9	−108.5	−82.6	−56.3
34"	−393.5	−369.7	−348.6	−321.1	−296.3	−271.2	−245.6	−219.9	−193.6	−167.1	−140.2	−113.0	−85.5
36"	−440.0	−414.8	−389.3	−363.4	−337.2	−310.7	−283.7	−256.5	−228.9	−200.9	−172.	−144.0	−115.0

Solution: In table 9.5, find 3 in. in the top boxheads. Follow this column down until you intersect the column whose stub is 18 in. In this case, the value is 65.0. Because 65.0 is a positive number, add it to 82.06 (the weight of the pipe) to obtain 147.06, which is the effective weight of the pipe in water, in lb/ft.

Pumping

Hydrostatic Head vs Psi

When dealing with centrifugal pumps, it is common to express pump pressure not in psi, but rather as the height of a column of water or other fluid (such as mercury) that the pump's pressure could support. This height is usually called hydrostatic head. Two equations are available for converting psi to hydrostatic head and hydrostatic head to psi. They are—

$$P = 0.433\, H \times sp\, gr \qquad \text{(Eq. 67)}$$

and

$$H = P \div (0.433 \times sp\, gr) \qquad \text{(Eq. 68)}$$

where

$\qquad P$ = pressure, psig
$\qquad H$ = height of fluid head, ft
$\qquad sp\, gr$ = specific gravity of fluid
$\qquad 0.433$ = hydrostatic pressure of a vertical column of water, psi/ft.

Example Problem: A pressure gauge on a pump reads 600 psig. What is the equivalent pressure expressed as hydraulic head of water? (The specific gravity of water is 1.0.)

Solution: Use equation 68 and substitute known values. Thus—

$\qquad H$ = $600 \div (0.433 \times 1)$
$\qquad H$ = 1,385.68 ft.

Example Problem: Express a head of 13 ft of mercury in psig. The specific gravity of mercury is 13.6.

Solution: Use equation 67 and substitute known values. Thus—

$\qquad P$ = $0.433 \times 13 \times 13.6$
$\qquad P$ = 76.6 psig.

Determining Pump Horsepower

The pump horsepower needed on a pipeline is found from the equation—

$$hp = \frac{R_f \times (P_d - P_s)}{2{,}450 \times E_p} \qquad \text{(Eq. 69)}$$

where

$\qquad R_f$ = rate of flow, in bbl/hr
$\qquad P_d$ = discharge, or maximum line, pressure, in psig
$\qquad P_s$ = suction pressure, in psig
$\qquad E_p$ = percentage of pump efficiency, expressed as a factor.

Oil Pipeline Design

Oil pipeline design considers many factors, including the terrain, the viscosity and gravity of the oil flowing through it, the size of the pipe, and the desired rate of flow. To illustrate the process by which these factors are evaluated, consider a pipeline to be built under the following conditions—

- elevation at origin: 400 ft
- elevation at destination: 3,875 ft
- distance: 100 mi
- pipe size: 12 in., ID
- oil viscosity: 70 Saybolt Seconds Universal (SSU); SSU is a measure of viscosity based on the amount of time, in seconds, a liquid of a given viscosity and temperature flows through an orifice of a specified size. A Dutch chemical engineer named Saybolt came up with the measurement.
- oil gravity: 40°API (sp gr = 0.825)
- desired rate of flow: 2,000 bbl/hr
- maximum line pressure: 700 psig
- suction pressure at each station: 25 psig
- pump efficiency: 79%.

The first step is to find the loss in pressure along the line due to friction in the pipe. Figure 9.18 plots rate of oil flow in bbl/hr against pressure loss in psi/mi (on the left) and pipeline size (along the top). Each curve represents oil viscosity in SSU. In figure 9.18 locate the curves dealing with 12-in. ID pipe, and the curve specifying a viscosity of 70 SSU. (The viscosity curves are not labeled SSU for clarity.) On the bottom of the chart, locate the line representing 2,000 bbl/hr. Follow this line up to the intersection with the 12-in., 70 SSU curve. From this intersection, move directly across to the left side of the chart and read 13.5 psi per mile on the pressure loss axis. (This frictional pressure loss is also called the hydraulic gradient.) By using equation 68, you can convert this loss to a head of 37.8 ft of oil per mile. Thus, the total pressure loss for the 100-mile line is—

$$37.8 \times 100 = 3,780 \text{ ft of head.}$$

Also, the destination of the oil is at elevation 3,875 ft, and the origin is at elevation 400 ft, which is a difference of 3,475 ft to be overcome. Thus, the total head to be overcome is (using equation 67)—

$$3,780 + 3,475 = 7.255 \text{ ft,}$$

which is equivalent to 2,590 psi. While one pumping station could supply all of the pressure, specifications state that the maximum line pressure cannot exceed 700 psi. In this case, four stations can supply the needed pressure with a little to spare, because 700 psi × 4 = 2,800 psi.

To determine the locations of these stations, plot a profile of ground elevations along the pipeline route (fig. 9.19). Convert 700 psi (the maximum line pressure) to head in feet of oil, which is 1,960 ft. This calculation is made using equation 67. Thus—

$$H = P \div (0.433 \times 56)$$
$$H = 700 \text{ psi} \div (0.433 \times 0.825)$$
$$H = 1959.5493 \text{ ft.}$$

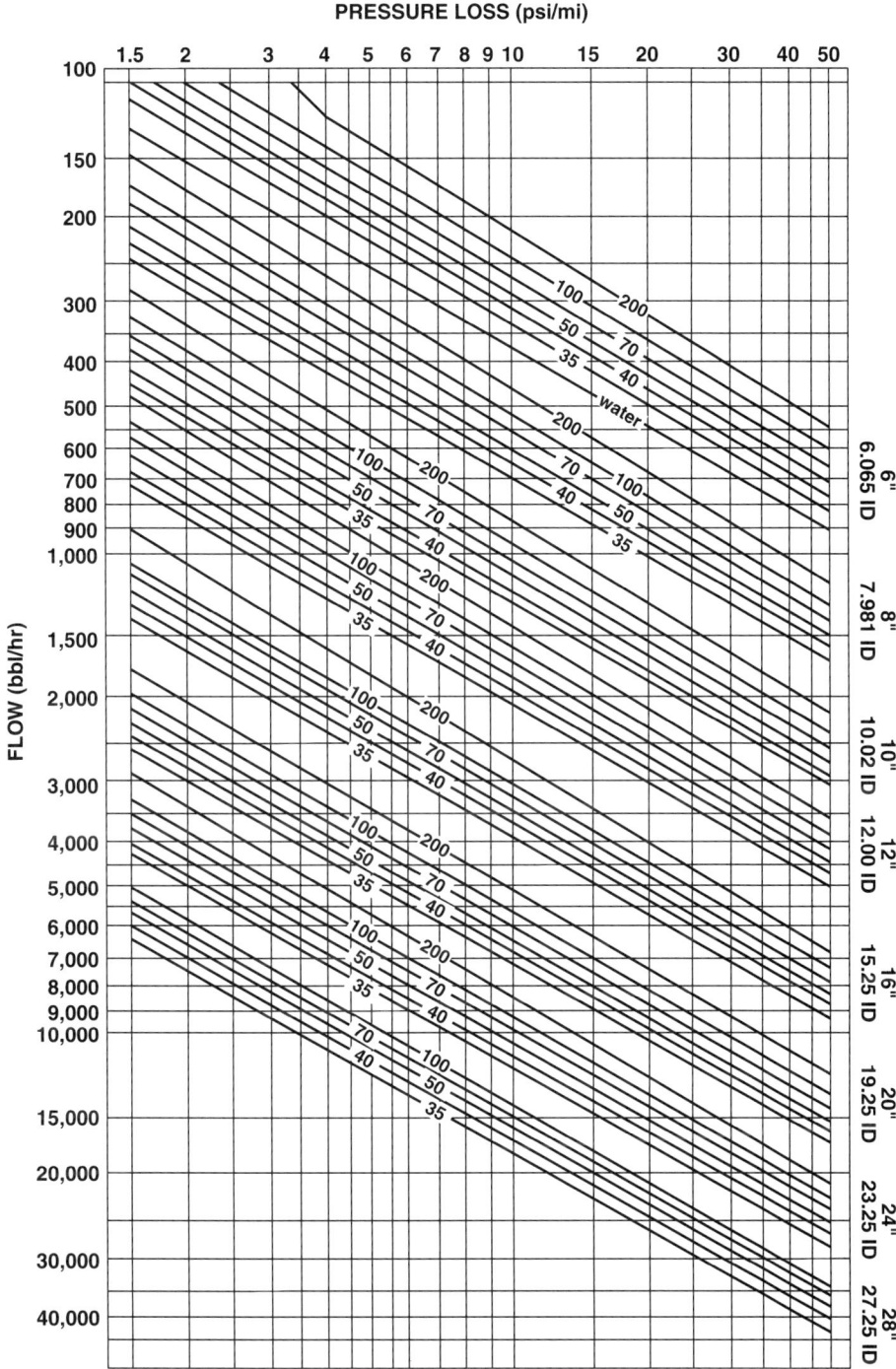

Figure 9.18 Flow of crude petroleum to pipelines

From the upper end of this line, draw a line sloping to the right at the rate of 37.8 ft per mile, the hydraulic gradient. The place where this line comes to 70 ft above the profile (70 ft is the head of oil equal to the desired suction pressure of 25 psi) is the projected location of station *B*. Repeat the procedure with a 1,960-ft vertical line from station *B* and a line sloping right to station *C*, and so on.

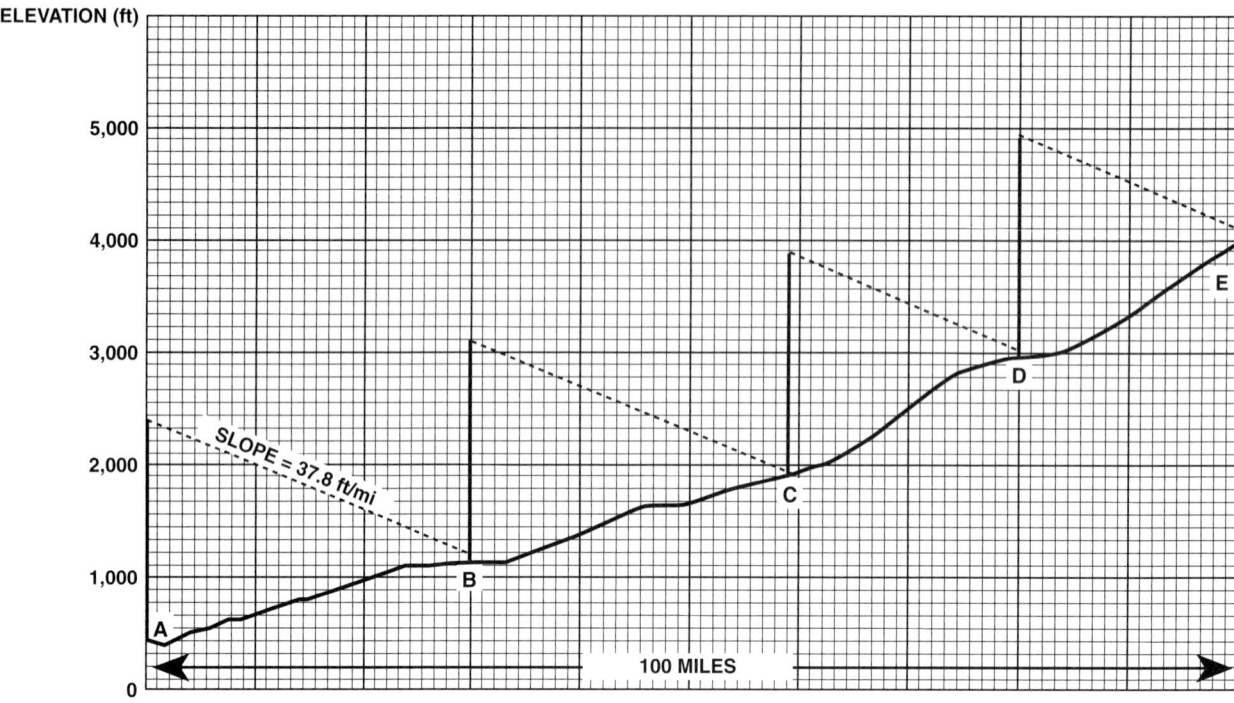

Figure 9.19   Profile method of calculating pipeline pump station spacing

The plot on figure 9.19 shows that the oil arriving at the tanks at station E, the destination, still has several hundred feet of head. Since this head is not necessary to fill the tanks, the pressure required from the pumps at station D can be reduced by this amount.

Using equation 69, the pump horsepower needed on this pipeline is—

$$hp = \frac{2,000 \times (700 - 250)}{2,450 \times 0.79}$$

$$hp = 697.49.$$

Practice Problems

1. What head of water is equivalent to a pressure of 750 psig?

2. Find the pump horsepower needed on a pipeline whose desired rate of flow is 1,500 bbl/hr, discharge pressure is 650 psig, suction pressure is 25 psig, and pump efficiency is 80%.

3. What is the negative buoyancy of a 20-in. pipe weighing 95 lb/ft if a 2-in. coating of concrete is used?

4. A 14" diameter pipe is being used in an underwater pipeline. The uncoated pipe weighs 65 lb/ft. How thick does the cement coating have to be to give the pipe an effective weight in water of 18 lb/ft?

5. A 200-mi pipeline loses 35.6 ft of head per mile because of friction. The elevation at its origin is 2,000 ft and at the destination, 3,685 ft. What is its total pressure loss in feet of head?

MATHEMATICS IN REFINING OPERATIONS

Weighted Averages

An understanding of weighted averages is necessary in refining problems. For instance, given the data on fuel gas in table 9.6, find the net heating value (*LHV*) of the fuel gas.

TABLE 9.6
Fuel Gas Data

Component	Heating Value of Component (Btu/ft³)	%Component in Fuel Gas
Methane	910	94
Ethane	1,630	5
Propane	2,350	1

To solve the problem, first recall the formula for weighted averages from chapter 3:

$$A_s = \frac{(N_a \times V_a) + (N_b \times V_b) + (N_c \times V_c)}{a + b + c \dots} \qquad \text{(Eq. 70)}$$

where

$$\begin{aligned} A &= \text{weighted average} \\ N &= \text{number of items} \\ V &= \text{value of each item} \\ a, b, c &= \text{items.} \end{aligned}$$

The weighted average, which in this case is the net heating value (*LHV*) of the fuel gas, is—

$$LHV = \frac{(94 \times 910) + (5 \times 1,630) + (1 \times 2,350) \dots}{94 + 5 + 1 \dots}$$

$$LHV = 960 \text{ Btu/ft}^3.$$

Heat Transfer

To find the amount of heat required to raise the temperature of a volume of liquid, use the formula—

$$Q = WC_p (T_2 - T_1) \qquad \text{(Eq. 71)}$$

where

$$\begin{aligned} Q &= \text{amount of heat, Btu} \\ W &= \text{weight of liquid, lb} \\ C_p &= \text{specific heat of liquid, Btu/lb/°F} \end{aligned}$$

T_2 = final temperature, °F
T_1 = initial temperature, °F.

Example Problem: How much heat is required to heat 100 gal of water from 80°F to 160°F? The specific heat of water is 1.0 Btu/lb/°F.

Solution: The weight of 100 gal of water is—

$$100 \text{ gal} \times 8.33 \text{ lb/gal} = 833 \text{ lb.}$$

Then, using equation 71 and substituting values for unknowns—

$$Q = 833 \times 1.0 \text{ Btu/lb/°F} \times (160 - 80)°\text{F}$$
$$Q = 66,700 \text{ Btu.}$$

If the rate of flow is 100 gpm, then the hourly heat requirement is 66,700 Btu × 60 = 4 MM Btu.

Example Problem: How much heat is required to heat 5,000 bbl of 40°API crude oil from 80°F to 250°F? The specific heat of the oil is 0.47 Btu/lb/°F at 80° and 0.57 Btu/lb/°F at 250°.

Solution: The specific gravity of 40°API oil is 0.825. The weight of 5,000 bbl is—

$$5,000 \text{ bbl} \times 0.825 \times 8.33 \text{ lb/gal} \times 42 \text{ gal/bbl} = 1,445,000 \text{ lb.}$$

The average C_p is (0.47 + 0.57) ÷ 2 = 0.52 and

$$Q = 1,445,000 \text{ lb} \times 0.52 \text{ Btu/lb/°F} \times (250 - 80)°\text{F}$$
$$Q = 127.7 \text{ MM Btu.}$$

Fuel Consumption

To find fuel costs, use the formula—

$$R = Q \div (E \times LHV) \qquad \text{(Eq. 72)}$$

where

R = fuel consumption, ft³
Q = heat requirement, Btu
E = heater efficiency, expressed as a decimal
LHV = net heating value of gas, Btu/ft³.

Then—

$$\text{fuel cost} = \text{consumption} \times \text{value.}$$

Example Problem: If a gas having an LHV of 960 Btu/ft³ costs $2.06 per Mcf, what is the daily cost of heating the oil in the previous problem (flowing at the rate of 5,000 bpd and requiring 5,350,000 Btu) with a 60% efficient heater?

Solution: Substitute the known values in equation 72. Thus—

$$R = 127.7 \text{ MM Btu} \div (0.60 \times 960 \text{ Btu/ft}^3) = 221,700 \text{ ft}^3.$$

Then, the fuel cost is—

$$221,700 \times \$2.06/1,000 = \$456.70.$$

Material Balance

In working with plant and refinery processes, it is often necessary to calculate the amount of product from a particular process, using known data on feeds and other products. The following example illustrates how this calculation is done.

Example Problem: Using the process and data given in figure 9.20, determine the daily production of bottoms.

Solution: From the data on input and output, convert all to pounds per hour.

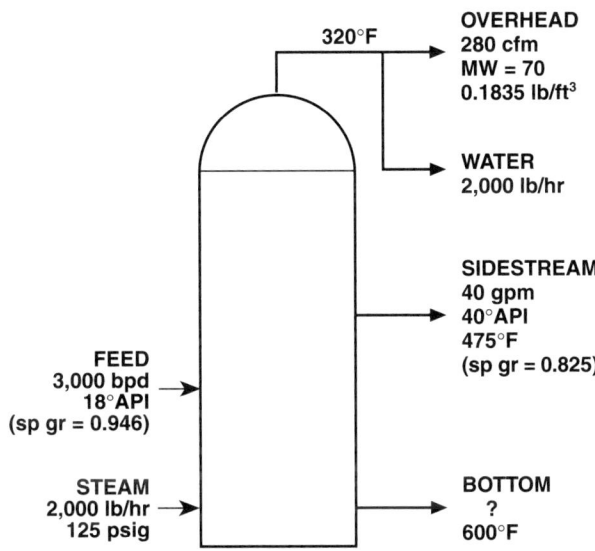

Figure 9.20   Material balance bottoms process

Feed: 3,000 bbl/day × (1 day ÷ 24 hr) × 42 gal/bbl × 8.33 lb/gal × 0.946 = 39,700 lb/hr (input)
Steam: 2,000 lb/hr (input)
Overhead: 280 ft³/min × 60 min/hr × 0.1835 lb/ft³ = 3,080 lb/hr (output)
Water: 2,000 lb/hr (output)
Sidestream: 40 gal/min × 60 min/hr × 8.33 lb/gal × 0.825 = 16,500 lb/hr (output).

Bottoms?

A balance sheet now shows—

Feed 39,700
Steam 2,000
Total Input .... 41,700 lb/hr
Overhead 3,080
Water.. 2,000
Sidestream 16,500
Total Output .... 21,580 lb/hr + bottoms.

Because the input and output should be equal,

$$\text{bottoms} + 21{,}580 = 41{,}700$$
$$\text{bottoms} = 41{,}700 - 21{,}580 = 21{,}120 \text{ lb/hr.}$$

And daily production is,

20,120 lb/hr × 24 hr × (1 ton ÷ 2,000 lb) = 241 tons/day.

Practice Problems

1. A barrel of liquid (42 gal) having a specific gravity of 0.9 required 32,000 Btu to raise the temperature from 75°F to 200°F. What is the average specific heat of the liquid?

_____

2. An oilfield heater has a throughput of 150 bbl/hr and consumes 3,500 ft^3 of gas per hour to increase the temperature from 65°F to 175°F. The gas has a heating value of 1,000 Btu/ft^3. The throughput product has a specific heat of 0.6 and specific gravity of 0.82. What is the efficiency of the heater in percent?

_____

9. Advanced Oil Industry Applications

*Multiply each problem answered correctly by five to arrive
at your percentage of competency.*

Round off your answers to two places. Use information given in the appendices as well as chapter 9 to solve these problems.

1. Solve for mud weight if the hydrostatic pressure is 2,330 psi and the true vertical depth of the well is 4,000 ft.

   _____

2. What is the input pressure to an electronic transmitter calibrated from 0-1000 psi if the signal current is 18 mA?

   _____

3. Air pressure is admitted into a closed container holding mercury as shown in the figure below. The distance between the mercury levels is 25 inches. What is the pressure in the container in psig? Mercury has a specific gravity of 13.6 (846 lb/ft³), and air has a specific gravity of 0.0012 (0.075 lb/ft³).

   _____

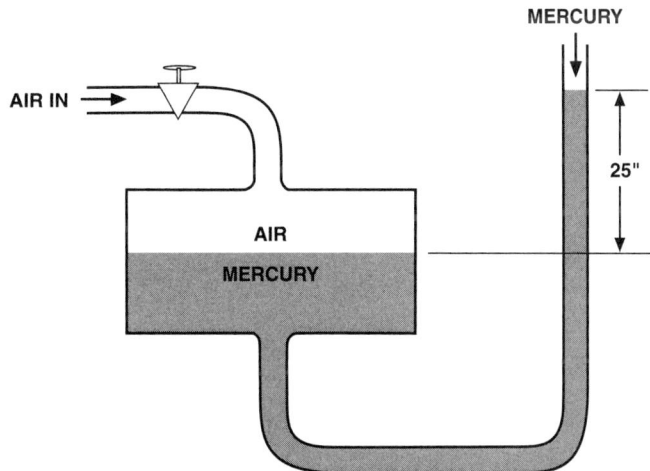

4. What is the hydrostatic pressure at the bottom of a well whose true vertical depth is 3,452 ft and is filled with mud that weighs 9 ppg?

   _____

5. Using an orifice flow constant of 1,000, find the volume in MMcf/day of a well whose differential pressure averages 100 in. of water and whose average static pressure is 150 psig. Use 14.65 rather than 14.7 as the psia factor.

   _____

6. The volume of a gasoline tank is 3,500 cubic feet. What is the weight of the gasoline it will contain when full?

   _____

7. In batching products through a pipeline, the rate of travel of the fluid through a 130-mi pipeline is 2.5 mph. If a new batch is started at 10:00 A.M. on Monday, when will it first arrive at the terminal end?

    _____

8. A pump is forcing water into an elevated reservoir tank. The pump must exert enough pressure to offset that of the column of water in the tank and in the pipe leading to it. Assume that the total vertical height from the pump to the surface of water in the tank is 150 ft. How much pressure must the pump produce to overcome the effects of hydrostatic pressure because of the water level height?

    _____

9. A turbine is delivering 200 hp to an alternator that has an efficiency of 95%. What is the output electrical power of the alternator in kW?

    _____

10. An electrical alternator is supplying three electrical loads of 100 kVA, 300 kVA, and 500 kVA. If the combined power factor of the loads is 0.7, what is the total kW being delivered by the alternator?

    _____

11. A tank contains 2,000 ft^3 of gas at 75 psig and 78°F. If the pressure changes to 80 psig and the temperature drops to 72°F, what is the volume of the gas?

    _____

12. A certain petroleum product has a specific heat capacity of 0.6 Btu/lb/°F and a specific gravity of 0.85. The product is flowing through a heater at 100 bbl/hr. Its temperature must be increased from 75°F to 180°F. How much heat energy must the heater supply in Btu/hr?

    _____

13. In problem 12, how much gas will be required for the heater if the gas has a heating value of 1,010 Btu/ft^3, and the heater has an efficiency of 65%? Give your answer in cubic feet.

    _____

14. An electronic process transmitter calibrated from 0 to 300 psi is delivering 8 mA of signal current to the 12-bit D/A converter of a PLC. What is the equivalent binary number produced by the D/A at its output?

    _____

15. An alternator has an efficiency of 85% and delivers 1,000 kVA with a pf of 0.85 to an electrical load. What is the horsepower required from the gas turbine prime mover?

    _____

16. A three-phase transformer is delivering 200 amperes at a voltage level of 480 VAC. What is the total kVA being provided?

    _____

17. What is the power factor of a three-phase electrical load requiring 200 amperes, 4,160 VAC, and 1 megawatt?

    _____

18. An electrical load rated at 800 kVA with a pf of 0.75 lagging is to have its power factor corrected to 0.9 lagging. What is the kVAR rating of the capacitor bank that produces the 0.9 pf?

_____

19. A diesel engine is delivering 1,000 pounds of force at a shaft that has a radius of 3 in. What is the torque being delivered by the engine?

_____

20. How much current does an electronic process transmitter deliver if the transmitter is calibrated from 0 to 100 in. of water and the differential pressure applied is 50 in. of water?

_____

APPENDIX A
Commonly Used Formulas

A-1. PROPORTION AND RATIO
Direct proportion:

$$a{:}b = c{:}d, \text{ or } ad = bc$$

Inverse proportion:

$$a{:}d = b{:}c, \text{ or } ac = bd$$

where

a = small, or first term
b = small, or second term
c = large, or third term
d = large, or fourth term.

Pulley ratio:

$$S_A/S_B = d_B/d_A$$

where

S_A = speed of pulley A
S_B = speed of pulley B
d_A = diameter of pulley A
d_B = diameter of pulley B.

Gear ratio:

$$S_A \times T_A = S_B \times T_B$$

where

S_A = speed of gear A
T_A = number of teeth in gear A
S_B = speed of gear B
T_B = number of teeth in gear B.

A-2. LENGTH, AREA, AND VOLUME
Board feet:

$$bd\,ft = \frac{t \times w \times 1}{12}$$

where

$bd\,ft$ = board feet in lumber
t = thickness of board, in inches
w = width of board, in inches
l = length of board, in feet.

Circle:

$$C = \pi d$$
$$C = \pi 2r$$
$$d = C/\pi$$
$$d = 2r$$

$$A = \pi r^2$$
$$r = \overline{A/\pi}$$

where

C = circumference
d = diameter
r = radius
π = 3.1416
A = area.

Cone, right circular:

$$L = \tfrac{1}{2}sC$$
$$V = \tfrac{1}{3}(BH)$$
$$V = \tfrac{1}{3}(\pi r^2 h)$$

where

L = lateral surface
s = slant height
V = volume
B = area of base
r = radius of base
h = height, or altitude
π = 3.1416
C = circumference of base.

Cube:

$$V = l^3$$

where

V = volume
l = length of one side.

Cylinder, right:

$$L = 2\pi rh$$
$$L = hC$$
$$T = 2\pi r(r + h)$$
$$V = Bh$$
$$V = \pi r^2 h$$

where

L = lateral surface
T = total surface area
r = radius of base
h = altitude
C = circumference of base
π = 3.1416
V = volume
B = area of base.

Ellipse:

$$A = \pi ab$$

where

π = 3.1416
a = one-half the major axis
b = one-half the minor axis.

Elliptical solid:

$$V = \pi abh$$

where

a = one-half the major axis
b = one-half the minor axis
h = height
π = 3.1416.

Frustum of cone or pyramid:

$$V = \tfrac{1}{3}\, h(B + B' + \overline{BB'})$$

where

V = volume
h = height of frustum
B = area of cone or pyramid base
B' = area of section base.

Parallelogram:

$$A = bh,\ \text{or}\ lw$$

where

b = base
h = height
l = length
w = width.

Polygon, regular:

$$A = \tfrac{1}{2}aP$$

where

A = area
a = altitude, or apothem
P = perimeter.

Pyramid, regular:

$$L = \tfrac{1}{2}sP$$
$$V = \tfrac{1}{3}(Bh)$$

where

L = lateral area
s = slant height
P = perimeter
V = volume
B = area of base
h = height, or altitude.

Rectangle:

$$A = bh,\ \text{or}\ lw$$
$$P = 2l + 2w$$

where

A = area
b = base
h = height
l = length
w = width
P = perimeter.

Rectangular solid:

$$V = lwh$$

where

V = volume
l = length
w = width
h = height

Sphere:

$$V = \tfrac{4}{3}\,\pi r^3$$

where

V = volume
r = radius
π = 3.1416.

Square:

$$P = 4 \times l$$
$$A = l^2,\ \text{or}\ bh$$

where

l = length of one side
b = base
h = height.

Trapezoid:

$$A = \tfrac{1}{2}h\,(b + b')$$

where

A = area
h = height
b = bottom base
b' = top base.

Triangle:

$$A = \tfrac{1}{2}bh$$

where

A = base
b = base
h = height.

A-3. TRIGONOMETRIC FUNCTIONS

In the following functions involving a right triangle,

A = angle opposite side a
B = angle opposite side b
C = right angle
c = hypotenuse, or side opposite right angle C
a = side adjacent to angle B, opposite angle A
b = side adjacent to angle A, opposite angle B.

Sine:

$$\sin A = a/c$$
$$\sin B = b/c.$$

Cosine:

$$\cos A = b/c$$
$$\cos B = a/c.$$

Tangent:

$$\tan A = a/b$$
$$\tan B = b/a.$$

Cosecant:

$$\csc A = c/b$$
$$\csc B = c/a.$$

Secant:

$$\sec A = c/a$$
$$\sec B = c/a.$$

Cotangent:

$$\cot A = b/a$$
$$\cot B = a/b.$$

A-4. ELECTRICITY

Ohm's law:

$$I = E/R$$
$$E = IR$$
$$R = E/I$$

where

I = current, in amperes
E = voltage, in volts
R = resistance, in ohms.

Series circuit current flow:

$$I = \frac{E}{R_1 + R_2 + R_3 + \ldots}$$

where

I = current, in amperes
E = voltage, in volts
$R_1, R_2, R_3, \ldots$ = resistance, in ohms

Parallel circuit total current:

$$I = E/R_1 + E/R_2 + E/R_3 + \ldots = I_1 + I_2 + I_3 + \ldots$$

Parallel circuit resistance:

$$R_p = \frac{1}{1/R_1 + 1/R_2 + 1/R_3 + \ldots}$$

Power (DC):

$$P = E \times I$$
$$P = I^2 R$$

where

P = power, in watts
E = voltage, in volts
I = current, in amperes.

Single-Phase AC Power:

$$VA_1 = V \times I$$
$$kVA_1 = \frac{V \times I}{1,000}$$
$$W_1 = VA_1 \times pf$$
$$kW_1 = kVA_1 \times pf$$
$$kVA_1 = \overline{kW^2 + kVARs}$$
$$kW_1 = \overline{kVAR^2 - kW^2}$$
$$kVARs_1 = \overline{kVA^2 + kW^2}$$

where

V = AC voltage, volts
I = line current, amperes
W_1 = single-phase real power in watts
kW_1 = single-phase real power in kilowatts
VA_1 = single-phase apparent power in voltamperes
kVA_1 = single-phase apparent power in kilovoltamperes
$kVARs_1$ = single-phase reactive power, kilovoltamperes-reactive
pf = power factor, ratio

Power-Three Phase AC:

$$VA_3 = V \times I \times \overline{3}$$
$$kVA_3 = \frac{V \times I \times \overline{3}}{1,000}$$
$$W_3 = VA_3 \times pf$$
$$kW_3 = kVA_3 \times pf$$
$$kVA_3 = \overline{kW^2 + kVARs^2}$$

$$kW_3 = \overline{kVA^2 + kVARs^2}$$
$$kVARs_3 = \overline{kVA^2 + kW^2}$$

where

VA_3 = three-phase apparent power, voltamperes

kVA_3 = three-phase apparent power, kilovoltamperes

W_3 = three-phase real power, watts

kW_3 = three-phase real power, kilowatts

$kVARs_3$ = three-phase reactive power, kilovoltamperes-reactive or kilovars

A-5. DENSITY AND SPECIFIC GRAVITY

Specific gravity, or relative density:

$$sp\ gr = D_s \div D_w$$
$$D_s = D_w \times sp\ gr$$

where

$sp\ gr$ = specific gravity

D_s = density of substance, or weight per volume of substance, in lb/ft^3, kg/m^3, or g/cm^3.

D_w = density of pure water, or weight per volume of water, in lb/ft^3, kg/m^3, or g/cm^3.

Baumé scale for heavy liquids:
$$^\circ Baumé = 145 - 145/sp\ gr.$$

Baumé scale of light liquids:
$$^\circ Baumé = (140/sp\ gr) - 130.$$

$^\circ$API gravity:
$$^\circ API = (141.5/sp\ gr) - 131.5.$$

A-6. MECHANICAL POWER

Pound-force:

$$lb_f = lb_m \times 32.15$$

where

lb_f = pound-force

lb_m = pound-mass

32.15 = a constant, in feet per second squared.

Horsepower:

$$hp = ft\text{-}lb/min \div 33,000$$
$$hp = kW/0.746$$
$$hp = \frac{T \times N}{12}$$

where

hp = horsepower

$ft\text{-}lb$ = foot-pounds of work

min = time, in minutes

kW = real power in kilowatts

T = torque in ft-lbs

N = speed in rpm

33,000 = a constant

0.746 = a constant

5,252 = a constant.

A-7. PRESSURE-VOLUME-TEMPERATURE

Converting gauge pressure to absolute pressure:

$$psia = psig + 14.65$$
$$psig = psia - 14.65$$

where

$psia$ = absolute pressure, in pounds per square inch

$psig$ = gauge pressure, in pounds per square inch

14.65 = a constant.

Charles's law:

$$P_2/P_1 = T_2/T_1$$

where

P_2 = absolute pressure after a change in temperature

P_1 = initial absolute pressure

T_2 = absolute temperature after a change in pressure

T_1 = initial absolute temperature.

Note: Volume of gas must remain the same.

Boyle's law:

$$P_1V_1 = P_2V_2$$

where

P_1 = initial absolute pressure

V_1 = initial volume of gas

P_2 = absolute pressure after a change in volume

V_2 = volume of gas after a change in pressure.

Note: Temperature must remain the same.

Ideal gas law:

$$P_1V_1 \div P_2V_2 = T_1 \div T_2$$

where

P_1 = original absolute pressure

P_2 = final absolute pressure

V_1 = original volume

V_2 = final volume

T_1 = original absolute temperature

T_2 = final absolute temperature.

APPENDIX B
Reference Tables

B-1. DECIMAL EQUIVALENTS

Fraction	Decimal Equivalent	Fraction	Decimal Equivalent
$1/64$	0.015625	$33/64$	0.515625
$1/32$	0.03125	$17/32$	0.53125
$3/64$	0.046875	$35/64$	0.546875
$1/16$	0.0625	$9/16$	0.5625
$5/64$	0.078125	$37/64$	0.578125
$3/32$	0.09375	$19/32$	0.59375
$7/64$	0.109375	$39/64$	0.609375
$1/8$	**0.125**	$5/8$	**0.625**
$9/64$	0.140625	$41/64$	0.640625
$5/32$	0.15625	$21/32$	0.65625
$11/64$	0.171875	$43/64$	0.671875
$3/17$	0.1875	$11/16$	0.6875
$13/64$	0.203125	$45/64$	0.703125
$7/32$	0.21875	$23/32$	0.71875
$15/64$	0.234375	$47/64$	0.734375
$1/4$	**0.25**	$3/4$	**0.75**
$17/64$	0.265625	$49/64$	0.765625
$9/32$	0.28125	$25/32$	0.78125
$19/64$	0.296875	$51/64$	0.796875
$5/16$	0.3125	$13/16$	0.8125
$21/64$	0.328125	$53/64$	0.828125
$11/32$	0.34375	$27/32$	0.84375
$23/64$	0.359375	$55/64$	0.857375
$3/8$	**0.375**	$7/8$	**0.875**
$25/64$	0.390625	$57/64$	0.890625
$13/32$	0.40625	$29/32$	0.90625
$27/64$	0.421875	$59/64$	0.921875
$7/16$	0.4375	$15/16$	0.9375
$29/64$	0.453125	$61/64$	0.953125
$15/32$	0.46875	$31/32$	0.96875
$31/64$	0.484375	$63/64$	0.984375
$1/2$	**0.5**	**1**	**1**

B-2. SQUARES, CUBES, SQUARE ROOTS, AND CUBE ROOTS OF NUMBERS

No.	Square	Cube	Square Root	Cube Root	No.	Square	Cube	Square Root	Cube Root
1	1	1	1.00	1.00	26	676	17,576	5.10	2.96
2	4	8	1.41	1.26	27	729	19,683	5.20	3.00
3	9	27	1.73	1.44	28	784	21,952	5.29	3.04
4	16	64	2.00	1.59	29	841	24,389	5.38	3.07
5	25	125	2.24	1.71	30	900	27,000	5.48	3.11
6	36	216	2.45	1.82	31	961	29,791	5.57	3.14
7	49	343	2.65	1.91	32	1,024	32,768	5.66	3.17
8	64	512	2.83	2.00	33	1,089	35,937	5.74	3.21
9	81	729	3.00	2.08	34	1,156	39,304	5.83	3.24
10	100	1,000	3.16	2.15	35	1,225	42,875	5.92	3.27
11	121	1,331	3.32	2.22	36	1,296	46,656	6.00	3.30
12	144	1,728	3.46	2.29	37	1,369	50,653	6.08	3.33
13	169	2,197	3.61	2.35	38	1,444	54,872	6.16	3.36
14	196	2,744	3.74	2.41	39	1,521	59,319	6.24	3.39
15	225	3,375	3.87	2.47	40	1,600	64,000	6.32	3.42
16	256	4,096	4.00	2.52	41	1,681	68,921	6.40	3.45
17	289	4,913	4.12	2.57	42	1,764	74,088	6.48	3.48
18	324	5,832	4.24	2.62	43	1,849	79,507	6.56	3.50
19	361	6,859	4.36	2.67	44	1,936	85,184	6.63	3.53
20	400	8,000	4.47	2.71	45	2,025	91,125	6.71	3.56
21	441	9,261	4.58	2.76	46	2,116	97,336	6.78	3.58
22	484	10,648	4.69	2.80	47	2,209	103,823	6.86	3.61
23	529	12,167	4.80	2.84	48	2,304	110,592	6.93	3.63
24	576	13,824	4.90	2.88	49	2,401	117,649	7.00	3.66
25	625	15,625	5.00	2.92	50	2,500	125,000	7.07	3.68

B-3. METRIC PREFIXES

Numerical Value	Name	Symbol	Numerical Value	Name	Symbol
$*10^1$	deca	da	$*10^{-1}$	deci	d
$*10^2$	hecto	h	$*10^{-2}$	centi	c
10^3	kilo	k	10^{-3}	milli	m
10^6	mega	M	10^{-6}	micro	μ
10^9	giga	G	10^{-9}	nano	n
10^{12}	tera	T	10^{-12}	pico	p
10^{15}	peta	P	10^{-15}	femto	f
10^{18}	exa	E	10^{-18}	atto	a

*Not encouraged for use in SI.

B-4. CONVENTIONAL AND METRIC (SI) UNITS OF MEASUREMENT

Quantity	Conventional		Metric (SI)	
	Unit	Symbol	Unit	Symbol
Fundamental Quantities				
Length	foot	ft	metre	m
Mass	pound	lb	kilogram	kg
Time	second	sec	second	s
Temperature	degree Fahrenheit	°F	degree Celsius	°C
	degree Rankine	°R	Kelvin	K
Electrical current	ampere	A	ampere	A
Amount of substance	mole	mol	mole	mol
Luminous intensity	candela	cd	candela	cd
Derived Quantities				
Acceleration	feet per second per second	ft/sec^2	metre per second per second	m/s^2
Area	square feet	ft^2	square metre	m^2
Density	pound per cubic foot	lb/ft^3	kilogram per cubic metre	kg/m^3
Electrical potential	volt	V	volt	V
Electrical resistance	ohm	Ω	ohm	Ω
Electrical capacitance	farad	F	farad	F
Energy	kilowatt-hour	kWh	kilowatt-hour	kWh
Force	pound-force	lb_f	newton	N
Frequency	hertz	Hz	hertz	Hz
Power	foot-pound per second	ft-lb/sec	joule per second	J/s
	horsepower	hp	watt	W
	watt	W	watt	W
Pressure	pound-force per square inch	psi	pascal	Pa
Velocity	foot per second	ft/sec	metre per second	m/s
Volume	cubic foot	ft^3	cubic metre	m^3
	gallon	gal		
	barrel	bbl		
Work	foot-pound	ft-lb	joule	J
			newton-metre	N•m

B-5. NATURAL TRIGONOMETRIC FUNCTIONS

Angle	Sine	Cosine	Tangent	Angle	Sine	Cosine	Tangent
1°	0.0175	0.9998	0.0175	46°	0.7193	0.6947	1.0355
2°	0.0349	0.9994	0.0349	47°	0.7314	0.6820	1.0724
3°	0.0523	0.9986	0.0524	48°	0.7431	0.6691	1.1106
4°	0.0698	0.9976	0.0699	49°	0.7547	0.6561	1.1504
5°	0.0872	0.9962	0.0875	50°	0.7660	0.6428	1.1918
6°	0.1045	0.9945	0.1051	51°	0.7771	0.6293	1.2349
7°	0.1219	0.9925	0.1228	52°	0.7880	0.6157	1.2799
8°	0.1392	0.9903	0.1405	53°	0.7986	0.6018	1.3270
9°	0.1564	0.9877	0.1584	54°	0.8090	0.5878	1.3764
10°	0.1736	0.9848	0.1763	55°	0.8192	0.5736	1.4281
11°	0.1908	0.9816	0.1944	56°	0.8290	0.5592	1.4826
12°	0.2079	0.9781	0.2126	57°	0.8387	0.5446	1.5399
13°	0.2250	0.9744	0.2309	58°	0.8480	0.5299	1.6003
14°	0.2419	0.9703	0.2493	59°	0.8572	0.5150	1.6643
15°	0.2588	0.9659	0.2479	60°	0.8660	0.5000	1.7321
16°	0.2756	0.9613	0.2867	61°	0.8746	0.4848	1.8040
17°	0.2924	0.9563	0.3057	62°	0.8829	0.4695	1.8807
18°	0.3090	0.9511	0.3249	63°	0.8910	0.4540	1.9626
19°	0.3256	0.9455	0.3443	64°	0.8988	0.4384	2.0503
20°	0.3420	0.9397	0.3640	65°	0.9063	0.4226	2.1445
21°	0.3584	0.9336	0.3839	66°	0.9135	0.4067	2.2460
22°	0.3746	0.9272	0.4040	67°	0.9205	0.3907	2.3559
23°	0.3907	0.9205	0.4245	68°	0.9272	0.3746	2.4751
24°	0.4067	0.9135	0.4452	69°	0.9336	0.3584	2.6051
25°	0.4226	0.9063	0.4663	70°	0.9397	0.3420	2.7475
26°	0.4384	0.8988	0.4877	71°	0.9455	0.3256	2.9042
27°	0.4540	0.8910	0.5095	72°	0.9511	0.3090	3.0777
28°	0.4695	0.8829	0.5317	73°	0.9563	0.2924	3.2709
29°	0.4848	0.8746	0.5543	74°	0.9613	0.2756	3.4874
30°	0.5000	0.8660	0.5774	75°	0.9659	0.2588	3.7321
31°	0.5150	0.8572	0.6009	76°	0.9703	0.2419	4.0108
32°	0.5299	0.8480	0.6249	77°	0.9744	0.2250	4.3315
33°	0.5446	0.8387	0.6494	78°	0.9781	0.2079	4.7046
34°	0.5592	0.8290	0.6745	79°	0.9816	0.1908	5.1446
35°	0.5736	0.8192	0.7002	80°	0.9848	0.1736	5.6713
36°	0.5878	0.8090	0.7265	81°	0.9877	0.1564	6.3138
37°	0.6018	0.7986	0.7536	82°	0.9903	0.1392	7.1154
38°	0.6157	0.7880	0.7813	83°	0.9925	0.1219	8.1443
39°	0.6293	0.7771	0.8098	84°	0.9945	0.1045	9.5144
40°	0.6428	0.7660	0.8391	85°	0.9962	0.0872	11.4301
41°	0.6561	0.7547	0.8693	86°	0.9976	0.0698	14.3007
42°	0.6691	0.7431	0.9004	87°	0.9986	0.0523	19.0811
43°	0.6820	0.7314	0.9325	88°	0.9994	0.0349	28.6363
44°	0.6947	0.7193	0.9657	89°	0.9998	0.0175	57.2900
45°	0.7071	0.7071	1.0000	90°	1.0000	0.0000	∞

B-6. DENSITY/SPECIFIC GRAVITY TABLES

Densities of Some Solids, Liquids, and Gases

Material	lb/ft^3	kg/m^3	g/cm^3 (sp gr)
Solids			
Gold	1,206.2	19,300	19.3
Mercury	846.0	13,500	13.5
Lead	712.5	11,400	11.4
Iron	485.0*	7,700*	7.7*
Aluminum	165.6	2,600	2.6
Wood	50.0*	800*	0.8*
Ice	56.9	900	0.9
Liquids			
Sulfuric acid	125.0	2,000	2.0
Seawater	64.3	1,030	1.03
Fresh (pure) water	62.5	1,000	1.00
Kerosene	50.0	800	0.80
Gasoline	46.8	750	0.75
Gases			
Air	0.075	1.20	0.0012
Oxygen	0.084	1.34	0.00134
Nitrogen	0.0737	1.18	0.0018
Carbon	0.0734	1.17	0.0017
Hydrogen	0.0053	0.085	0.000085

*Wood and iron have variable densities. The values shown are approximate figures.

Specific Gravity Based on Air

Gas	Sp gr
Air	1.00
Oxygen	1.120
Nitrogen	0.0983
Carbon monoxide	0.979
Hydrogen	0.071
Butane	2.004

APPENDIX C
Conversion Factors

C-1. CONVERSION FACTORS FOR CONVENTIONAL AND METRIC (SI) MEASUREMENTS

To use this table, multiply the number of known units of measure (find in left-hand column) by the conversion factor for the desired unit of measure; for example, to convert 50 barrels to gallons, multiply: 50 × 42 = 2,100 gallons.

Unit of Measure	Conventional Units		Metric Units	
To convert:	to:	multiply by:	to:	multiply by:
acre	square foot	43,560	square metre	4,046.875
	square yard	4,840	square kilometre	0.004046875
	square rod	160		
	square mile	0.0015625		
	section	0.0015625		
acre-foot	cubic foot	43,560	cubic metre	1,233.48766
bar	inch of water @ 60°F	407.229	millimetre of water @ 60°F	10,343.6
	inch of mercury @ 32°F	29.5282	millimetre of mercury @ 0°C	750.0187
barrel, U.S. petroleum (bbl)	cubic foot	5.6146	cubic metre	0.1589873
	gallon	42	litre	158.9873
British thermal unit @ 60°F (Btu)	foot-pound	777.97265	joule	1,054.68
	kilowatt-hour	0.00029283	kilowatt-hour	0.00029283
	calorie	251.98		
	horsepower-hour	0.0003927		
calorie (cal)	British thermal unit	0.003968	joule	4.1868
centimetre (cm)	mil	393.70	millimetre	10
	inch	0.3937	metre	0.01
	foot	0.032808		
centipoise (cp)	poise	100	pascal-second	0.001
cubic centimetre (cm³)	cubic inch	0.061023	cubic millimetre	1,000
	cubic foot	0.000035314	millilitre	1
	gallon	0.00026417		
cubic foot	cubic inch	1,729.98829	cubic centimetre	28,317
	cubic yard	0.037036	cubic metre	0.028317
	gallon	7.48050		
	barrel	0.17811		
cubic foot of water (ft³ of H₂O)	pound of water	62.3		
cubic foot/second (ft³/sec)	gallon/minute	488.883		
cubic inch (in.³)	cubic foot	0.0005787	cubic centimetre	16.38716
	cubic yard	0.000021434	litre	0.016387
	gallon	0.0043290	millilitre	16.38716

| Unit of Measure | Conventional Units | | Metric Units | |
To convert:	to:	multiply by:	to:	multiply by:
cubic metre (m³)	cubic foot	35.314445	kilolitre	1
	cubic inch	61,022.9388	litre	1,000
	cubic yard	1.307943	cubic centimetre	1,000,000
	gallon	264.17762		
	barrel	6.28994		
cubic yard (yd³)	cubic foot	4.80897	cubic metre	0.76456
	gallon	201.97350	kilolitre	0.76456
	barrel	4.80897		
day	minute	0.0006944		
	hour	0.0041666		
	week	7		
	year	365		
degree, angle	minute	60	radian	0.01745
	second	3,600		
degree, temperature (See Appendix C-2.)				
foot/second (ft/sec)	miles/hour	0.68182	metre/second	0.3048
foot-pound force (ft-lb_f)	inch-pound force	12	joule	1.35582
	Btu	0.001286		
gallon of water	pound of water	8.328		
gallon, U.S. liquid (gal)	cubic inch	231	cubic metre	0.0037854
	cubic foot	0.133681		
	barrel	0.0128095	cubic metre	0.0037854
gram (g)	oz, avoir.	0.0352739	kilogram	1,000
hectare	acre	2,471	square metre	10,000
horsepower (hp)	foot-pound/minute	33,000	watt	745.6999
	foot/pound/second	550		
horsepower-hour	Btu	2,546.473	megajoule	2.68452
hour (hr)	second	0.0002777	second	0.0002777
	minute	0.016666		
	day	24		
	year	8,766		
inch (in.)	foot mil	0.083333	metre	0.0254
	mil	1,000	centimetre	254
			millimetre	25.4
inch of mercury @ 32°F	inch of water @ 60°F	13.595326	pascal	3,386.389
	pound/square inch	0.491161	millimetre of mercury @ 0°C	
	bar	0.0338659		
joule (J)	foot-pound	0.73756	kilowatt-hour	0.0000002778
	watt-second	1	watt-second	1

Unit of Measure	Conventional Units		Metric Units	
To convert:	*to:*	*multiply by:*	*to:*	*multiply by:*
kilogram (kg)	pound mass, avoir.	2.204622	gram	1,000
			tonne	0.001
kilometre (km)	mile, U.S. statute	0.62137	metre	1,000
kilometre/hour .... (km/h)	mile/hour	0.62137	metre/second	0.2777778
kilowatt (kW)	foot-pound/second	737.56	joule/second	3,600,000
	horsepower	1.341		
kilowatt-hour (kWh)	horsepower-hour	1.341	joule	3,600,000
litre	gallon	0.264178	cubic metre	0.001
	cubic inch	61.025		
megajoule (MJ)	horsepower-hour	0.3725	joule	1,000,000
metre (m)	inch	39.370	centimetre	100
	foot	3.280833	kilometre	0.001
mil	inch	0.001	millimetre	0.0254
mile, U.S. statute (mi)	rod	320	kilometre	1.60935
	foot	5,280	metre	1,609.35
	yard	1,760		
miles per hour (mph)	foot/second	1.467	metre/second	0.44704
millibar	bar	1000	pascal	101.356
millimetre of mercury .... @ 0°C (mm of Hg)	inch of mercury @ 32°F	0.03937	pascal	133.3224
minute (min)	second	0.016666	second	0.016666
	hour	60		
	day	1,440		
newton (N)	pound-force	0.2248		
newton-metre (N•m)	pound-force-foot	0.7375621	joule	1
ohm (Ω)	megohm	0.000001	megohm	0.000001
ounce, U.S. fluid (oz)	cubic inch	1.80469	cubic metre	0.00002957
ounce-force (oz_f)	pound-force	16	newton	0.278
ounce-mass (oz_m)	pound-mass	16	kilogram	0.4535
pascal (Pa)	pound-force/ square inch	0.000145038	newton/square metre	1
pascal-second (Pa•s)	centipoise	0.001	gram/centimetre-second	0.10
poise	centipoise	100	pascal-second	0.10
pound-force (lb_f)			newton	4.448
pound-force/foot			newton/metre	14.59390
pound-force/square .... inch (psi)	foot of water @ 60°F	2.306787	pascal	6.894757
	inch of mercury @ 32°F	2,0360	kilopascal	6.894757
	bar	0.068947		

Unit of Measure *To convert:*	Conventional Units *to:*	*multiply by:*	Metric Units *to:*	*multiply by:*
pound-mass (lb_m)	ounce-mass	16	kilogram	0.4536
pound of water	gallon of water cubic foot	0.120 0.016		
radian, angle (rad)	degree degree, minute, second	57.29578 57° 17′44.8″		
rod (rd)	foot yard mile, U.S. statute	16.5 5.5 0.003125	metre centimetre	5.0292188 502.921875
section	square mile acre	1 640	square metre hectare	2,589,988 258.9988
square foot (ft²)	square inch square yard	144 0.11111	square metre square centimetre	0.0929034 929.0341
square inch (in.²)	square foot square mil circular mil	0.00694444 1,000,000 1,273,240	square centimetre square millimetre	6.451626 645.16258
square kilometre (km²)	square mile acres	0.383101 247.1044	hectare square metre	100 1,000,000
square metre (m²)	square foot acres	10.7639 0.000247104	square centimetre hectare	10,000 0.0001
square mil (mil²)	circular mil square inch	1.273224 0.000001	square millimetre	0.000645163
square mile, statute (mi²)	section acres square foot	1 640 27,878,400	square metre square kilometre hectare	2,589,988 2.589988 258.9988
square millimetre (mm²)	square inch square mil	0.00155 1,550.0	square metre square centimetre	0.000001 0.01
square rod (rd²)	square foot	272.25	square metre	25.2929
square vara, Texas	square foot	7.716	square metre	0.716843
square yard (yd²)	square foot	9	square metre	0.836131
ton, U.S. long (tn)	pound-mass barrels of oil, 36°API	2,240 7.33627	kilogram tonne	1,016.047 1.016047
ton, U.S. short (tn)	pound-mass	2,000	tonne	0.90718
tonne, metric SI (t)	pound-mass ton, U.S. short barrel of oil, 36°API	2,204.6 1.01606 7.454	kilogram gram	1,000 1,000,000
township	square mile	36	square kilometre	93.24
vara, Texas	foot	2.777778	metre	0.84667

Unit of Measure	Conventional Units		Metric Units	
To convert:	*to:*	*multiply by:*	*to:*	*multiply by:*
watt (W)	kilowatt	0.001	kilowatt	0.001
	British thermal unit/hour	3.413772	kilogram-metre/hour	367.1
	foot-pound/second	0.73756	joule/second	1
	horsepower	0.001341		
watt-hour (Wh)	foot-pound	2,655.22	kilogram-metre	367.1
	kilowatt-hour	0.001	kilowatt-hour	0.001
	horsepower-hour	0.001341	joule	3,600
week (wk)	day	0.1428571		
	year	52		
	hour	0.0059523		
yard (yd)	mile, statute	0.00568182	metre	0.914404
	foot	3	millimetre	914.40360
	inch	36		
	mil	36,000		
year (yr)	month	0.083333		
	day	0.0027397		

C-2. TEMPERATURE CONVERSION FACTORS

To convert:	Conventional		Metric	
	to:	*factor:*	*to:*	*factor:*
degree Fahrenheit (°F)	°R	°F + 459.67	°C	(°F − 32)/1.8
			K	(°F + 459.67)/1.8
degree Celsius (°C)	°F	°C/1.8 + 32	K	°C + 273.15
	°R	(°R − 491.67)/1.8		
degree Rankine (°R)	°F	°R − 459.67	°C	(°R − 491.67)/1.8
			K	°R/1.8
Kelvin (K)	°F	1.8(K − 273.15) + 32	°C	K − 273.15
	°R	1.8K		

ANSWER KEY

Problems involving decimals may have discrepancies in their answers—for example, the problem involving the number of fans needed to change the air every 3 minutes in a given area. Possible answers are 4.053, 4.05, 4, or 5 fans, depending upon the practical application. Four fans will not quite do the job. Also, with 4 fans, there is no standby for emergency. The engineer would probably call for 5 fans. If the specification for a change of air each 3 minutes included a factor for safety, perhaps 4 fans would suffice. Obviously, it is not possible to install 4.053 fans, but that is the solution to the problem to three decimal places. At this point, judgment enters the picture.

Problems involving only whole numbers are precise in the results obtained. Many problems in fractions cannot be worked to a final answer. Rounding off fractions to whole numbers, to one place, two places, and so forth will change the end result. For example, using pi on a calculator instead of 3.1416 will slightly alter an answer. Using rounded-off conversion factors as given in the chapter tables will result in answers different from those solved by using the exact conversion factors given in Appendix C. These answers may vary considerably if large quantities are involved.

As a user of this manual you should compare your answer with that given, and if there is a difference, check your work. If it is not in error, then you might ask, "How can this difference be accounted for?" or "Is my answer satisfactory although perhaps not exact?" If it is satisfactory for the purpose you had in mind from the beginning, then it should be acceptable. However, you should always be able to explain how the difference arose.

Answers to Practice Problems

1. THE NUMBER SYSTEM

Whole Numbers, pp. 13–14

1. a. 5,000
    b. 110
    c. 30
    d. 10
2. a. 42
    b. 361,000,000
    c. 27,051,289,000
    d. 20,400,502,000,000
3. a. four thousand, six hundred
    b. seventy-eight thousand
    c. eight million, six hundred thousand
    d. eighty billion, seven hundred forty million dollars
4. a. 953
    b. 236
    c. 1,281,150
    d. 6,485,381
5. a. $3,520 per month
       $42,240 per year
    b. 17 hr
    c. 1,090 ft
    d. 4,811 ft

Common Fractions, pp. 19–20

1. a. $^6/_{64}$
    b. $^{12}/_{16}$
    c. $^{14}/_{16}$
    d. $^7/_8$
2. a. $^4/_8, ^1/_8, ^6/_8$
    b. $^4/_{32}, ^2/_{32}, ^7/_{32}$
    c. $^{56}/_{64}, ^{19}/_{64}, ^{22}/_{64}, ^{17}/_{64}, ^4/_{64}, ^{62}/_{64}$
3. a. $^1/_2$
    b. $^7/_8$
    c. $^1/_8$
    d. $^5/_6$
4. a. $1^1/_2$
    b. $3^5/_{16}$
    c. $^7/_{32}$
    d. $^{43}/_{64}$
5. a. $5^5/_8$
    b. $^{35}/_{160}$ or $^7/_{32}$
    c. $1^1/_{10}$
    d. $^5/_{18}$
6. 30 bbl
7. Jones: $28, 125
    Smith: $14,062.50
    White: $32,812.50
8. $^5/_{16}$"

9. $7'5^9/_{16}$"
10. $^{21}/_{64}$"

Mixed Numbers, pp. 23–24

1. 40 hr
2. $17^{15}/_{16}$"
3. $20^{29}/_{32}$"
4. $12^1/_4$ hr
5. $^5/_{16}$"
6. $45^3/_8$"
7. $764^4/_{13}$ bbl
8. 10
9. 21
10. 1,100 bbl

Decimal Fractions, pp. 31–32

1. a. three thousandths
    b. six hundred twenty-five thousandths
    c. one hundred twenty and four hundredths
    d. eight and three thousand, seven hundred, forty-five ten thousandths

299

2. a. 0.115; $^{115}/_{1,000}$
   b. 76.76; $76^{76}/_{100}$ or $76^{19}/_{25}$
   c. 5,000.6; $5000^{6}/_{10}$ or $5,000^{3}/_{5}$
   d. 3,125.8; $3,125^{8}/_{10}$ or $3,125^{4}/_{5}$
3. $216.16
4. $570.15
5. 8.928 gal
6. 42.48 hp
7. 53,400 lb
8. $23.03
9. 1.53 in.
10. 1.25 in.

Roots and Powers, p. 35

1. 28
2. 2,744
3. 36 mi²
4. 3.2
5. 13,839

2. THE CALCULATOR, p. 56

1. 117.8
2. 1,391.84
3. 5,035.24
4. 1,548.55
5. 1,872
6. 0.50
7. 548.17
8. 625
9. 408
10. 1,610,069.92
11. 59.45
12. 308 mi
13. $1.29
14. $14.70
15. 1,276.66

3. NUMBER RELATIONS

Percentage, pp. 61–62

1. $54.60
2. 17
3. $116.40
4. a. Tin—416.50 lb
   b. Copper—28 lb
   c. Antimony—55.50 lb
5. 200
6. $294.40
7. 15%
8. a. Lathe—53.70%

b. Drill press—8.33%
   c. Shaper—25%
   d. Welding—12.96%
9. 1,282.05 bbl
10. a. 22
    b. $400/wk
    c. 99%
    d. 87½%
    e. 50%

Ratio and Proportion, pp. 66–67

1. 25 to 1
2. 5 to 3
3. 14 to 1
4. 816 sacks
5. 40 ft
6. 50 men
7. 80 mph
8. 435 rpm
9. 450 rpm
10. 123.75 rpm

Mean (Average), Median, and Mode, p. 70

1. 110 rigs
2. 43.5¢ or 44¢/copy
3. 40.91
4. $1,860
5. 12", 1', or 30.48 cm

Tables and Graphs, pp. 76–78

1.

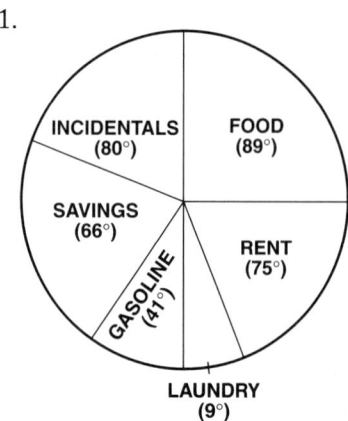

2. a. September, 1984
   b. Increased
   c. 8 million bbl/day
   d. 44 million bbl/day
   e. 1981

3.

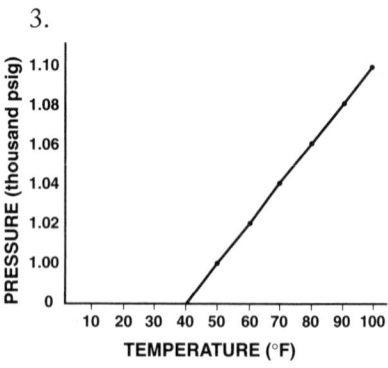

4.

GAS PIPE INSTALLED BY TYPE

YEAR	PLASTIC PIPE (mi)	STEEL PIPE (mi)
1975	15,985	12,663
1976	11,640	9,718
1977	15,991	6,157
1978	14,676	7,212
1979	18,826	8,145
1980	18,912	8,334

5.

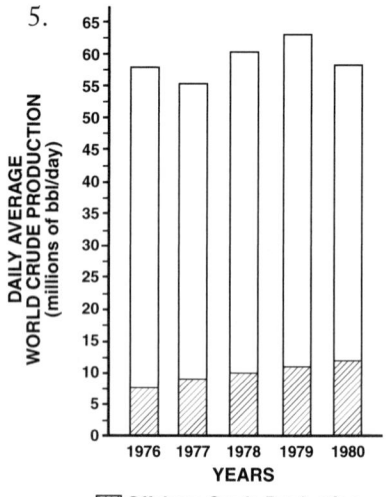

4. SOME PHYSICAL QUANTITIES AND THEIR MEASUREMENT

Length, Area, and Volume, pp. 94–96

1. a. 3.8 in.
   b. 5¾ in.
   c. 1½ in.
   d. 3¹/₁₆ in.
   e. 3¼ in.
2. a. 7.4 cm
   b. 3.8 cm
   c. 5.1 cm

d. 1.2 cm

e. 7.9 cm

3. 91.44 m

4. 30.48 cm

5. 51.49 km

6. 1.93 m

7. a. 2.40 m

b. 2,400 mm

c. 0.0024 km

d. 2,400,000 mm

e. 0.00000240 Mm

8. 1,196.17 yd^2

9. 14 hr

10. 10,857 in.^{3}

11. 144 in.^{3}

12. 4 bd ft

13. 1.48 yd^3

14. 0.00156 mi^2

15. 320

16. 1.323 yd^3

17. 158.928 litres

18. 7.0861 ft^3

19. 1,308.64 acres

20. 2.396 mi^2

Time and Temperature, pp. 100–101

1. a. 530 mph

b. 8.833 mi/min

2. 21 min

3. 1,350 rpm

4. a. 161.6°F

b. 68°F

c. 32°F

d. 22°F

e. −108.4°F

5. a. −67.2°C

b. 0°C

c. 4°C

d. −33.3°C

e. −263°C

6. a. 51°C

b. −70°C

c. 20°C

d. −23°C

e. -224°C

7. a. 1,140°R

b. 463°R

c. 503°R

d. 436°R

e. 347°R

8. 12:47 P.M., Wed.

9. 99 min

10. 180.6 K

Mass and Related Derived Quantities, pp. 109–110

1. 907.185 kg

2. 53,381 daN; 533.8kN

3. 631.579 ft

4. 11.023 lb

5. 1.8 lb/ft^3

6. 40.7669°API

7. 3,200 ft-lb

8. 51.6 psia

9. 15.92 lb

10. 6.928 psi

Electricity, pp. 115–116

1. 0.12 A

2. a. 60 V

b. 3 A

c. 12 V

d. 3 A

3. 172.5 kWh

4. a. 1.96 Ω

b. 0.385 A

c. 0.25 A

d. 19.2 V

e. 148 Ω

5. 180 W

6. 240 Wh

7. 4 Ω

8. a. 7.5 A

b. 225 W

9. 0.5 A

10. 500 W

5. PRINCIPLES OF ALGEBRA

Algebraic Expressions, pp. 127–128

1. $3a - 3b$

2. $11a + 3b - bc + 6c$

3. $2a - 4ab + 26b - 2c$

4. $-3x + xy + y$

5. $15a - 8ab + 8b + 9c$

6. $a^2 - 4b + c$

7. $4x^2 - 6x - 6xy - 7y$

8. $-4a + 3ab - 15ac - 12b - 8c$

9. $5a^2 - 3ab - 14c^2$

10. $6x^2 - 10x - 4xy - 6y$

11. $4ax^2 + 8axy + 4ay^2$

12. $20a^2b + 30a^2b^2 + 15abc$

13. $3a^2 + ab - 4b^2$

14. $-4a^3 + 4a^2 - 8a^2b + 12ab - 4ab^2$

15. $x^2 + 2xy - 2x + 3y + y^2$

16. $-4a - 2b$

17. $9a^2 + 3a + 4b + 2b/a$

18. $m^2 - 3m + 14$

19. $x^2 - 2xy + y^2$

20. $m^2 - 4m + 8$

Equations, pp. 130–131

1. a. 36

b. 22

c. 10

2. 2

3. a. $a + c + b = e + a + b + d$

b. $b + c + e = b + d + 2e$

c. $c + a + d = c + e + b + d$

4. $71.50 per day

5. 7.34 lb

6. Greater number: 20

Lesser number: 15

7. 1,225 ft^2

8. $147.69

9. 90

10. $5.00

Formulas, pp. 139–141

1. 0.4 darcys

2. a. 0.2364 in.

b. 94.24 mi

c. 6.56 yd

d. 11.82 in.

e. 6.56 ft

3. $130.00

4. 3.68 gal

5. a. $D = ST$

b. $S = D/T$

c. $T = D/S$

6. 55,063, or approx. 55,000 bbl

7. $sp\ gr = 141.5 \div (API + 131.5)$

8. 38.76 °API

9. $P_1 = P_2V_2 \div V_1$

$V_1 = P_2V_2 \div P_1$

$P_2 = P_1V_1 \div V_2$

$V_2 = P_1V_1 \div P_2$

10. 1.651.26 ft^3

6. PRACTICAL GEOMETRY

Plane Figures, pp. 156–159

1. 330 ft²
2. 4.5 ft²
3. 42 in.
4. 37.19 acres
5. 1,111.11 yd²; 136.66 yd
6. 25 lb
7. ¾ of an acre
8. a. 11,200 ft²
   b. 1,400 boards
9. 73.9, or 74 ft²
10. a. 38.0431, or 38 ft²
    b. 9 in.²
    c. 65.96, or 66 in.²
    b = 8.25"
11. 37.70 in.²
12. 179.93, or 180
13. 0.0865 of an acre
14. 15 in.²
15. a. 212.06 in.²
    b. 805.82 in.²

Solid Figures, pp. 167–170

1. 10,378.125 lb
2. 138.88 yd³
3. 7.48 gal
4. 1,468.698, or 1,469 gal
5. 204.75 bbl
6. 7.83 gal
7. 24 lb
8. 10 ft³
9. 3⅑ yd³
10. 10.728 ft³
11. 11.13 lb
12. a. 18.96 ft³
    b. 26.4 times/min

13. 4.14 or 5 fans
14. 20.23 lb
15. 67.5 ft²

Geometric Constructions

Since these are construction exercises, no answers are given.

7. TRIGONOMETRY

Right Triangle Trigonometry, pp. 195–196

1. 75°
2. 8.49 in.
3. 42.43 ft
4. 8 in.
5. 6.93 in.
6. 20.42 in.
7. 17.54 ft²
8. a. 22.29 ft
   b. 20.07 ft
9. 41°24"
10. 4.29°

Oblique Triangle Calculations, pp. 199–201

1. 81.91 ft
2. 9.25 in.
3. 58.40°
4. 5.63 in.
5. X = 0.88 in.
   Y = 1.52 in.
6. 2.39 in.
7. 15.06 ft ²
8. Approximately 300 ft²
9. 13.5 ft
10. 3.53 in./side

9. ADVANCED OIL INDUSTRY APPLICATIONS

Mathematics in Drilling Operations, pp. 258–259

1. 1,209 sacks
2. 173 bbl
3. 444 hp
4. 299 ton-miles
5. 285 ton-miles
6. 4,212 psig
7. 11.7 ppg
8. 1,325 psi
9. 10.9 ppg
10. 0.52
11. 625 psi
12. 0.78

Mathematics in Production Operations, pp. 268–269

1. 0.74 bpd/psi
2. 14.06 hp
3. 4.03 pt
4. 38.59 Mcf/hr
5. 1.54 MMcf
6. 175°F

Mathematics in Pipeline Operations, pp. 274–275

1. 1,732 ft
2. 478.32 hp
3. 81.1 lb/ft
4. ½"
5. 8,805 ft

Mathematics in Refining Operations, p. 278

1. 0.813 Btu/lb/°F
2. 81%

Answers to Self-Tests

1. THE NUMBER SYSTEM, pp. 36–38

1. 2,242,000,000
2. $\frac{1}{4}$
3. $^{27}/_{32}$"
4. $4^{1}/_{8}$"
5. $^{27}/_{32}$"
6. Royalty owner: $19,880
   Operator: $139,160
7. 1,837 gal/hr
   44,100 gal/day
8. a. 0.5625
   b. 0.0938
   c. 0.4844
   d. $2^{21}/_{32}$
   e. $^{1}/_{16}$
9. 610.3 mi
10. 152.4 bbl
11. a. 266.4 lb
    b. 177.6 lb
12. $2.11
13. 64 pins
14. 4.7430 in.
15. a. $190,000
    b. 6,552 bbl
    c. 26,207 bbl
    d. $464
    e. 409 or 410 days
16. 533 1/3 bbl
17. a. 11
    b. 348
    c. 71,639,296
    d. 6.346
    e. 10,000
18. a. 0.109 in.
    b. 0.133 in.
    c. 0.154 in.
    d. 0.203 in.
    e. 0.216 in.
19. a. 11,250 gal
    b. 236,250 gal
    c. 337,500 gal
    d. $6,851.25
    e. $2,500,706.25
20. a. 3,063
    b. $8.50

c. 2.75
d. $9.27
e. 6,400

2. THE CALCULATOR, pp. 57–58

1. 45
2. 33.4411
3. 31.88
4. 8.33
5. 1
6. 365.6
7. 14,300
8. $2.00
9. 2,329.6
10. 1.12
11. 14.213986
12. 436.8
13. 2,496.7091
14. 3.898885
15. 59.16567
16. 12,016
17. 74.934007
18. 14.3175
19. $71.70
20. 2,303.818

3. NUMBER RELATIONS, pp. 79–82

1. 1 to 4
2. 312.5 ft
3. 1:3
4. 14.28 gal
5. $576.00
6. 3.66 hr
7. 480 rpm
8. 294.54 rpm
9. 200 rpm
10. $40.71/bbl
11. 128 ft
12. Mean: 51.62°F
    Median: 52°F
    Mode: 52°F
13. $2.65/Mcf
14. 7.39¢/kWh
15. 39 qt

16.

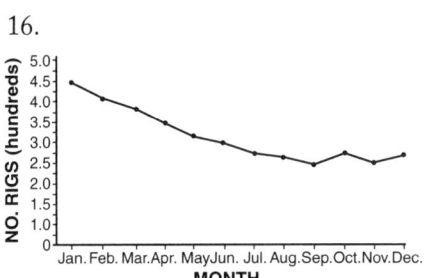

17.

U.S. WHOLESALE PRODUCT PRICES ($/bbl)

	OCTOBER 1982	NOVEMBER 1982	DECEMBER 1982
Gasoline	$39.68	$38.87	$36.88
Kerosene	$42.20	$42.97	$36.88
Fuel oil	$40.06	$40.79	$37.84
Residual fuel	$24.78	$26.06	$24.57

18.

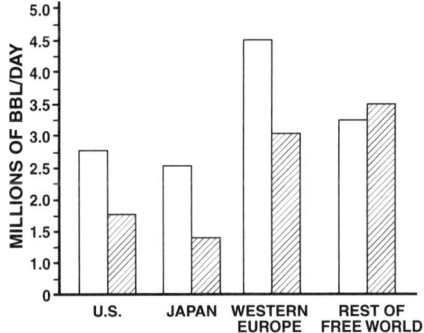

19.

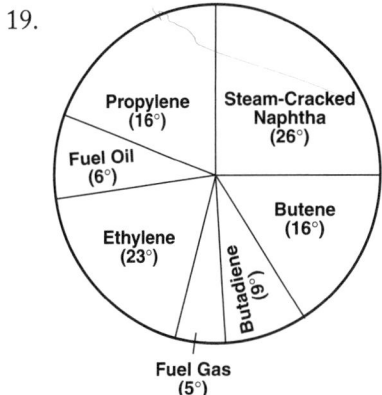

20.

COAL PRODUCED IN U.S.
1970–1980

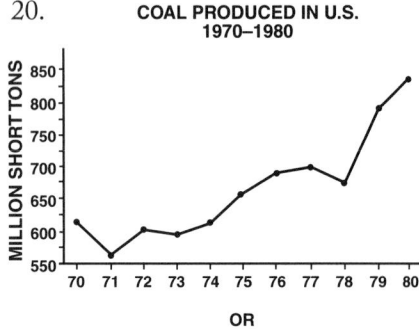

OR

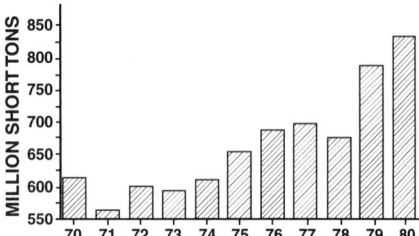

4. SOME PHYSICAL QUANTITIES AND THEIR MEASUREMENT, pp. 117–119

1. 6.21 mi
2. 79.45 m^3
3. a. a fundamental quantity
4. a. mi
   b. lb/gal, or ppg
   c. ft^3
   d. m^2
   e. lb/in.^{2}, or psi
   f. g/cm^3, or g/cc
   g. lb_f
   h. W
   i. ft/sec
   j. rpm
   k. kWh
5. a. 10^3, or 1,000
   b. 10^{-1}, or 0.10
   c. 10^{-3}, or 0.001
   d. 10^{-6}, or 0.000001
   e. 10^6, or 1,000,000
6. a. 29.03 kg
   b. 3.5816 tn
   c. 17.984 lb_f
   d. 165.345 lb_m
   e. 2.222 oz
7. 26.25 lb/m
8. 268.56 MJ
9. a. 21.1°C
   b. 51°C
   c. −18.4°F
   d. 365.4°R

e. 612°R
10. 7.52
11. 335,000 lb
12. 12.08 ft^3
13. 21,600 ft-lb
14. 279.75 kW
15. 10 A
16. 30 V
17. 6.93 psi
18. a. 579°R
    b. 296 K
    c. 53.7 psia
    d. 30.7 psia
    e. 1,760°R
19. 347°R or − 112.2°F
20. 2.4 kW

5. PRINCIPLES OF ALGEBRA, pp. 142–146

1. a. $x = 4$
   b. $x = 9$
   c. $x = 30$
   d. $x = 20$
   e. $x = 16$
2. a. 45
   b. 66
   c. −78
   d. 54
   e. 107
3. 1,500 lb
4. a. $9x − 3y$
   b. $−x − 4y$
   c. $20x − 5y − 7z$
   d. $a + 5b + 8c + 21d$
   e. $x^2 + 11x^2y + 8xy^2$
5. a. $−a − 4c$
   b. $−3b − c$
   c. $x − y −4z − 6$
   d. $3a^2 − 7a^2b + 2ab^2 + b^3$
   e. $8x^2z + 9z + 2x^3 − xz^2$
6. a. $7a + 2b − 4$
   b. $2x^2 − 3x + 12$
   c. $3x − 12y$
   d. $9a − 7b + 6c$
   e. $−2y^2 + 4a^2$
7. a. $−5ab^2$
   b. $6x^5$
   c. $−24x^8$
   d. $4x^3y + 8x^2y^2 + 4xy^2$
   e. $−x^5ay + 2x^4a^2y − x^3ay^3$
8. a. $8a$

b. $−3x^3$
   c. $x/5$
   d. $1 − 2(a + x)^2 = 1 − 2a^2 − 4ax − 2x^2$
   e. $a^2 + 2ab − b^2$
9. 140 acres
10. 20 ohms
11. 150°F
12. a. $A = 118.30$
    b. $h = 16.77$
    c. $d = 10.69$
    d. $A = 31.82$
    e. $P = 23.55$
13. Larger part: 36
    Smaller part: 19
14. 84 ft/sec
15. 57.25 mph
16. a. Plan A: 33¢; plan B: 25¢
    b. Plan A: 2,500; plan B: 3,750
    c. Plan A: 12,500; plan B: 26,250
    d. Plan A: $25,000; plan B: $52,500
    e. Plan B costs $2,000 less.
17. °R = ⅘°C + 492
18. $w = A/l$
19. 8 ft
20. 289.15 lb
21. 108 tiles
22. 48 kW
23. 8 ohms
24. 16.8 mA
25. 4 amperes

6. PRACTICAL GEOMETRY, pp. 182–186

1. 12.5664 mm
2. 0.3314, or 0.33 in.^{2}
3. 1,368.84 gal
4. 150.8 in.^{2}
5. 1,436.76 in.^{3}
6. 1,024.43 bbl
7. 18 in.^{2}
8. 16.98 acres on right; 144.63 acres on left
9. 1.2, or 1⅕ gal
10. 27,073.54 gal
11. 563.98 gal
12. a. 97.23 ft
    b. 7.57 ft

c. 1.55 ft

d. 11.66 mm

e. 4.69 in., or 4$^{7}/_{10}$ in.

13. 131.9472 + 1, or 133 bolts

14. 534.38 ft^2

15. 1.2, or 1$^{1}/_{5}$ gal

16. 105 in.^{2}

17. 2.25 ft^3

18. 448.18 rpm

19. 37.70 in.^{2}

20. 4.97 in.^{3}

7. TRIGONOMETRY, pp. 202–206

1. 17.3 ft

2. a. 39.3 ft

 b. 22.6 ft

3. 46.7 ft

4. 5.6 in.

5. 12.1 ft

6. 591 ft

7. a. 619.3 ft

 b. 385.5 ft

 c. 143 ft

8. Angle A: 10°

 Angle B: 80°

 Angle C: 90°

9. 7.3 ft

10. 8.7 ft

11. 17'1$^{7}/_{8}$"

12. Side a: 156.2 ft

 Side c: 294.8 ft

13. 22'3.5"

14. a. 1.2349

 b. 0.8098

15. 3 in./side

16. 2.3 in./side

17. 13.1 in.

18. 2.5 in.

19. 2.3 in.

20. 9.9 in.

8. ADVANCED MATH CONCEPTS, pp. 223–225

1. a. 11

 b. 55

 c. 14

 d. 731

 e. 255

2. a. 1010

 b. 1100100

 c. 1111101000

 d. 100110011000

 e. 100000000

3. a. 100110

 b. 1001100

 c. 101100101

4. a. 10011

 b. 001000

 c. 00010001

5. a. 10110000 (decimal equivalent = 176)

 b. 10110010100 (decimal equivalent = 1428)

 c. 111110000101110 (decimal equivalent = 31790)

6. a. 10 (decimal equivalent = 2)

 b. 100 (decimal equivalent = 4)

 c. 10100 (decimal equivalent = 20)

7. a. 103652$_8$

 b. 73414$_8$

 c. 12525$_8$

 d. 1306563$_8$

8. a. 87AA$_{16}$

 b. 770C$_{16}$

 c. 1555$_{16}$

 d. 58D73$_{16}$

9. a. 1000001

 b. 1011001

 c. 1011100

 d. 1011110

 e. 0111111

 f. 0101111

 g. 0100110

10. a. A•B=C

 b. (A+B) + (C+D) = E

 c. (A • B) + C = D

9. ADVANCED OIL INDUSTRY APPLICATIONS, pp. 279–281

1. 11.2 ppg

2. 875 psi

3. 12.27 psig

4. 1,615.54 psi

5. 3.08 MMcf/d

6. 163,800 lb

7. 2 P.M. on Wednesday

8. 65 psi

9. 142 kW

10. 630 kW

11. 1,873 ft^3

12. 1,873,500 Btu/hr

13. 2,853.76 ft^3

14. 111001100110

15. 1000 hp

16. 166 kVA

17. 0.69

18. 238 kVARs capacitive

19. 250 ft-lbs

20. 12 mA